. . . .

Neuro

The New Brain Sciences and the Management of the Mind

Nikolas Rose and Joelle M. Abi-Rached

PRINCETON UNIVERSITY PRESS

Princeton and Oxford

Copyright © 2013 by Princeton University Press

Published by Princeton University Press, 41 William Street,
Princeton, New Jersey 08540

In the United Kingdom: Princeton University Press, 6 Oxford Street,
Woodstock, Oxfordshire OX20 1TW

press.princeton.edu

Library of Congress Cataloging-in-Publication Data

Rose, Nikolas S.
    Neuro : the new brain sciences and the management of the mind / Nikolas Rose
and Joelle M. Abi-Rached.
        p.   cm.
    Includes bibliographical references and index.
    ISBN 978-0-691-14960-8 (hardcover : alk. paper) — ISBN 978-0-691-14961-5
(pbk. : alk. paper) 1. Neuropsychology.   I. Abi-Rached, Joelle M., 1979–   II. Title.
    QP360.R655 2013
    612.8—dc23   2012023222

British Library Cataloging-in-Publication Data is available

This book has been composed in Minion Pro and Ideal Sans

Printed on acid-free paper. ∞

Printed in the United States of America

10 9 8 7 6 5 4 3 2 1

For Diana, as always.
NR

For my parents, May and Maroun.
JAR

Contents

. . . .

            The "Social Brain Hypothesis"   *143*
            Pathologies of the Social Brain   *148*
            Social Neuroscience   *151*
            Social Neuroscience beyond Neuroscience   *156*
            Governing Social Brains   *160*

Six   **The Antisocial Brain**   *164*
            Embodied Criminals   *167*
            Inside the Living Brain   *173*
            Neurolaw?   *177*
            The Genetics of Control   *180*
            Nipping Budding Psychopaths in the Bud   *190*
            Sculpting the Brain in Those Incredible Years   *192*
            Governing Antisocial Brains   *196*

Seven   **Personhood in a Neurobiological Age**   *199*
            The Challenged Self   *202*
            From the Pathological to the Normal   *204*
            The Self: From Soul to Brain   *213*
            A Mutation in Ethics and Self-Technologies?   *219*
            Caring for the Neurobiological Self   *223*

Conclusion **Managing Brains, Minds, and Selves**   *225*
            A Neurobiological Complex   *225*
            Brains In Situ?   *227*
            Coda: The Human Sciences in a Neurobiological Age   *232*

Appendix   **How We Wrote This Book**   *235*

            Notes   *237*

            References   *277*

            Index   *325*

Acknowledgments

· ·

This book arises from research funded by the United Kingdom's Economic and Social Research Council in the form of a three-year professorial research fellowship to Nikolas Rose (grant number RES-051-27-0194), and we are happy to acknowledge this support and the opportunities that it provided. We have also benefited greatly from the work with our colleagues in the European Neuroscience and Society Network (ENSN), funded by the European Science Foundation (ESF), and the numerous events organized by this network that bring researchers from the human sciences together with leading neuroscience researchers in the spirit of critical friendship within which we have written this book: thanks to Trudy Dehue, Giovanni Frazzetto, Cornelius Gross, Ilpo Helén, Kenneth Hugdahl, Ilse Kryspin-Exner, Klaus-Peter Lesch, Linsey McGoey, Cordula Nitsch, João Arriscado Nunes, Andreas Roepstorff, Ilina Singh, and Scott Vrecko. We would also like to acknowledge our lively conversations with Francisco Ortega and Fernando Vidal, and to thank as well the many members of the international network of social scientists working in the field of neuroscience for so many productive discussions on its social dimensions. The research for this book was carried out within the wonderfully supportive community of the BIOS Research Centre at the London School of Economics, who also helped with organization of the final "Brain, Self, and Society" conference and follow-up workshop, "Personhood in a Neurobiological Age," held in London on September 13–14, 2010: special thanks go to Btihaj Ajana, Valentina Amorese, Rachel Bell, Astrid Christoffersen-Deb, Megan Clinch, Caitlin Connors, Des Fitzgerald, Amy Hinterberger, John MacArtney, Sara Tocchetti, and our excellent administrator Victoria Dyas and outstanding manager, Sabrina Fernandez. We also wish to thank our editorial team at Princeton University Press for their support for this book and their careful work on our text.

Abbreviations

. . . .

AD: Alzheimer's disease
ADHD: attention deficit hyperactivity disorder
APA: American Psychiatric Association
APIRE: American Psychiatric Institute for Research and Education
BNA: British Neuroscience Association
BOLD: blood-oxygen-level-dependent (contrast imaging)
BPD: bipolar disorder (sometimes termed "manic depression")
BRA: Brain Research Association
Caltech: California Intitute of Technology
CNRS: Centre National de la Recherche Scientifique (France)
CNVs: copy number variations
CT scan: Computerized tomography (formerly computerized axial
                tomography or CAT scan)
DALYs: disability adjusted life years
DSM: *Diagnostic and Statistical Manual of Mental Disorders*
DSPD: dangerous and severe personality disorder
EBC: European Brain Council
ECNP: European College of Neuropsychopharmacology
EEG: electroencephalography
ENSN: European Neuroscience and Society Network
ESF: European Science Foundation
ESRC: Economic and Social Research Council (U.K.)
FDA: Food and Drug Administration (U.S.)
fMRI: functional magnetic resonance imaging
GWAS: genome-wide association studies
IBRO: International Brain Research Organization
ICD: International Classification of Diseases
LSD: lysergic acid diethylamide
MAOA: monoamine oxidase A
MCI: mild cognitive impairment
MIT: Massachusetts Institute of Technology
MRC: Medical Research Council (U.K.)
MRI: magnetic resonance imaging

NESARC: National Epidemiologic Survey on Alcohol and Related Conditions (U.S.)
NGBRI: not guilty by reason of insanity
NHS: National Health Service (U.K.)
NIDA: National Institute on Drug Abuse (U.S.)
NIH: National Institutes of Health (U.S.)
NIMH: National Institute of Mental Health (U.S.)
NMR: nuclear magnetic resonance
NRP: Neurosciences Research Program (U.S.)
OED: *Oxford English Dictionary*
PET: positron emission tomography
RSA: Royal Society for the Encouragement of Arts, Manufactures and Commerce (U.K.)
SfN: Society for Neuroscience
SNPs: single nucleotide polymorphisms (pronounced "snips")
SPECT: single photon emission computed tomography
SSRI: selective serotonin receptor inhibitor
UC Berkeley: University of California, Berkeley
UCLA: University of California, Los Angeles
WHO: World Health Organization
YLDs: years lived with disability

Neuro

. . . .

Introduction

What kind of beings do we think we are? This may seem a philosophical question. In part it is, but it is far from abstract. It is at the core of the philosophies we live by. It goes to the heart of how we bring up our children, run our schools, organize our social policies, manage economic affairs, treat those who commit crimes or whom we deem mentally ill, and perhaps even how we value beauty in art and life. It bears on the ways we understand our own feelings and desires, narrate our biographies, think about our futures, and formulate our ethics. Are we spiritual creatures, inhabited by an immaterial soul? Are we driven by instincts and passions that must be trained and civilized by discipline and the inculcation of habits? Are we unique among the animals, blessed or cursed with minds, language, consciousness, and conscience? Are we psychological persons, inhabited by a deep, interior psyche that is shaped by experience, symbols and signs, meaning and culture? Is our very nature as human beings shaped by the structure and functions of our brains?

Over the past half century, some have come to believe that the last of these answers is the truest—that our brains hold the key to whom we are. They suggest that developments in the sciences of the brain are, at last, beginning to map the processes that make our humanity possible—as individuals, as societies, and as a species. These references to the brain do not efface all the other answers that contemporary culture gives to the question of who we are. But it seems that these other ways of thinking of ourselves—of our psychological lives, our habitual activities, our social relations, our ethical values and commitments, our perceptions of others—are being reshaped. They must now be grounded in one organ of our bodies—that spongy mass of the human brain, encapsulated by the skull, which weighs about three pounds in an adult and makes up about 2% of his or her body weight. This 'materialist' belief has taken a very material form. There has been a rapid growth in investment of money and human effort in neurobiological research, a remarkable increase in the numbers of papers published in neuroscience journals, a spate of books about the brain for lay readers, and many well-publicized claims that key aspects of human affairs can and should be governed in the light of neuroscientific knowledge. A host of neurotechnologies have been invented—drugs, devices, techniques—that seem to open ourselves up to new strategies of intervention through the brain. In the societies of what we used to call the West, our brains are becoming central to understanding who we are as human beings.[1]

For many in the social and human sciences, these developments are profoundly threatening. Their unspoken premise, for at least the past century, has been that human beings are *freed* from their biology *by virtue of that biology*—that we come into the world unfinished and that our individual capacities, mores, values, thoughts, desires, emotions (in short, our mental lives), as well as our group identities, family structures, loyalties to others, and so forth are shaped by upbringing, culture, society, and history. Practitioners of these disciplines can point, with good reason, to the disastrous sociopolitical consequences of the biologization of human beings: from eighteenth-century racial science to twentieth-century eugenics, and more recently to the reductionist simplifications of sociobiology and evolutionary psychology. To give priority to the biological in human affairs, it seems, is not only to ignore what we have learned from more than two centuries of social, historical, and cultural research, but also to cede many of the hard-won disciplinary and institutional achievements of the social and human sciences to others, and to risk their sociopolitical credibility.

We are sympathetic to these anxieties, and there is much truth in them. One only has to glance at the wild overstatements made by many of the popular writers in this field, the recurring overinterpretation of the findings from animal research in neurobiology, the sometimes willful misrepresentation of the significance of the images generated by brain scanning, not to mention the marketing on the Internet of many dubious products for improving brain power, to realize that there is much scope for critical sociological analysis and for cultural investigation of the contemporary lure of the brain sciences. But in this book, while we certainly seek to develop tools for a critical relation to many of the claims made in both serious and popular presentations of neuroscience, we also seek to trace out some directions for a more affirmative relation to the new sciences of brain and mind.

We do this for two reasons. On the one hand, there is no reason for those from the social and human sciences to fear reference to the role of the human brain in human affairs, or to regard these new images of the human being, these new ontologies, as fundamentally threatening. These disciplines have managed to live happily with the claims of psychoanalysis and dynamic psychologies, even though these disciplines also see much of human individual mental life and conduct as grounded in processes unavailable to our consciousness. Perhaps, then, we should not be so wary of the reminder that we humans are, after all, animals—very remarkable ones, indeed, but nonetheless not the beneficiaries of some special creation that sets us in principle apart from our forebears.[2] And, on the other hand, the new brain sciences share much with more general shifts within contemporary biological and biomedical sciences: at their most sophisticated, they are struggling toward a way of thinking in which our corporeality is in constant transaction with its milieu, and the biological and the social are not distinct but intertwined. Many of the assumptions and extrapolations that are built into the heterogeneous endeavor of neurobiology are ripe for critique. But at a time when neurobiology, however hesitantly, is opening its explanatory systems to arguments and evidence from the social sciences, perhaps there is a relation beyond commentary and critique that might be more productive. We will return to these issues in our conclusion. For now, though, it is sufficient to say that it is in the spirit of critical friendship between the human sciences and the neurosciences that we have written this book.

Beyond Cartesianism?

Mind is what brain does. This little phrase seems to encapsulate the premise of contemporary neurobiology. For many, it now merely states the obvious. But it was not always so. In 1950, the BBC broadcast a series of talks titled *The Physical Basis of Mind*, introduced by the eminent neurophysiologist Sir Charles Sherrington, with contributions from leading philosophers, psychiatrists, and neurologists (Laslett 1950). In his opening remarks, Sherrington pointed out that half a century earlier he had written, "We have to regard the relation of mind to brain as still not merely unsolved, but still devoid of a basis for its very beginning"; in 1950, at the age of ninety-two, he saw no reason to change his view: "Aristotle, two thousand years ago, was asking how the mind is attached to the body. We are asking that question still."[3] Whatever their differences, all the distinguished contributors to this series agreed that this debate over the relation between mind and body, between mind and brain, had lasted many centuries, and that it was unlikely that a consensus would soon be reached as to whether there *was* a physical basis for the mind in the brain, let alone what that basis was, or where it was, or how such a basis should be conceptualized.[4]

A half century later, Vernon Mountcastle, celebrated for his fundamental discoveries about the structure of the cerebral cortex, contributed the introductory essay on "Brain Science at the Century's Ebb" to a special issue of *Daedalus* (the journal of the American Academy of Arts and Sciences) devoted to the state of brain research. "The half-century's accumulation of knowledge of brain function," he wrote, "has brought us face to face with the question of what it means to be human. We make no pretention that solutions are at hand, but assert that what makes man human is his brain. . . . Things mental, indeed minds, are emergent properties of brains" (Mountcastle 1998, 1).[5] Minds are properties of that organ of the body that we term the brain. And brains makes humans human, because the minds that constitute their humanity emerge from their brains. Mountcastle spoke here for most of those working in the field that had come to call itself neuroscience. In a way that they did not quite understand and yet that they could not doubt, the human mind did indeed have a physical basis in the human brain. And that brain, however remarkable and complex, was an organ like any other organ, in principle open to neuroscientific knowledge: the 'explanatory gap' between the processes of brain and the processes of mind they somehow produce was beginning to narrow and would, in time, be closed.

Of course, many would say, there is nothing much new here. Have we not known that the brain is the seat of consciousness, will, emotion, and cognition since the Greeks? Closer to our present, from the nineteenth century onward, especially in Europe and North America, there was an intense focus on the significance of the brain to human character, to human mental pathologies, and to the management of the moral order of society: while many may now scoff at the attempts of the phrenologists to read intellectual and moral dispositions in the shape and contours of the skull, few would dispute the pioneering work on brain anatomy and function memorialized in the brain areas designated by the names of Wernicke, Broca, Flechsig, and their colleagues (Hagner 1997, 2001; Hagner and Borck 2001). If we wanted further evidence that there was nothing new about the salience long accorded to brain research, we could point to the fact that in the first six decades of the twentieth century more than twenty scientists were honored by the award of a Nobel prize for discoveries concerning the nervous system—from Santiago Ramon y Cajal in 1906 to John Eccles, Alan Hodgkin, and Andrew Huxley in 1963.

Nor is there anything particularly novel in the challenge that contemporary neuroscientists mount to dualism. For example, in the early decades of the twentieth century, Charles Sherrington sought to develop an integrated theory of brain and mind, and this was the prelude to a host of neurological, psychological, and philosophical attempts to clarify the mind-body relation; it also led to a host of worries about the implications for the higher human values of morality, autonomy, wholeness, and individuality (R. Smith 2002). Like their contemporaries today, neurologists and brain researchers in the first half of the twentieth century certainly believed—and claimed— that their research had uncovered mechanisms of the brain that would have major social implications. The gradual acceptance of the usefulness of the electroencephalograph in the 1930s, and the image of the electrical brain that it seemed to embody, appeared to some (notably William Grey Walter) to offer the possibility of objective diagnoses of psychiatric conditions, and indeed of normal characteristics; it was thought of "as a kind of truth machine or electrical confessional" that would reveal the workings of the human mind and enable public access to private mental life, and also have implications for the management of everything from child rearing to the choice of marriage partners (R. Hayward 2002, 620–21ff.). Perhaps, then, when neuroscientist Michael Gazzaniga titled his recent book *Human: The Science behind What*

Makes Us Unique (Gazzaniga 2008a), claiming that advances in research on the brain will reshape our understandings of who we are, he was only the latest in a long tradition.

Yet, as the twenty-first century began, there was a pervasive sense, among the neuroscientific researchers themselves, among clinicians, commentators, writers of popular science books, and policymakers that advances in our understanding of the human brain had implications that were nothing short of revolutionary. Much had indeed changed in the fifty years between Sherrington's pessimism and Mountcastle's optimism. By the end of the twentieth century, the term *neuroscience* slipped easily off the tongue, yet it dates only to 1962.[6] The Society for Neuroscience (SfN) was formed in 1969 and held its first major conference in 1979, which was attended by about 1,300 people; by 1980 it attracted about 5,800 people; by 1990 this number had grown to more than 13,000; and by 2000 it reached more than 24,000.[7] Alongside this annual event there were now dozens of other conferences and workshops organized by more specialist associations of brain researchers, each with its own membership, websites, and newsletters, along with undergraduate and graduate programs in neuroscience, 'boot camps' for those who sought a rapid immersion in the field, and much more. These activities were not confined to the United States but spanned Europe, Japan, China, and many other countries. This was not a unified field: there were many different formulations of the problems, concepts, experimental practices, professional allegiances, and so forth. But nonetheless, by the start of the twenty-first century, a truly global infrastructure for neuroscience research had taken shape.

These organizational changes were accompanied by a remarkable burgeoning of research and publishing. While in 1958 there were only some 650 papers published in the brain sciences, by 1978 there were more than 6,500. By 1998 this figure had risen to more than 17,000, and in 2008 alone more than 26,500 refereed papers were published on the neurosciences in more than four hundred journals.[8] In the wake of the decade of the 1990s, which U.S. President George Bush designated "the decade of the brain," things seemed to shift into a new phase, with discussions of the crucial role of the brain for individuals and society in the light of advances in neuroscience moving from the specialized literature into a wider domain. In the subsequent ten years, dozens of books were published suggesting that we have witnessed the birth of new sciences of brain and/or mind, and drawing on research findings to illustrate these claims (Andreasen 2001; Kandel 2006; Restak 2006; Iacoboni 2008; Rizzolatti and Sinigaglia 2008; Begley 2009; Lynch and Laursen 2009).[9] These books, and the regular newspaper articles and television programs about these discoveries and their importance, are now almost always accompanied by vibrant visual illustrations derived from brain imaging of the living brain in action as it thinks, feels, decides, and desires (Beaulieu 2002): the brain has entered popular culture, and mind seems visible in the brain itself.

Governing through the Brain

By the turn of the century, it seemed difficult to deny that the neurosciences had, or should have, something to say about the ways we should understand, manage, and treat human beings—for practices of cure, reform, and individual and social improvement. Across the first half of the twentieth century, the prefix *psy-* was attached to a great many fields of investigation of human behavior, seeming to link expertise and authority to a body of objective knowledge about human beings (Rose 1989); now the prefix *neuro-* was being invoked in the same way. Psychiatry was an obvious niche, not least because of the belief, since the 1950s, that new pharmacological treatments had been discovered that were effective because they acted directly on the neurobiology underpinning mental disorder. While the term *neuropsychiatry* had been used as early as the 1920s (for example, Schaller 1922) and gained popularity in the European literature in the 1950s (for example, Davini 1950; Garrard 1950; Hecaen 1950), by the start of the twenty-first century the term was being used in a very specific sense—to argue that the future of psychiatry lay in the integration of insights from genetics and neurobiology into clinical practice (Healy 1997; J. Martin 2002; Sachdev 2002, 2005; Yudofsky and Hales 2002; Lee, Ng, and Lee 2008).

But while psychiatry might seem an obvious niche for neuroscience, it was not alone in using the *neuro-* prefix to designate a novel explanatory framework for investigating phenomena previously understood in social, psychological, philosophical, or even spiritual terms.[10] Thus we now find *neurolaw*, which, especially in its U.S. version, claims that neuroscientific discoveries will have profound consequences for the legal system, from witness interrogation to ideas about free will and programs of reform and prevention; the first papers proposing this term appeared in the mid-1990s (Taylor, Harp, and Elliott 1991; Taylor 1995, 1996). We encounter *neuroeconomics*, which argues for the importance of studying the neurobiological underpinnings of economic behavior such as decision making.[11] We read of *neuromarketing* (*Lancet* 2004; Lee, Broderick, and Chamberlain 2007; Renvoisé and Morin 2007; Senior, Smythe, Cooke, et al. 2007); *neuroaesthetics* (Zeki 1993, 1999), which concerns the neuronal basis of creativity and of perceptions of beauty; *neuroergonomics*, which studies brain and behavior at work; *neurophilosophy* (Churchland 1986, 1995); and *neurotheology*, or the neuroscience of belief and spirituality (Trimble 2007).

For some, even the capacity to think ethically, to make moral judgments, is a brain kind of thing (see for example, Greene, Sommerville, Nystrom, et al. 2001; Tancredi 2005; Koenigs, Young, Adolphs, et al. 2007). And, reuniting apparent rivals for a knowledge of the human mind, we find *neuropsychoanalysis*; proponents always remind their readers that Freud was a neurologist and hoped for just such an integration in his early *Project for*

a Scientific Psychology (Bilder and LeFever 1998). We see *neuroeducation*—
thus Johns Hopkins University established an initiative bringing together
educators and brain science researchers to "magnify the potential for current findings to enrich educational practice," and the University of London
launched its own platform on "educational neuroscience."[12] We see *social
neuroscience*, which, in the words of the journal of that name, "examine[s]
how the brain mediates social cognition, interpersonal exchanges, affective/
cognitive group interactions [and] the role of the central nervous system
in the development and maintenance of social behaviors."[13] Centers, institutes, and laboratories in social neuroscience were established at the Max
Planck Institute, New York University, the University of Chicago, UCLA,
Columbia University, and elsewhere.

For others, the new brain sciences had important implications for the
reform of social policy and welfare (cf. Blank 1999). In 2009, the United
Kingdom's Institute for Government was commissioned by the Cabinet
Office to produce a report on the implication of neuroscience for public
policy.[14] The following year the nation's Royal Society—the oldest scientific
academy in the world—launched a project called Brain Waves to investigate developments in neuroscience and their implications for policy and
for society.[15] In 2009, the French government's *Centre d'Analyse Stratégique*
launched a new program dedicated to inform public policy based on neuroscientific research.[16] And there is much more of the same. Hence, perhaps inevitably given the contemporary ethicalization of biomedical matters, we have seen the rise of a new professional enterprise for worrying
about all this: *neuroethics* (Marcus 2002; Moreno 2003; Kennedy 2004; Illes
2006; Farah 2007; Levy 2007). It appears that to understand what is going
on when people engage in social interactions with one another, when they
feel empathy or hostility, when they desire products and buy goods, when
they obey rules or violate laws, when they are affected by poverty or child
abuse, when they do violence to others or themselves, and indeed when
they fall in love or are moved by works of art, we should turn to the brain.

What are we to make of all this? How has it come about that in the
space of half a century, the neurosciences have become such a repository
of hope and anticipation? How have they emerged from the laboratory and
the clinic, and have not only entered popular culture, but have become
practicable, amenable to being utilized in practices of government? And
with what consequences? We know that there are close linkages between
the ways in which human beings are understood by authorities, and the
ways in which they are governed. The various psychological conceptions
of the human being in the twentieth century had a major impact on many
practices: on understanding and treatment of distress; on conceptions of
normality and abnormality; on techniques of regulation, normalization,
reformation, and correction; on child rearing and education; on advertising, marketing, and consumption technologies; and on the management of

human behavior in practices from the factory to the military. Psychological languages entered common usage across Europe and North America, in Australasia, in Latin America, and in many other countries. Psychological training affected professionals from child guidance counselors and social workers to human resource managers. In the process, our very ideas of our selves, identity, autonomy, freedom, and self-fulfillment were reshaped in psychological terms (Rose 1979, 1985, 1988, 1989, 1996, 1999; Miller and Rose 1988, 1990, 1994, 1995a, 1995b, 1997; K. Danziger 1990; Hacking 1995).

Will these developments in the neurosciences have as significant a social, political, and personal impact? If we now consign the Cartesian split of mind and body to history and accept that mind is nothing more than what the brain does, does that mean that neuroscientists, after years of toiling in relative obscurity, are poised to become nothing less than "engineers of the human soul"?[17] It is undoubtedly too early for a considered judgment; it is far from clear what we would see if we were to look back on these events from the twenty-second century. Despite all the grand promises and expectations generated by neuroentrepreneurs, we cannot know for certain whether any lasting new bodies of expertise will emerge, nor can we foretell the role of neurobiology in the government of conduct across the next decades.

In this book, we abstain from speculation wherever possible. We also seek to distance ourselves from the overgeneralized critiques of 'neuromania' and other fundamentally defensive reactions from the social and human sciences. For while we raise many technical and conceptual problems with these new ways of thinking and acting, and point to many premature claims and failed promises of translation from laboratory findings to treatments and policies for managing human miseries and ailments, we also find much to appreciate in many of these attempts to render human mental life amenable to explanation and even to intervention.

And, unlike many of our disciplinary colleagues, we do not think that the social and human sciences have anything to fear, provided that they maintain an appropriate critical awareness, from our new knowledge of the brain. The *Oxford English Dictionary* (*OED*) provides a definition of criticism that matches our aims. Rather than fault finding or passing censorious judgment, we are critical here in the sense of "exercising careful judgement or observation; nice, exact, accurate, precise, punctual." It is in that critical spirit that we aim to describe the new ways of thinking about the nature of the human brain and its role in human affairs that are taking shape, to consider the problems around which these have formed and the conceptual and technical conditions that have made it possible to think in these new ways, and to analyze the ways in which these developments have been bound up with the invention of novel technologies for intervening upon human beings—governing conduct though the brain, and in the name of the brain.

Our Argument

In the course of the intertwined investigations that make up this book, we argue that a number of key mutations—conceptual, technological, economic, and biopolitical—have enabled the neurosciences to leave the enclosed space of the laboratory and gain traction in the world outside. It may be helpful to summarize these here, to help guide readers through the rather detailed analyses contained in the following chapters.[18]

Concepts and Technologies

Over the course of the half century that we focus on in this book, the human brain has come to be anatomized at a molecular level, understood as plastic and mutable across the life-course, exquisitely adapted to human interaction and sociality, and open to investigation at both the molecular and systemic levels in a range of experimental setups, notably those involving animal models and those utilizing visualization technologies. This has generated a sense of human neurobiology as setting the conditions for the mental lives of humans in societies and shaping their conduct in all manner of ways, many of which are not amenable to consciousness. Each of the major conceptual shifts that led to the idea of the neuromolecular, plastic, and visible brain was intrinsically linked to the invention of new ways of intervening on the brain, making possible new ways of governing through, and in the name of, the brain. Yet despite the ontological changes entailed, and the emerging belief that so much of what structures human thoughts, feelings, desires, and actions is shaped by nonconscious neurobiological processes, few of those who work in this area believe that humans are mere puppets of their brains, and the emerging neurobiologically informed strategies for managing human conduct are rarely if ever grounded in such a belief. Neurobiological conceptions of personhood are *not* effacing other conceptions of who we are as human beings, notably those derived from psychology. On the contrary, they have latched on to them in the many sites and practices that were colonized by psychology across the twentieth century—from child rearing to marketing, and transformed them in significant ways. In this way, and through these processes, our contemporary 'neurobiological complex' has taken shape.[19] Let us say a little more about some of these developments.

The central conceptual shift that we chart in the chapters that follow is the emergence of a neuromolecular vision of the brain. By this we mean a new scale at which the brain and nervous system was conceptualized, and a new way in which their activities were understood. At this molecular level, the structure and processes of the brain and central nervous system were made understandable as material processes of interaction among molecules in neurons and the synapses between them. These were conceived in terms of the biophysical, chemical, and electrical properties of their constituent parts. At this scale, in a profoundly reductionist approach, despite the recognition that there was much that could not yet be explained, there seemed nothing mysterious about the operations of the nervous system. Mental processes— cognition, emotion, volition—could be explained in entirely material ways, as the outcome of biological processes in the brain, understood as an organ that was, in principle, like any other, even if, in the case of humans and many other animals, it was far more complex than any other organ. While the explanatory gap still remained, and the move from the molecular level to that of mental processes was highly challenging, the dualism that had haunted philosophy and the sciences of mental life increasingly seemed anachronistic.

The project of neuroscience—for it was indeed an explicit project to create interactions between researchers from the whole range of disciplines that focused on the brain, from mathematics to psychology—had as its aim to revolutionize our knowledge of the brain, and in so doing, radically to transform our capacities to intervene in it. One key transactional point was psychiatric pharmacology—that is to say, the development of pharmaceuticals to treat mental disorder. The emergence of this neuromolecular gaze was intrinsically intertwined with the development of psychopharmacology and the increasing resort to drugs for treating people diagnosed with mental illness, first within, and then outside the asylum walls. Many key findings about molecular mechanisms were made in the course of trying to identify the mode of action of those drugs, almost always using animal models. Indeed, we argue that animal models were epistemologically, ontologically, and technologically crucial to the rise of neuroscience. Research using such models focused on the molecular properties of drugs that appeared to act on mental states and behavior, and hence almost inescapably led to the belief that the anomalies in those mental states could and should be understood in terms of specific disturbances, disruptions, or malfunctions in neuromolecular processes. Since the drugs seemed to affect the components of neurotransmission, this led both to the triumph of the chemical view of neurotransmission over the electrical view that had previously been dominant, and to the belief that malfunctions in neurotransmission underpinned most if not all mental disorders.

The two founding myths of psychopharmacology—the monoamine hypothesis of depression and the dopamine hypothesis of schizophrenia provided ways of organizing these linkages conceptually and technologically. Both have proved mistaken, perhaps fundamentally wrong. However, this 'psychopharmacological imaginary' enabled the growth of novel transactions between laboratory, clinic, commerce, and everyday life. In particular, it was linked to the growing associations between the pharmaceutical companies, the neurobiological research community, and the profession of psychiatry. It was associated with many inflated statements about the therapeutic potency of the compounds being produced and marketed, with the growing routinization of the use of psychoactive drugs that claimed to be able to manage the travails of everyday life by acting on the brain, and with the reshaping of distress in ways that might best accord with the vicissitudes of an increasingly competitive and profitable market for pharmaceuticals.

There were, of course, many who were critical of these new relationships. Critics denounced the medicalization of social problems, linked this to an analysis of psychiatry—and in particular of biological psychiatry—as an apparatus of social control, and argued that the profession, its explanatory claims, its diagnostic categories, and its preference for drugs as a first line of intervention, resulted from its capture by the pharmaceutical industry.[20] For many of these critics, aware of the doleful history of eugenics, genetic explanations of mental disorders were particularly distasteful. Despite the certainty of psychiatric geneticists that mental disorders had a genetic basis, critics correctly pointed out that the repeated claims to have discovered 'the gene for' schizophrenia, manic depression, and many other conditions were always followed by failures of replication. However as the twentieth century came to a close, a radical transformation in the styles of thought that characterized genetics made a different approach possible.[21] This focused on variations at a different level—at the level of changes in single bases in the DNA sequences themselves, and the ways in which such small variations in the sequence might affect the nature of the protein synthesized or the activity of the enzyme in question, with consequences for susceptibility to certain diseases or response to particular drugs.

This molecular vision of genomic complexity thus mapped onto the vision of the neuromolecular brain. It thus became possible to move beyond studies of heritability in lineages and families to seek the specific genomic variants and anomalies that had consequences for susceptibility to certain diseases or pathological conditions such as impulsive behavior. One now looks for the variations that increase or decrease the activity of an enzyme, the operation of an ion channel, or the sensitivity of a receptor site, and which, in all their multiple combinations, underpin all differences in human mental functioning, whether these be deemed normal variations or pathologies. Further, one tries to locate these within the environmental or other conditions that provoke or inhibit the onset of such conditions. As we have moved to such a neurogenomics of susceptibilities and resiliencies, new translational possibilities appear to emerge for neuroscience to engage with strategies of preventive intervention in the real world, whether via early identification and treatment of mental disorder or of neurodegenerative diseases, or in enabling preventive intervention to steer children from a pathway that will lead to antisocial behavior and crime.

Alongside psychopharmacology and psychiatric genomics, there was a third pathway, equally significant in our view, for the transactions between the knowledge of the brain and interventions in human lives—the growing belief that, at least when it comes to the human brain, neither structure nor function is inscribed in the genes or fixed at birth. One term has come to designate this new way of thinking—*plasticity*. The neural architecture of the brain was now located in the dimension of time—not just the time of development from fertilization to birth and into the early years of life, but also throughout the life-course, through adolescence, into adulthood, and indeed across the decades. While it had long been recognized that plasticity existed at the level of the synapse—that synaptic connections constantly formed or were pruned in response to experience—new ideas of plasticity were taken to mean that a wider 'rewiring' was also possible. Notable, here, were the results of work on rehabilitation after stroke in humans, and related work with animal models, which showed that the primate brain could remap itself after injury and that this process could be accelerated by neurobiologically informed practices of rehabilitation—an argument that was commercialized in the development of a number of therapeutic methods, often patented by the neuroscientific researchers themselves.

At the other end of life, researchers argued that experience in the very early days and months following birth, perhaps even in utero, shaped the brain in fundamental ways through modifying gene methylation. Epigenetic arguments sought to establish the ways in which experience 'gets under the skin' at the level of the genome itself. In particular, it seemed, early maternal behavior toward offspring might so shape their neural development to affect not only the behavior of offspring over their whole life span, but also their own maternal behavior. There now seemed to be a mechanism to pass these environmentally acquired characteristics of the brain down the generations. And finally, the long-held dogma that no new neurons were produced after the first years of life was itself overturned with the finding that in humans, neurogenesis or the growth of new nerve cells in the brain, was possible throughout adult life and might be stimulated or inhibited by environmental factors from nutrition to cognitive activity. No matter that many doubts remained about the translation of these findings from animals to humans, and the interpretation of these results. The brain now appeared as an organ that was *open* to environmental inputs at the level of the molecular processes of the genome, shaping its neural architecture and its functional organization, with consequences that might flow down the generations. The implications were clear: those who were concerned about the future of our children, and the conduct and welfare of the adults they would become, needed to recognize, and to govern, these processes of shaping and reshaping our plastic brains.

If these three imaginaries—of pharmacology, neurogenomics, and neuroplasticity—provided pathways linking the work of brain labs to interventions in the everyday world, so too did a fourth: the visual imaginary, associated in particular with the development of powerful technologies of brain imaging. While the skull initially proved an impenetrable barrier to techniques of medical imaging such as X-rays, there were early attempts—notably by Edgar Adrian—to explore the electrical activity of the living brain using electroencephalography (Adrian and Matthews 1934). However, the fundamental shift in the visibility of the living brain was linked to the development of computerized tomography (CT) scanning in the 1970s and magnetic resonance imaging (MRI) in the 1980s. These produced images of the structure and tissues of the brain that were, to all intents and purposes, equivalent to the images produced of any other bodily tissues. They were simulations, of course, not simple photographs, but they were open to confirmation by physical interventions into the imaged tissues.

Two further developments, positron emission tomography (PET) and functional magnetic resonance imaging (fMRI), seemed to produce identical images of something with a very different ontological status—not the *structure* of the brain but its *functioning*, its activity as its human host engaged in certain tasks or experienced certain emotions. We seemed to be able to see the neural correlates of the activities of mind itself in real time. And once we did, it seemed impossible to doubt that mind is what brain does. As these technologies became more widely available to researchers, thousands of papers were published claiming to identify the neural correlates of every human mental state from love to hate, from responses to literature to political allegiances; by 2011, such publications were appearing at a rate of about six hundred a month. Despite the well-known technical problems, assumptions, and limitations of these technologies, and the fact that they do not speak for themselves and must be interpreted by experts, the images have undoubted powers of persuasion, and their apparent ability to track mental processes objectively, often processes outside the awareness of the individual themselves, have proved persuasive in areas from neuromarketing to policies on child development. The belief that we can see the mind in the living brain, can observe the passions and its desires that seemingly underlie normal and pathological beliefs, emotions, and behaviors, has been a key element in the claim that neuroscience can provide useful information about the government of human beings, the conduct of their conduct in the everyday world.

Governing the Future—through the Brain

We should beware of scientific or technological determinism. Truths and technologies make some things possible, but they do not make them inevitable or determine the sites in which they find a niche. Different societies, cultures, and sociopolitical configurations offer different opportunities for the new brain sciences. Nonetheless, there is one feature of contemporary biopolitics that has proved particularly welcoming to the image of the molecular, visible, and plastic brain—that which concerns the future. Contemporary biopolitics is infused with futurity, saturated with anticipations of imaged futures, with hope, expectation, desire, anxiety, even dread. The future seems to place a demand not just on those who govern us but also on all those who would live a responsible life in the present (O'Malley 1996).

No doubt this widespread sense of obligation to take responsibility for the future is not unique to advanced liberal democracies in the early twenty-first century. Biopolitics, since at least the eighteenth century, has been future-oriented. From earliest politics of the population, governing vitality operates on axis of time and orients to the future, and images of the future are intrinsic to biopolitical thought and strategies from the politics of health in the eighteenth century, to concerns with the degeneracy of the population in the nineteenth century, through the rise of eugenics and the birth of strategies of social insurance in the first half of the twentieth century. Today we are surrounded by multiple experts of the future, utilizing a range of technologies of anticipation—horizon scanning, foresight, scenario planning, cost-benefit analyses, and many more—that imagine those possible futures in different ways, seeking to bring some aspects about and to avoid others. It would not be too much of an exaggeration to say that we have moved from the risk management of almost everything to a general regime of futurity. The future now presents us neither with ignorance nor with fate, but with probabilities, possibilities, a spectrum of uncertainties, and the potential for the unseen and the unexpected and the untoward. In the face of such futures, authorities have now the obligation, not merely to 'govern the present' but to 'govern the future.'[22] Such futurity is central to contemporary problematizations of the brain.

One feature of these imagined futures is the growing burden of brain disorder. Public funding for research in the new brain sciences, not just in the so-called Decade of the Brain but from at least the 1960s, when initiatives such as the Neurosciences Research Program (NRP) were established,[23] has almost always been linked the belief that conquering this new frontier of the brain will, in an unspecified time line, lead to major advances in tackling that burden.[24] The idea of *burden*, here, has disturbing resonances for those who know their history (Proctor 1988; Burleigh 1994). Nonetheless, in a very different sociopolitical context than that of eugenics, psychiatrists, lobby groups, and international organizations make dire predictions about the rising numbers of those in the general population who suffer from depression and other mental disorders, not to mention the "dementia time bomb": a recent estimate was that, in any one year, more than one-third of the population of the European Union could be diagnosed with such a brain disorder (Wittchen, Jacobi, et al. 2011). In the emerging style of thought that we trace in this book, brain disorders encompass everything from anxiety to Alzheimer's disease, and often include both addictions and obesity—all, it seems, have their origin in the brain. These disorders, the majority of which are undiagnosed, lead to many days lost from work and many demands on medical and other services, costing European economies hundreds of billions of euros.[25] The corollary seems obvious: to reduce the economic burden of mental disorder, one should focus not on cure but on prevention. And prevention means early intervention, for the sake of the brain and of the state.

Children are the key—children who are at risk of mental health problems as they grow up. Many pathologies—ADHD, autism, schizophrenia, bipolar disorder, dementias—are now reframed as *developmental* and hence amenable to early detection and ideally to preventive intervention. This logic can then be extended from mental disorders to antisocial conduct, resulting in the attempt to discover biomarkers in the brain or in the genes of young children that might predict future antisocial personality or psychopathy. In this logic, one first identifies susceptibility, and one then intervenes to minimize the chances of that unwanted eventuality coming about, in order to maximize both individual and collective well-being and to reduce the future costs of mental health problems. Earlier is almost always better—as the mantra has it. Earlier usually means during childhood, because the brain of the developing child is more 'plastic,' believed to be at its most open to influences for the good (and for the bad)—and hence leads to intensive interventions in the parenting of those thought to be potentially at risk. This is the rationale of "screen and intervene" (Singh and Rose 2009; Rose 2010b). Neuroscientifically based social policy thus aims to identify those at risk—both those liable to show antisocial, delinquent, pathological, or criminal behavior and those at risk of developing a mental health problem—as early as possible and intervene presymptomatically in order to divert them from that undesirable path.[26]

At the other end of life, many argue for early intervention to forestall the development of Alzheimer's disease and other dementias, which has led to regular announcements of tests claiming to identify those at risk, the rise of the prodromal category of "mild cognitive impairment," the growing number of "memory clinics" to diagnose such brain states and prescribe interventions to ameliorate them, and much research, so far largely unsuccessful, to find effective forms of intervention into the dementing brain (discussed in Whitehouse and George 2008).

In the era of the neuromolecular and plastic brain, those who advocate such strategies think of neurobiology not as destiny but *opportunity*. Many believe that to discover the seeds of problematic conduct in the brain will reduce stigma rather than increase it, despite research showing the reverse (Phelan 2002, 2006). Those seeking biomarkers for psychopathy, even when they believe that there is a clear, genetically based, neurobiological basis for antisocial conduct, argue that neurobiology informs us about susceptibility but not inevitability. Their wish to identify the gene-environment interactions, which provoke vulnerability into frank psychopathy, is linked to a hope for protective strategies, for "the goal of early identification is successful intervention" (Caspi, McClay, Moffitt, et al. 2002; Kim-Cohen, Moffitt, Taylor, et al. 2005; Odgers, Moffitt, Poulton, et al. 2008).[27]

Interventions sometimes involve behavior therapy, cognitive therapy, and psychopharmaceuticals. But the preferred route to the problematic child—as so often in the past—is through the parents. In the age of the plastic brain, many undesirable neurobiological traits appear to be malleable by changing the ways parents deal with their vulnerable children (Dadds, Fraser, Frost, et al. 2005; Hawes and Dadds 2005, 2007). Such arguments for early intervention have been strengthened over recent years by the proliferation of brain images seeming to show the consequences of early adverse environments on the developing brain of the child (Perry 2008). On the one hand, these images provide powerful rhetorical support for early intervention into the lives of the most disadvantaged families, in the name of the individual, familial, and social costs of the developing brain, and hence future lives, of their children. On the other hand, in situating the origins of all manner of social problems and undesirable forms of conduct so firmly in neurobiology, even in a neurobiology that is itself shaped by environment, we see a repetition of a strategy that we have seen innumerable times since the nineteenth century—to prevent social ills by acting on the child through the medium of the family: a neurobiological explanation for the persistence of social exclusion in terms of a 'cycle of deprivation' grounded in the inadequate parenting provided by the socially deprived.

Economies of the Brain

These arguments about the burden—in this case the economic costs—of brain disorders and the increasing faith in the economic benefits to be gained through strategies of prediction and preventive medicine, have been one important factor for the growth of public investment in neuroscientific research.[28] The National Institutes for Health (NIH) in the United States, their equivalent in the U.K. research councils, the European Commission's Framework Programmes, the European Research Council, and the European Science Foundation have all invested in this work, as have private foundations and charitable bodies, such as the MacArthur Institute in the United States and the Wellcome Trust in the United Kingdom. Given the diversity of these funding sources, the total sums involved are hard to estimate. However, in the United States as of 2006, it was estimated that the combined commitment to neuroscience research of the NIH, the pharmaceutical industry, large biotech firms, and large medical device firms increased from $4.8 billion in 1995 to $14.1 billion in 2005—adjusted for inflation, this meant a doubling of investment, which was more or less in line with that for biomedical research as a whole.

In the case of neuroscience, over half of this investment came from industry, a proportion that remained around that level over that decade, despite the fact that investment in this area was not matched by an increase in new pharmaceuticals coming on to the market: the U.S. Food and Drug Administration (FDA) approved only "40 new molecular entities for indications within the neurosciences from 1995 to 2005, with the annual number of approvals remaining relatively stagnant during this period" (Dorsey, Vitticore, De Roulet, et al. 2006).[29] However, commenting on developments over the decade from 2000 to 2010, and into the near future, leading neuroscientists have expressed their fears that public funding for research in neuroscience in many major industrialized countries—notably the United States, the United Kingdom, and Japan—has not kept up with inflation and has been affected by the recession; they argue that this is producing something of a crisis and does not match the scale of the social costs of what they variously refer to as brain diseases, mental disorders, or brain disorders (Amara, Grillner, Insel, et al. 2011). In the United Kingdom, in early 2011, the British Neuroscience Association (BNA) estimated overall U.K. neuroscience research funding from public sources to be in the region of £200 million per year. In the United States, despite a one-time supplement to the NIH of $10 billion as a result of the American Recovery and Reinvestment Act of 2009, funding in the second decade of the century was expected to be flat.

Commercial investment in neuroscience is, of course, linked to the belief that there is a growing global market for the drugs, diagnostics, and devices that are being produced. Zack Lynch of NeuroInsights, perhaps the leading organization appraising the neurotechnology market, has estimated that the global value generated by the neurotechnology industry in 2009 was around $140 billion, slightly down from that in previous years (a fall attributed to the drop in cost of neuroimaging devices), but that over the previous decade the average annual growth had been on the order of 9%.[30] As we have already remarked, many critics have viewed the business strategies of pharmaceutical companies as a key factor in reshaping psychiatry toward neuropsychiatry, influencing its diagnostic systems and the use of psychiatric drugs as the main form of therapy. In fact, global neuropharmaceutical revenue actually slowed over 2009, largely as a result of the failure of new drugs to come onto the market and the replacement of patented products by generics, although there is now some evidence that the pharmaceutical companies are withdrawing investment for the development of novel psychiatric compounds that are based on the hypotheses of specificity of etiology and treatment that had informed their strategies since the 1960s.[31] But other key neurotechnology areas showed growth, including neurodiagnostic technologies ranging from brain scanners to biomarkers, and neurodevices ranging from cochlear implants, through brain stimulation technologies, to neuroprosthetics purporting to arrest memory decline and optimize attention.

We can see immediately that it would be misleading to separate the academic and the industrial components of this neuroeconomy: what Stephen Shapin has termed "the new scientific life" entails an entrepreneurial spirit on the part of both researchers and universities, and the search for intellectual property and the rhetoric of knowledge transfer are endemic (Shapin 2008). Thus, for example, David Nutt, president-elect of the BNA, commenting on the effects of the decisions by GlaxoSmithKline and Merck in 2011 to close their U.K. neuroscience research sites, with a loss of about one thousand jobs in neuroscience, remarked that it not only "removes the only major site of job opportunities for neuroscience graduates and postdoctoral researchers outside academia" but "the company pull-out directly impacts research spend[ing] in neuroscience as many PhD students and postdoctoral researchers were funded by, or in partnership with, this industry."[32]

Indeed, such partnership funding has been a major element in the strategy of the U.K. Medical Research Council. Many of the start-up companies in the neuroindustry are initiated by research scientists with the assistance of venture capitalists, and in turn, their investments are fueled by expectations about future market growth and market opportunities generated by organizations such as IMS Health and NeuroInsights ("the neurotech market authority") whose own business models depend on their effective servicing of these expectations.[33] Running across all actors in this new configuration is a 'translational imperative': an ethos that demands and invests in research that is predicted to generate returns of investment—in therapies and in products. This translational imperative now inflects every serious research proposal or grant application in the United States, the United Kingdom, continental Europe, and many other regions. It has arisen, at least in part, because of failure: the failure of many hopes that an increased knowledge of basic biological processes of body and brain would lead inevitably to better therapies for individuals and valuable products for national bioeconomies. If that return in health and wealth, rather than advances in knowledge for its own sake, was the quid pro quo for public investment in basic research, it seemed that the researchers had not kept their side of the bargain. The response of the funding organizations—the NIH in the United States (Zerhouni 2003, 2005) and the MRC in the United Kingdom (Cooksey 2006; National Institute for Health Research 2008)—was to argue for a radical reshaping of the pathways from research to application, and a specific focus, organizationally and financially, on 'translational medicine.'[34]

Of course, there is no simple path from 'bench to bedside' or vice versa. It is rare for a single piece of research, or even the research program of a single group or lab, to translate on its own. The time frames over which research findings are integrated into novel therapies or products are often of the order of decades; the acceptance and utilization of research findings often depends more on social, political, and institutional factors than on the inherent productiveness of the research itself.[35] Nonetheless, we have seen the formation of a number of 'translational platforms'—sites of diverse material exchanges, from knowledge production to decision making and commercial transactions, from innovation to technical assemblages of material entities, where diverse agents and agencies, practices and styles of thought, discourses and apparatuses, converge in the name of the promissory benefits of translational neuroscientific research.[36] This is, of course, an agonistic space, especially when it comes to the role of pressure groups. Some such groups are strongly opposed to the neuro-biomedicalization of conditions such as autism or depression, while others seek to shape biomedical funding toward their own translational demands as a sign of the new democratization of biomedical research (Heath, Rapp, and Taussig 2004). Many activist pressure groups of patients or their families are funded in part by donations from those very

commercial corporations and pharmaceutical companies that stand to benefit from increased public awareness and concern about the disorders in question (Moynihan 2002, 2008; Moynihan and Cassels 2005). These new vital
economies are tangled webs and permit of no easy ethical judgments.

In the analyses that follow, we do not argue that there is something essentially malign in the intertwining of researchers' hopes for academic success,
hopes for a cure for one's loved ones, hopes for private financial advantage
for individual scientists and for companies, and hopes for public economic
benefits in terms of health. Indeed, these intertwinings characterize contemporary biomedicine in what Carlos Novas has termed our contemporary
"political economy of hope" (Novas 2006). But we do point to zones where
such entanglements may be highly contentious, for example, where there are
powerful but hidden financial links between researchers and those who stand
to benefit from the claims made in their research.[37]

Even where frank corruption is not entailed, we have seen significant
changes in the economics of research in the life sciences generally, which certainly extend to neuroscience. Steven Shapin quotes a writer in *Science* magazine in 1953: "The American scientist is not properly concerned with hours of
work, wages, fame or fortune. For him an adequate salary is one that provides
decent living without frills or furbelows. . . . To boil it down, he is primarily
interested in what he can do for science, not in what science can do for him."[38]
This description certainly seems to fit some of the key figures in the early
years of neuroscience, and it undoubtedly fits some of those working in the
field today. But it is doubtful if it characterizes those many others who would
be scientific entrepreneurs; even for those for whom the research itself is the
prize—the much-cited article, recognition by peers, career advancement—
the economy of neuroscience as a whole, and hence the comportment of most
of those who work in this field, has had to change.

In the United States, the passage of the Bayh-Dole Act in 1980 gave up federal rights to intellectual property that arose from research supported from
government funds, allowing those property rights and the value that might
flow from them to be claimed by universities or by individual scientists (Kennedy 2001).[39] The ostensible aim was to avoid important scientific discoveries
lying idle, and thus not contributing to national wealth and well-being. But
this act, coupled with other legal changes, opened biomedical research to a
flood of private investment and venture capital. Trends in the United States
were followed across much of the rest of the world, and the new opportunities were embraced enthusiastically by research universities, which set up
technology transfer offices, participated in start-up companies, and entered
into the murky realm of patenting, licensing, distribution of royalties, and
complex commercial relations with the corporate world.[40]

But many argued that the principles that should govern scientific knowledge were being compromised by the drive for intellectual property and
the growing entanglements between universities, researchers, industry, and

knowledge claims. For example, in 2002, the Royal Society of London set up a study to ask "whether the use of laws which encourage the commercial exploitation of scientific research is helping or hindering progress in fields such as genetics." The conclusion of its study was that "[i]ntellectual property rights (IPRs) can stimulate innovation by protecting creative work and investment, and by encouraging the ordered exploitation of scientific discoveries for the good of society. . . . [But] the fact that they are monopolies can cause a tension between private profit and public good. Not least, they can hinder the free exchange of ideas and information on which science thrives."[41]

While questions of patenting and intellectual property have been crucial in reshaping the neuroeconomy, we would argue that this intense capitalization of scientific knowledge, coupled with the other pressures on researchers to focus on, and to maximize, the impact of their research, has additional consequences. It exacerbates tendencies to make inflated claims as to the translational potential of research findings, and, where those potentials are to be realized in commercial products, to a rush to the market to ensure that maximum financial returns are achieved during the period of a patent. It produces many perverse incentives (Triggle 2004). These include the possibilities of 'corporate capture,' where universities, departments, or research centers are significantly funded by commercial companies that gain priority rights to patent and commercialize discoveries that are made.[42] They increase the already existing incentives on researchers selectively to report positive findings of research—an issue that is particularly relevant where the studies are funded by commercial companies. And, as we show in the chapters that follow, they can lead researchers (or the press releases issued by their university communications departments) to overclaim the generalizability of studies carried out with small samples and to imply that studies with animals will, very soon, lead to therapeutic developments for humans.[43]

But from our point of view, above and beyond these specific problems generated by the new forms of scientific life, there is a more general question about truth. For if one has a path-dependent theory of truth—if you believe as we do that it is difficult to make things become true in science, and among the necessities for making things become true today are the funds to enable the research to proceed—then the decisions by public, private, and commercial bodies as to which areas of research to fund, and the often unacknowledged intertwining of promises, hopes, anticipations, expectations, and speculations that are involved, play a key role in shaping our contemporary regimes of truth about persons and their mental lives.

Brains and Persons

Many critics have suggested that the rise of neurobiology is leading to a kind of reductionism in which mental states are reduced to brain states, human actions are generated by brains rather than conscious individuals, and the key dimensions of our humanness—language, culture, history, society—are ignored (Ehrenberg 2004; Vidal 2005; Vidal 2009; Tallis 2011). For some, this rests on a philosophical error: attributing to brains capacities that can only properly be attributed to people (Bennett and Hacker 2003). For others, it is the apotheosis of contemporary individualism—a turn away from social context to a vision of society as an aggregate of isolated individuals (Martin 2000, 2004). There is an analogy here with the concerns that social scientists expressed about 'geneticization' before and after the Human Genome Project. Dire warnings of genetic reductionism, genetic discrimination, and hereditary fatalism proved wide of the mark. And as with genomics, so with neuroscience. This is partly because of the findings of the research itself. But it is also because of the way such arguments are aligned with existing conceptions of personhood and regimes of self-fashioning in advanced liberal societies (Novas and Rose 2000; Rose and Novas 2004; Rose 2007c).

We argue that despite their apparent contradictions, neurobiological research emphasizing the role of nonconscious neural processes and habits in our decisions and actions can—and does—happily coexist with long-standing ideas about choice, responsibility, and consciousness that are so crucial to contemporary advanced liberal societies (Rose 1999). Such societies are premised on the belief that adult human beings, whatever the role of biology and biography, are creatures with minds, who have the capacity to choose and to intend on the basis of their mental states. Humans can be held accountable for the outcomes of those decisions and intentions, even when they are shaped by those nonconscious forces, except in specific circumstances (compulsion, mental disorder, brain injury, dementia, automatism, etc.).[44] Indeed, it is hardly radical to suggest that human beings are swayed by forces that come from beyond their consciousness. In most cultures and most human practices, individuals have believed in the importance of fates, passions, instincts and drives, unconscious dynamics, and the like. And, as a result, they have been urged, and taught, to govern these forces in the name of self-control, whether by spiritual exercises, by prayer and mortification, by the inculcation of habits, by learning how to govern one's will through inhibition,[45] by understanding the dynamics of projection and denial, by consciousness raising, or a multitude of other techniques.

Similarly, the recognition of nonconscious neurobiological factors in our mental lives does not lead policymakers to propose that we should resign ourselves before these neural forces. On the one hand, it leads them to believe that those who govern us should base their strategies on knowledge of these nonconscious neurobiological mechanisms, for example, by creating settings that make it easier for human beings to make the right decisions, with nonconscious cues that steer them in the desired directions (Thaler and Sunstein 2008). Likewise, they urge our authorities to nurture the nonconscious bonds of fraternity and good citizenship and to minimize those that weaken culture and character (Brooks 2011).[46] They suggest that our governors need to recognize the nonconscious roots of valued moral understandings such as empathy and the sense of fairness, and to reinforce them—to build social policies on the neurobiological evidence that our beliefs, actions, and aspirations are shaped more by our evolved emotional judgments and the habits that have become ingrained in our nonconscious neural pathways, than by conscious, rational deliberations. On the other hand, they argue that we as individuals should seek to manage these underlying processes, to become conscious of them, to develop mindfulness, to work responsibly to improve ourselves as persons, to reform our habits through acting on our brains: we have entered, they claim, the age of neurological reflexivity (Rowson 2011).

In this emerging neuro-ontology, it is not that human beings *are brains*, but that we *have brains*. And it is in this form—that our selves are shaped by our brains but can also shape those brains—that neuroscientific arguments are affecting conceptions of personhood and practices of self-fashioning. In the final decades of the twentieth century in the West, we saw the rise of a "somatic ethic" in which many human beings came to identify and interpret much of their unease in terms of the health, vitality, or morbidity of their bodies, and to judge and act upon their soma in their attempts to make themselves not just physically better but also to make themselves better persons (Rose 2007c). The contemporary cultural salience of the brain does not mark the emergence of a new conception of personhood as "brainhood" (as suggested by Vidal 2009), in which persons become somehow conceived of as identical to, or nothing more than, their brains. Rather, we are seeing this somatic ethic gradually extending from the body to the embodied mind—the brain.

The pedagogies of 'brain awareness' and the rise of practices and devices for working on the brain in the service of self-improvement thus find their locus within a more general array of techniques for working on the somatic self in the name of maximizing our well-being.[47] While it is true that much neuroscience presents a picture of brains as isolated and individualized, an alternative image is also taking shape. In this image, the human brain is evolved for sociality, for the capacity and necessity of living in groups, for the ability to grasp and respond to the mental states of others: human brains are both shaped by, and shape, their sociality. As various aspects of sociality are attributed to human brains, neurobiological self-fashioning can no longer so simply be criticized as individualistic and asocial. It is now for the *social* good that parents need to understand the ways in which their earliest interactions with their children shape their brains at the time when they are most plastic, to recognize the ways in which they learn to understand other minds, to enhance their capacities for empathy and the inherent emotional ability of their brains to respond positively to fairness and commitment to others, to maximize the mental capital and moral order of society as a whole. And it is in the name of improving the well-being of our societies that each of us is now urged to develop a reflexive understanding of the powers of these nonconscious determinants of our choices, our affections, our commitments:

in doing so, we will no longer be passive subjects of those determinants, but learn the techniques to act on them in order to live a responsible life. Once more, now in neural form, we are obliged to take responsibility for our biology, to manage our brains in order to bear the responsibilities of freedom.

Human Science?

What, then, of the relations between neuroscience and the human and social sciences? There is much to be critical of in the rise of these technologies of the neurobiological self. Experimental practices in laboratory settings still fail adequately to address the fact that neither animal nor human brains exist in isolation or can be understood outside their milieu and form of life. Conceptions of sociality in social neuroscience are frequently impoverished, reducing social relations to those of interactions between individuals, and ignore decades of research from the social sciences on the social shaping and distributed character of human cognitive, affective, and volitional capacities. Strategies of intervention, of governing through the brain, are based on many dubious assumptions and often blithely ignorant of the likely social consequences of their endeavors.

Arguments claiming that neuroscience offers a radical and challenging reconceptualization of human personhood and selfhood are often confused and are based on the very culturally and historically specific premises that that they claim to explain. Many, if not all, of the claims made for *neuro* products claiming to improve mental capacities by brain stimulation of various types have no basis in research, although they appeal to long-standing cultural beliefs that mental capacities are like physical capacities and can be improved by exercise, training, and rigorous ascetic control of the body. Practices such as 'mindfulness' have swiftly migrated from being self-managed radical alternatives to other forms of 'governing the soul' to become yet another element in the armory of the psychological, psychiatric, and lifestyle experts trying to persuade their clients to improve themselves by becoming mindful. And we can see how the practices of self-improvement, focusing on enhancing each person's capacity to manage themselves flexibly and adaptably in a world of constantly changing demands, do aim to produce the forms of subjectivity that might be able to survive in the new patterns of work and consumption that have taken shape over the past twenty years.

Yet, as we argue throughout this book, new styles of thought are beginning to emerge in neuroscience that recognize the need to move beyond reductionism as an explanatory tool, to address questions of complexity and emergence, and to locate neural processes firmly in the dimensions of time, development, and transactions within a milieu. These offer the possibilities of a more positive role for the social and human sciences, an opportunity to seize on the new openness provided by conceptions of the neuromolecular, plastic, and social brain, and to move beyond critique to find some rapprochement. The human sciences have nothing to fear from the argument that much of what makes us human occurs beneath the level of consciousness; indeed, many have long embraced such a view of our ontology (Ellenberger 1970). If we take seriously the renewed assault on human narcissism from contemporary neurobiology, we may find the basis of a radical way of moving beyond notions of human beings as individualized, discrete, autonomous, coherent subjects, free to choose. Neuroscience may seem an unlikely ally of progressive social thought, but its truth effects could surprise us.

Chapter One

· · · · · · · · · · · · · · · · ·

The Neuromolecular Brain

Those of us who have lived through the period in which
molecular biology grew up . . . realized that here was an enormous
opportunity for a new synthesis . . . [an] approach to understanding
the mechanisms and phenomena of the human mind that applies
and adapts the revolutionary advances in molecular biology
achieved during the postwar period. The breakthrough to precise
knowledge in molecular genetics and immunology—"breaking
the molecular code" resulted from the productive interaction
of physical and chemical sciences with the life sciences. It now
seems possible to achieve similar revolutionary advances in
understanding the human mind.

—**Francis O. Schmitt**, "Progress Report on the Neurosciences
Research Program," 1963[1]

In the summer of 1961, at the Massachusetts Institute of Technology (MIT),
Francis O. Schmitt and a small group of collaborators began to lay the plans
for a project that was as simple to state as it was ambitious to imagine: to
do for the brain what Watson and Crick had so recently done for the gene
when they discovered the structure of DNA and "cracked the genetic code."[2]
In the process they were to invent what they would come to term *neuroscience*
(Swazey 1975, 531). "To this end, new devices are being forged," wrote Schmitt
in 1967, "with one of which I am closely identified. This is the Neurosciences
Research Program (familiarly known as NRP), being a group of 34 gifted
scientists—mathematicians, physicists, chemists, neurobiologists, neurologists, psychologists, and psychiatrists—who meet regularly to study the problem and who sponsor a center and a center staff which arranges work sessions
on relevant topics of conceptual importance and which communicates the
harvest of the work sessions in the program's Bulletin" (Schmitt 1967, 562).

The NRP had grown out of the first Intensive Study Program (ISP) that
took place in 1966 at the University of Colorado at Boulder, where more
than a hundred participants from diverse professional backgrounds gathered to "unify the disparate neurosciences" and shed the light on the emerg-

ing trends, issues, hypotheses, and findings that could further advance our knowledge of the human brain (Quarton, Melnechuk, Schmitt et al. 1967). The community brought together by this first study program defined itself by its set of objects of study: the brain, the nervous system, and their relation to their arising "products" (for example, learning, memory, sleep, arousal, reflex, etc.). The challenge, as they saw it, was to bring together the disparate fields of neurologists, behavioral scientists, physicists, chemists, and mathematicians into meaningful communication with one another under the rubric of what Schmitt (1970)—in one of the first journal articles to mention *neuroscience* in its title—was to call "brain and behavior." In other words, the intention was to integrate not just the *neuro* disciplines—neurophysiology, neuroanatomy, neurochemistry, and neurology—but also the *psy* disciplines of psychology, psychoanalysis, and psychiatry, as well as some other less obvious fields such as immunology (Schmitt 1990). Their approach was characterized by the attempt to combine analysis at many levels: cellular, molecular, anatomical, physiological, and behavioral (Quarton, Melnechuk, Schmitt, et al. 1967). Twenty years later it seemed easy to state the program's aim: it was to answer the question "What is neuroscience?" (Schmitt 1985, 1990).

The terms *neuroscience* and *neuroscientist* now seem the obvious ones for this kind of research on the brain. But neither the name nor its implications were immediately evident to those visionaries at the time. Judith Swazey tells us that Schmitt first used the terms "mental biophysics" and "biophysics of the mind" to encapsulate his attempt to bring together the "wet, the moist and the dry" disciplines that addressed central nervous system functioning (Swazey 1975).[3] He and his colleagues later worked under the title "the Mens Project" (adopting the Latin term for mind) and initially applied to the NIH for a grant for a "Neurophysical Sciences Study Program"—which was approved in just two weeks. The project only became the "Neurosciences Research Program" in a subsequent grant application to NASA and the term *neuroscience* made its first public appearance (in the plural) in 1963 in the title of the *Neurosciences Research Program Bulletin* of that year.

Schmitt was a biophysicist who was recruited to MIT in 1941 at the request of MIT's president, Karl Compton, and its then vice-president, Vannevar Bush, author of the classic text on the endless frontier of science published in 1945 (Bush 1945). Compton and Bush were keen on establishing a "new kind of biology" or "biological engineering" that would combine physics, mathematics, and chemistry (Bloom 1997). This vision, as we have suggested, was explicitly informed by the rapid developments in molecular biology during the 1950s (Schmitt 1990). The new kind of biology that Schmitt was called upon to establish at MIT was a molecular biology heavily rooted in biophysics, biochemistry, and biophysical chemistry (Schmitt 1990). Swazey tells us that when Schmitt was sketching out his plans in 1961, "he listed nine 'basic disciplines' for mental biophysics: solid state physics, quantum chemistry, chemical physics, biochemistry, ultrastructure (electron microscopy and x-ray diffraction), molecular electronics, computer science, biomathematics, and literature research" (Swazey 1975, 531). Physical chemistry, bioelectric studies and computer science seemed to be essential to study the "physical basis" of memory, learning, consciousness, and cognitive behavior.

In a later paper the list was expanded to twenty-five areas of study, and in its progress report to the NIH in 1963, the NRP enumerated fourteen disciplines represented in their program: chemistry, biochemistry, neurochemistry, physical chemistry, mathematics, physics, biophysics, molecular biology, electrophysiology, neurobiology, neuroanatomy, psychology, psychiatry, and clinical medicine. This was in danger of becoming an enterprise without limits! By 1967, however, Schmitt created a simpler diagram that divided the neurosciences into "three intellectual areas": *molecular neurobiology*, working at the molecular, subcellular, and cellular level; *neural science*, dealing with "net characteristics," including neuroanatomy, neurophysiology, and neuropathology; and *behavioral or psychological science*, or what he called "molecular psychology" and which he—like so many of his successors—found the most challenging to integrate (Schmitt 1967, 562).

Swazey argued that the NRP had a "catalytic role in creating and promoting the growth of a *neuroscience* community. . . . [T]he term and concepts of 'neuroscience,' of a new type of integrated multidisciplinary and multilevel approach to the study of the brain and behavior, came into being with the NRP" (Swazey 1975). While those working in this area would retain their specific disciplinary allegiances—molecular biologists, neuroanatomists, biophysicists, psychologists, and so on—they would also locate themselves within this emerging scientific project with a common goal: they would think of themselves as *neuroscientists*. By 1964, this aspect of the project had achieved some success. The term *neuroscientist* was first used publicly in an article in the journal *Technological Review* titled "A neuroscientist sees 'invisible colleges' as a means of meeting the information problem," with specific reference to the "invisible college" of the recently created Neuroscience Program "activated and made 'visible' in Boston."[4] To create the neuroscientist, then, required, first and foremost, the creation of a matrix—the invisible college—that could bind those who identified with this term into some kind of community.

For Schmitt, as for other prominent scientists of the time, it would be through the unification or "synthesis" of these diverse and separate disciplines and through the reductionist molecular approach (i.e., a bottom-up approach) that substantial breakthroughs could be delivered (cf. Kay 1993). He based this belief on his perception that the success of molecular biology, genetics, immunology, and biophysics arose from the multidisciplinary dialogue between three traditionally separate disciplines: physics, chemistry, and biology. "It now seems possible," Schmitt wrote in a 1963 progress report, "to achieve similar revolutionary advances in understanding the human mind. . . . By making full use of [the approaches of physiology and the behavioral sciences] and by coupling them with the conceptual and technical strengths of physics, chemistry, and molecular biology, great advances are foreseeable" (quoted in Swazey 1975, 530). This was not to be simply a natural evolution of scientific knowledge; it was an explicit project to anatomize the workings of mind in terms of brain, to be undertaken at a novel, neuromolecular level: "[M]olecular neuropsychology is becoming ripe for study by the methods and concepts that proved significant in molecular genetics and molecular immunology. It seems inevitable that the problem of the mechanisms of brain function will soon become the primary intellectual salient, offering the greatest promise for the life sciences—indeed, as Schroedinger has suggested, of all science" (Schmitt 1967, 561–62). This, then, was to be the promise, and the mission, of the new molecular vision of the brain that was born in 1962.

How Should One Do the History of the Neurosciences?

But surely this claim, that the neurosciences were born in 1962 is absurd, or at the least, purely semantic. As we have already mentioned, human beings have been concerned with the brain, the nerves, let alone the mind, since time immemorial, and rigorous reflections, often based on empirical examples, go back at least to the Greeks. So why focus on 1962? To explain this, it is necessary to reflect for a moment on different ways of writing the history of the neurosciences.

Consider, for example, one of the first articles published in French to use the term *neuroscience*, written by André Holley in 1984: its title was "Les neurosciences: Unité et diversité" (Holley 1984). Holley argued that there is one object of study, and one objective of these new sciences of the brain, namely, "the knowledge of the nervous system, its function and the function of the diverse phenomena that emerge from it" (Holley 1984, 12; our translation). Yet, says Holley, on the other hand there are a plurality of methods and concepts. Thus, for him, the neurosciences were born with the convergence of all these sciences around one object of study and concern: the brain. This account of unity in the face of conceptual and methodological diversity seems to fit the account given in most 'authorized' versions of the history, which usually take the form of chronologies of the great thinkers, establishing a lineage of precursors for the present (Garrison and McHenry 1969; DeJong 1982; Brazier 1984, 1988; Finger 1994). These are often supplemented, for the modern period, by compilations of autobiographical accounts of key figures in the manner of oral history (Healy 1996, 1998, 2000; Squire 1996; Tansey 2006).

There are also a number of institutional versions of this history: many university departments of neuroscience have created their own accounts or time lines of the milestones—key discoveries, key texts, key persons. Some versions are presented from the perspective of the various disciplinary trajectories— from the perspective of neurology, neurochemistry, or neuroimaging. For example, the Society for Neuroscience (SfN), has its own history and has produced eleven DVDs of interviews with prominent neuroscientists.[5] Along the same lines, accounts of other societies and organizations have been published as well, for example, the International Neuropsychological Symposium, founded in 1951 (Boller 1999); the International Brain Research Organization (IBRO), founded in 1961 (Marshall, Rosenblith, Gloor, et al. 1996); the International Society for Neurochemistry, founded in 1965 (McIlwain 1985); and the European Brain and Behavior Society, founded in 1968 (Marzi and Sagvolden 1997).

In addition, numerous memoirs have been written by famous neuroscientists of the twentieth century (Ramón y Cajal 1937; Schmitt 1990; Hodgkin 1992; Kandel 2006), as well as abundant biographies and articles on the contributions and achievements of particular individuals, either in the form of "neuro-anniversaries" (Eling 2001)[6] or as part of the development of a particular concept or a discovery, for example, the discovery of acetylcholine by Henry Dale (Tansey 2006) or the discovery of dopamine by Arvid Carlsson (Benes 2001). Other published papers name this or that historical figure as the true founder of neuroscience (Changeux 1985, 1997; Wills 1999). Alongside these accounts of founding fathers—and they were mostly men—we find accounts of the major discoveries, such as the synapse (Rapport 2005), and of major controversies, such as the famous "war of the soups and the sparks" between the proponents of electrical transmission and neurotransmission in the central nervous system (Valenstein 2005). And, starting in the 1990s, the history of neuroscience becomes a kind of subdiscipline in its own right, with the founding of the *Journal of the History of the Neurosciences*, and then of the International Society for the History of the Neurosciences (Rosner 1999).

What are they for, all these histories? What role do they play for the neurosciences themselves? As in other truth-claiming disciplines, a historical sensibility is internal to the science itself. It is performative: it establishes a respectable past for neuroscientific knowledge, thus securing a basis for its future. Knowingly or not, such histories solidify the scientificity of the present state of the science whose tale they tell by presenting it as the culmination of a long trajectory of the progress of knowledge. Georges Canguilhem terms such accounts "recurrent history"—the history written and rewritten by almost every science, viewing the past from the perspective of its own regime of truth.[7] These histories establish a division between the sanctioned and the lapsed. The sanctioned: a more or less continuous sequence that has led to the present—a past of genius, of precursors, of influences, of obstacles overcome, crucial experiments, discoveries, and the like. The lapsed: a history of false paths, of errors and illusions, of prejudice and mystification, and all those books, theories, arguments, and explanations associated with the past of a system of thought but incongruous with its present. To point to this use of history within science is not to denounce it but to characterize its constitutive role: using the past to help demarcate a contemporary regime of truth, to police the present, and to try to shape the future.

Central to such a history is a succession of predecessors. This quest for precursors fabricates "an artifact, a counterfeit historical object" and ignores the fact that the precursor is the creature of a certain history of science, and not an agent of scientific progress; Canguilhem quotes Koyré: "It is quite obvious (or should be) that no-one has ever regarded himself as the 'forerunner' of someone else. . . . [T]o regard anyone in this light is the best way of preventing oneself from understanding him" (Koyré 1973, 77; quoted in Canguilhem 1994, 51). Indeed, as Canguilhem points out (1968, 21): "The precursor is a thinker who the historian believes he can extract from one cultural frame in order to insert him into another, which amounts to considering concepts, discourses and speculative experimental acts as capable of displacement or replacement in an intellectual space in which reversibility of relations has been obtained by forgetting the historical aspect of the object dealt with."

For such histories, we can reinterpret the arguments of those precursors in terms of our own concepts and problems. The objects of study—the brain, the nerves—remain static, intact, brute entities in a world external to thought, which are merely grasped more or less adequately by a succession of thinkers. But the kind of historical epistemology practiced by Canguilhem recognizes the historicity of the very object of explanation, the set of problems, issues, phenomena that an explanation is attempting to account for. In a way, André Holley is right when he says that the neurosciences (in the plural form) are centered on one object of study: "la connaissance du système nerveux, de son fonctionnement et des phénomènes qui emergent de ce fonctionnement" (Holley 1984, 12). He insists on this 'unity': "pluralité des methods, diversité des concepts mais unité de l'objet" (ibid.).

Yet the contemporary brain is not the object that formed the focus of nineteenth-century neurologists, nor that explored by Sherrington and Adrian in the first half of the twentieth century. In this chapter we will suggest that the new style of thought that has taken shape in the form of neuroscience has so modified the object of thought, the brain, that that it appears in a new way, with new properties and new relations and distinctions with other objects. And yet, as pointed out by Hans-Jörg Rheinberger in another context, the research object—what he terms the "epistemic thing"—has a "characteristic, irreducible vagueness" (Rheinberger 1997, 28). That is because this object embodies what is not yet known. The object, in our case, the neuromolecular brain, is both there and not there. Rheinberger quotes Latour: "[T]he new object, at the time of its inception, is still undefined. [At] the time of its emergence, you cannot do better than explain what the new object is by repeating the list of its constitutive actions. [The] proof is that if you add an item to the list you *redefine the object*, that is, you give it a new shape" (Latour 1987, 87–88, quoted in Rheinberger 1997, 29). That is no doubt true, yet we will suggest in this chapter that while the object of the new brain sciences is elusive, vague, shifting, and while it constantly refers back, and gains some of its power, from that brain that we think we have always known, that mass inside the skull, a new style of thought has taken shape, which has provided the possibility for the neurosciences to achieve their current status, to move out of the labora-

tory and the clinic, to become a kind of expertise in understanding and intervening in human conduct in many different practices.

It is, of course, a mistake to think of historical time as continuous: the time
of history is not that of a calendar, it is intercalated by discontinuous events
(Canguilhem, 1968; Foucault 1972) or "turbulences" (Serres 1977). The birth of
neuroscience marks something like an event. Yet even a discontinuous genealogy cannot do without dates, symbolic landmarks to alert us to something
like an 'event.' It is in that sense that we mark this event with a date—1962.
This is the point at which a number of distinct lines of thought and practice
seem to come together to create a difference. It is beyond our capacity to undertake a genealogy of this event here. So, at the risk of simplification, let us
try to orient ourselves by distinguishing, purely for heuristic purposes, three
intertwining pathways that seem to intersect in that event—the path through
the nerves, the path through the brain, and the path through madness.

The Path through the Nerves

Neuroscience, neurology, neuroanatomy, neuroembryology, neurophysiology,
neurochemistry, and the burgeoning *neuro* fields, from neuroeducation and
neuroeconomics, to neurophilosophy and neuromarketing all have in common
the prefix *neuro-*, which is taken from the ancient Greek *neûron* or the Latin
nervus, meaning "of or relating to nerves or the nervous system." The *OED*
dates the first use of *neuro-* in words of Greek origin to the seventeenth century
with the introduction of the term *neurology* and in the eighteenth and early
nineteenth centuries in loanwords from post-classical Latin, as with the terms
neurotomy (the anatomical dissection of nerves), *neurography* (the anatomical
description of nerves), and *neuralgia* (pain in the area served by a nerve). In
French, the prefixes *névro-* and *neuro-* both occur from the late seventeenth
century onward (1690, in *névrologie*; 1691, in *neurologie*). In German, the prefix
makes its first appearance probably in 1856 with the term *Neuroglia*. In Italian,
the *OED* dates the first use of *neurologia* in 1764 and *nevralgia* in 1828.

Thomas Willis is thought to have coined the term *neurology* to refer to the
"Doctrine of Nerves": the first entry in the *OED* for the term *neurology* is 1681,
the date of publication of Samuel Pordage's translation of Willis's 1664 influential *Cerebri anatome*. The use of that term to describe a treatise on the nervous
system dates back to 1704, and the word gradually came to refer to the branch
of science or medicine dealing with the nervous system and its diseases and
disorders. Wilhelm Griesinger, in Berlin, founded the *Archiv für Psychiatrie
und Nervenkrankheiten* in 1867, and the American Neurological Association
was formed in 1875. The *OED* tells us that term *neuropath*—for both the doctor
who attributes diseases to disorders of the nerves, and to those suffering from
or susceptible to nervous diseases—dates from the 1880s.

All histories of these matters credit Golgi with inventing the silver salt
staining method that enabled the visualization of the elements of the nervous

system, although he believed that they formed a single continuous network, and Cajal with establishing the doctrine that neurons were discrete cells that were the elementary units of the brain. The *OED* finds the first English use of the term *neuron* in a paper by Alex Hill, published in *Brain* in 1891, summarizing the research of Camillo Golgi, Santiago Ramón y Cajal, Albert von Kölliker and Wilhelm His Sr. But Jean-Pierre Changeux tells us: "As to the term 'neuron,' we owe it neither to Ramón y Cajal nor to His, but to Wilhelm Waldeyer (1891) about whom Ramón y Cajal wrote that all he had done was to 'publish in a daily paper a résumé of his research and invent the term neuron'" (Changeux 1997, 28). In 1906, despite their bitter disagreements, and Cajal's attacks on Golgi's reticular or neural net theory, Golgi and Cajal jointly received the Nobel Prize for Physiology or Medicine for their discoveries of the structure of the nervous system.[8]

The French Société de Neurologie was formed in 1899, and its German equivalent in 1907. We find *neuroanatomy* used in 1900, the first reference to *neurobiology* in 1906, and *neuroembryology* emerged in the 1930s. The first chair in *neurochemistry* was established in 1945, the *Journal of Neurochemistry* started publishing in 1956, the International Society for Neurochemistry was founded in 1965, and the American Society for Neurochemistry was inaugurated in 1969.[9] The mode of action of the neuron was gradually unraveled over this period: in 1926, Edgar Adrian published his recordings of impulses for individual sensory nerve fibers; in 1939, Hodgkin and Huxley published their classic paper reporting the recording of action potentials from inside a nerve fiber; and in 1944, Joseph Erlanger and Herbert Gasser were awarded the Nobel Prize in Physiology or Medicine "for their discoveries relating to the highly differentiated function of single nerve fibres." *Neuropharmacology* appears in the late 1940s, used in relation to the introduction and discovery of more effective antiepileptics (Toman 1949). In 1951, the International Neuropsychological Symposium was formed under the leadership of cofounder Henry Hécaen,[10] and it met almost every year from 1951 until 1998 for the purpose of promoting knowledge and understanding of brain functions and cognate issues on the borderland of neurology, psychology, and psychiatry.[11] The 1950s saw the famous "war of soups and sparks"—the disputes between the believers in electrical neurotransmission and those who believed that neurotransmission in the central nervous system was chemical—and the gradual victory of the chemical view of neurotransmission (cf. Valenstein 2005). And thus we arrive, more or less, at 1962.[12]

The Path via the Brain

Most historians agree that contemporary research in neuroscience builds on studies of the brain that flourished in Europe in the nineteenth century. It was at this point, it seems, that the brain began to be truly understood as an organ analogous to others, and that specific areas and structures in the brain came to be seen as responsible for certain functions—such as those that are still known by the names of their discoverers—Broca, Wernicke, Flechsig, for example. The journal *Brain* was first published in England in 1878, calling itself "A Journal of Neurology"—thus making it clear that our separation of nerves and brain is purely heuristic! It had four editors. J. C. Bucknill and James Crichton-Browne both had backgrounds as medical superintendents of insane asylums; both were prolific authors and advocates for what was then often termed psychological medicine. David Ferrier and John Hughlings Jackson were neurologists working on the localization of brain functions, work that we discuss in more detail in chapter 2.

In the half-century that followed, work on brain structure and function continued, notably by Charles Sherrington and his group in the United Kingdom, using technologies of stimulation or ablation of specific of brain areas. Others, notably Sherrington's colleague Edgar Adrian, used technologies of visualization, such as the electroencephalogram. A meeting of electroencephalographers in London in 1947 led to the establishment of an International Federation of Electroencephalography and Clinical Neurophysiology, and the proposal for an International Brain Research Organization arose from a meeting of this group in Moscow in 1958 initiated by the French clinical neurophysiologist Henri Gastaut.[13] IBRO was established as a component of the United Nations Educational, Scientific, and Cultural Organization (UNESCO), which was founded in 1951. Initial proposals for UNESCO had included an "Institut du Cerveau" (Institute for the Brain), which was not accepted (McIlwain 1985). IBRO was constituted with seven panels—in neuroanatomy, neurochemistry, neuroendocrinology, neuropharmacology, neurophysiology, behavioral sciences and neurocommunications, and biophysics.[14] This multidisciplinary structure of collaboration between approaches and across national borders was retained when it became an independent body: again note the year of its foundation—1961.[15]

The Path via Insanity

We arrive at roughly the same point in the 1960s if we follow the path of those dealing with those deemed insane or otherwise mentally afflicted. In fact, the journal *Brain*, not surprisingly, given the interests of two of its founding editors, was much concerned with the ways in which brain damage or disorder could produce symptoms like those of hysteria or mania, as well as loss of speech, hearing, memory, or other functions. Broca, Wernicke, Flechsig, and others studied the brains of those who died in lunatic asylums in their attempts to discover the basis of mental afflictions in the brain. Many also studied those who had suffered brain injuries, for example, those who drew conclusions from emblematic figures like Phineas Gage, who had suffered changes in character after an injury to specific regions of his brain caused by that famous tamping iron (Macmillan 2000). There were those like the

French psychiatrist Bénédict Morel, whose understanding of madness as a physical condition was closely linked to their ideas of inheritance and the problem of degeneration (Morel 1852–53, 1860). The birth of psychiatric genetics was intrinsically bound up with the eugenic style of thought that was prevalent in Europe and North America in the first half of the twentieth century, although the pathways for the action of the supposed inherited diathesis were unknown.

It is true that this period was also that of the early development of dynamic psychotherapies, but most of those who worked in this tradition were—like Freud himself—quite open to the belief that the propensities to the disorders they addressed were organic and inherited, although once more the locus of these disorders in the nerves or the brain was a matter of conjecture. The first half of the twentieth century saw the birth of movements to address the 'maladaption' and 'maladjustment' of the instincts that underpinned the troubles of everyday existence at work or in the family, such as the mental hygiene movement. But in the asylums, physical treatments were flourishing: shock therapies using electroshock- or insulin-induced coma, malaria therapies (for neurosyphilis), and psychosurgery (Sargant and Slater 1944; Sargant, Hill, and Slater, 1948). Somehow, it seemed, the brain was the locus of mental pathology, and potentially therefore the target for therapy, but just how and where remained elusive.

In 1936, Egas Moniz reported the first results of his technique of frontal lobotomy.[16] These were apparently inspired by a report of the calming effects of destruction of the prefrontal brain area of monkeys and chimpanzees; Moniz's version of this procedure involved the destruction of a portion of the frontal lobes, first by injection of alcohol, and later by cutting nerve fibers with a surgical leucotome (Moniz 1948, 1954; Valenstein 1986; Pressman 1998).[17] These methods were taken up by Walter Freeman and James Watts in the United States, who invented a technique to enter the brain through the eye socket using an instrument modeled on the ice pick (Freeman and Watts 1942, 1950). By 1948, lobotomies had been performed on about twenty thousand patients worldwide. In 1949, Moniz was awarded the Nobel Prize in Physiology or Medicine for this work.

Joseph Zubin, in his article on "Abnormalities of Behavior" for *Annual Reviews of Psychology* in 1952, notes: "The tremendous impact that psychosurgery has made on current investigations is evidenced by the following books and monographs which were published between 1949 [*sic*] and 1951" (Halstead 1947; Mettler 1949; Freeman and Watts 1950; Partridge 1950; Fulton 1951; Greenblatt, Arnot, and Solomon 1951) and the number of symposia that focused on this topic (Association for Research in Nervous and Mental Disease 1948; Rees, Hill, and Sargeant 1949; Congrès International de Psychiatrie 1950; Halstead 1950; U.S. National Institute of Mental Health 1951), as well as the studies of the brain that showed the impact of this research (Hebb 1949; Golla and Richter 1950; Penfield and Rasmussen 1950; Porteus

1950). Psychosurgery, he remarks, acted as a "shot in the arm" for psychiatry in the 1940s, and while "the mechanisms whereby the operation brings about improvement remain unknown. . . . By the end of the third month, most of the psychological functions have returned to their preoperative level, or have even exceeded it" (Zubin 1952, 262–63). Subsequent experience was to prove him wrong.

The enthusiasm for physical treatments in the 1940s and 1950s, along with the conviction that they had a scientific basis and major therapeutic implications for psychiatry, led to the formation in 1945 of the Society for Biological Psychiatry, whose mission was to encourage "the study of the biological causes of and treatments for psychiatric disorders." Edward Shorter credits Stanley Cobb, one of the originators of this society, with being the founder of biological psychiatry in the United States, not only building the Department of Psychiatry at Massachusetts General Hospital beginning in 1934 (Shorter 1997, 263) but also promoting psychoanalysis.[18] The opposition between the biological and the psychoanalytic does not emerge until the 1960s. When the first International Congress on Psychiatry was held in Paris in 1950, sessions on Psychotherapy and Psychoanalysis (chaired by Franz Alexander) were presented alongside those on Genetic and Eugenic Psychiatry (chaired by Torsten Sjögren), on Shock Therapy (chaired by Josef Handesman), and on Cerebral Anatomy and Physiology in the Light of Lobotomy and Topectomies (chaired by Frederick Golla) (Congrès International de Psychiatrie, 1950). Present at the meeting were Denis Hill, who talked on the use of the EEG in psychiatry, and Max Reiss and Derek Richter, both of whom were then working on the psychiatric implications of hormones and endocrinology.[19] The overall chairman of the Congress was Jean Delay, who was the emblem of the new bridge between psychiatry and neurochemistry via psychiatric drugs that would rapidly come to replace most, if not all, of the more direct physical interventions into the brain (whose promises were only to be enthusiastically explored again half a century later).

This is not the place to recount the history of psychopharmacology in any detail; a rapid sketch will serve our purposes. The effects of chlorpromazine were first reported by Henri Laborit working at Val-de-Grâce military hospital in Paris in 1951, and in 1952, Jean Delay and Pierre Deniker published the apparently dramatic results of administering this drug to psychotic patients at the Saint Anne Hospital in Paris (Delay 1953). In a series of transactions between France and North America, the drug was widely taken up by those working in psychiatric hospitals in North America during the 1950s (Horwitz 2002). The first results of another drug, reserpine, on psychotic patients were reported in 1954, and by 1955, when Jean Delay organized an international colloquium on psychopharmacology in Paris, it seemed as if at last drugs had been developed that acted on the symptoms of mental disorder by virtue of their specific effects in the brain. Once more, however, this was not framed as an opposition between physical and psychological treatments, for it was argued that one of the main advantages of the drugs was that it made the patients who had been treated more amenable to psychotherapy.

Throughout the 1950s, a number of key papers begin to discuss the effects of other drugs, notably isoniazid, iproniazid, reserpine, and imipramine on disturbed or depressed patients.[20] Gradually the theory took shape that all these novel psychotherapeutic agents produced their effects on mood and conduct by acting on the levels of monamines in the brain (Salzer and Lurie 1953; Kline 1954; Delay and Buissoj 1958; Kuhn 1958). Bernard ("Steve") Brodie, who decided to enter psychopharmacology "after hearing about the sensational clinical actions of the new antipsychotic drugs and the ability of the hallucinogenic LSD to block the effects of serotonin on various peripheral organs" (Carlsson 1990, 486), developed new analytical methods that enabled him and his colleagues to demonstrate that the effects of reserpine were linked to its ability to deplete serotonin in the brain (Brodie, Pletscher, and Shore 1955; Pletscher, Shore, and Brodie 1955; Shore, Silver, and Brodie 1955). Arvid Carlsson, who had worked with Brodie, generalized this to other catecholamines, notably to dopamine (Carlsson, Lindqvist, Magnusson, et al. 1958; Carlsson 1959; Carlsson and Lindqvist 1963; Dahlstrom and Carlsson 1986), and began to formulate the link between the depletion of dopamine by reserpine and the symptoms of Parkinson's disease.

Although by the 1950s it was accepted that neurotransmission in the *peripheral* nervous system was chemical, as late as 1960, according to Arvid Carlsson, there was still considerable resistance to the argument that chemical neurotransmission happened in the brain and was probably the dominant form of neurotransmission.[21] However, the work of Carlsson and his collaborators over the next few years, involving many technical developments, established the neuronal localization of dopamine, noradrenaline, and serotonin in the central and peripheral nervous system, and mapped the major monoaminergic pathways and the sites of action of the major psychotropic drugs. According to Carlsson, by 1965 a "paradigm shift" had taken place, and it was now almost universally accepted that the monoamines played the key role in neurotransmission in both peripheral and central nervous systems and that their presence and actions were clearly linked to psychiatric symptoms (Dahlstrom and Carlsson 1986; Carlsson 1990, 2001).

A kind of informal consensus began to develop in which each of the major classes of neurotransmitters then known was allocated to a specific disorder—catecholamines to mood disorders, serotonin to anxiety, dopamine to psychosis, and acetylcholine to dementia (Healy 1997). In 1963, Carlsson and Lindqvist published what is usually taken as the first formulation of the dopamine hypothesis of schizophrenia, describing the interaction between the antipsychotic drugs chlorpromazine and haloperidol, and the level of the dopamine metabolite, 3-methoxytyramine, in the mouse brain—thus suggesting that high levels of dopamine might be linked to the symptoms and that reduction in dopamine might be the mode of effect of the drug (Carlsson and Lindqvist 1963).

In 1965 Joseph Schildkraut published his paper on the catecholamine hypothesis of depression: "what has been designated 'the catecholamine hypothesis of affective disorders,' proposes that some, if not all, depressions are associated with an absolute or relative deficiency of catecholamines, particularly norepinephrine, at functionally important adrenergic receptor sites in the brain. Elation, conversely, may be associated with an excess of such amines" (Schildkraut 1965, 509). This paper, which hedged its conclusions with many reservations, marked the moment when a new language was assembled together, one that would come to shape the style of thinking and research in psychopharmacology and the forms of clinical practice in psychiatry. Indeed, Schildkraut claimed that he decided to publish first in the *American Journal of Psychiatry*, a clinical journal, because he saw his work as more than about catecholamines—it was about the "notion that pharmacology can become a bridge, linking neuroscience and clinical psychiatry"— though he later published a version in a journal that would be read by the psychopharmacologists, when he discovered that they didn't read the clinical journals (Schildkraut 2000, 124).

The significance of the hypothesis was not limited either to the catecholamines or to depression. What one had to do, from this point onward, was to understand mental disorders in terms of their characteristic patterns of biogenic amines in the brain and to understand the action of actual or potential psychiatric drugs in terms of their actions on the secretion, breakdown, depletion, and reuptake of these amines in the synapses or the receptors. A whole continent of potential experiments, clinical trials, and drug development was opened for exploration. One might examine the amines in the brain directly, if one could. One might examine them in experimental animals. One might examine the behavior of synapses or receptors by using in vitro cultures. One might examine them or their breakdown products in the blood or urine of psychiatric patients. One might, in a reverse move, seek to refine psychiatric diagnoses in terms of amine malfunctions. This, then, was the path to a future neuropsychiatry (Healy 1997).

The dopamine hypothesis of schizophrenia and the monoamine hypothesis of depression were the two founding myths of the psychopharmacological imaginary. Both have proved to be wrong, perhaps fundamentally so. However, for the biological psychiatrists, who would gradually come to dominate the field, the biogenic amine system in the brain would increasingly become the obligatory passage point of all accounts of mental disorder. If there were environmental or biographical factors in the etiology of disorder, they would have their effects by acting through this mechanism. If there were genetic factors, they would work through their impact on this mechanism. If there were psychodynamic factors, well, they too would work via this mechanism. Even evidence that seemed to conflict with the argument would have to be explained in analogous neurochemical terms. Once more, note the date when this neuromolecular vision of the mode of action of psychiatric drugs was formulated—the early 1960s.

Infrastructure

Despite so many histories of neuroscience, most social historians working in this area agree that we do not have anything approaching a sociological account of the development of the neurosciences—that is, at a minimum, something that would link these developments to their social, cultural, perhaps even political conditions of emergence.[22] We have not attempted such an account here. Our claim is a limited one: that the event here was a new way of seeing the brain—a 'neuromolecular gaze.' And in the chapters that follow, we will argue that it is this new way of envisaging the brain and nervous system, and the style of thought and intervention to which it is linked, that has established the conditions that allow for the burgeoning of the neurosciences over the past five decades, and hence for the birth of our contemporary neurobiological complex. This is not to say that Schmitt was the founder of the neurosciences or of the new complex of sciences of mind and behavior that takes shape after 1962. Rather, his NRP is emblematic; it exemplifies a moment, dispersed across different sites, cultures, and countries, in which something new was coming into existence—a neuromolecular vision of the brain as an intelligible organ that was open to knowledge.

But a vision itself is seldom enough. A vision needs some kind of infrastructure.[23] Many different infrastructural forms can be found in scientific research. We have already encountered a number in our rapid overview of pathways to neuroscience. Many involve a kind of discipleship, where a charismatic scientist infuses certain beliefs and commitments into those who are trained at his or her feet. Others involve the founding of learned journals developed to promulgate a particular kind of identity for those who publish, contribute to, and read them. In some, a university has played a key role, for example where a university department employs those who share and transmit the vision of the professor. Another ancient form of infrastructure is the scientific society or the academy. And one more form, that of an 'invisible college,' refers us back to the phrase used by Robert Boyle in the 1640s to designate a network of intellectuals without an institutional form, exchanging ideas, information, and speculation on specific topics (Crane 1972). And a framework has also been provided by scientific institutes—usually located within or attached to universities, with a physical form, an organizational structure, a group of faculty, students, visitors, and so forth.[24]

Schmitt was certainly aware of the significance of infrastructure. While he did sometimes speak of the NRP as an invisible college, he spent much time and energy pondering and organizing the rather material conditions under which his project might best flourish. Beginning with the idea of a bricks-and-mortar research center or institute, Schmitt concluded that this was not the best form to foster the novel mode of scientific communication and interaction that he sought—he wanted to establish a different kind of center "oriented towards probing and evaluating the state of the art, seeking to map new research and conceptual directions, and providing new modes of education for the communication among a worldwide network of scientists" (Swazey 1975, 533). He took the idea of a study program from those he already knew from biophysics, and chose a structure in which a core group of individuals from the different disciplines would sponsor a series of such programs—conferences and workshops—with a relatively small central physical base independent of any university for the co-location of staff and for meetings of the associates.

This was indeed the form the NRP took, with premises located within the American Academy of Arts and Sciences in Brookline, Massachusetts, undertaking a set of activities—study programs, focused conferences, teaching activities to introduce those with no knowledge of neuroscience to the fundamentals of brain structure and function—plus the *NRP Bulletin* which would disseminate the results of all these endeavors. The NRP thus involved many of the key elements of infrastructure: a charismatic leader, a network of loyal students, a shared language and nomenclature, a shared moral economy among its members, and, perhaps more than anything, a certain way of seeing the object of investigation, rendering it thinkable and actionable via a neuromolecular gaze. But the NRP was no 'origin'; rather, it was exemplary of something happening at several other sites, around other key individuals. For Schmitt was not alone in his endeavor to forge a community among those who were addressing this organ of the brain from diverse disciplinary perspectives.

In their own 'recurrent history' of neuroscience, Eric Kandel and his colleagues choose David Rioch at the Walter Reed Army Institute of Research as the key early figure, and they single out Stephen Kuffler, who created the first Department of Neurobiology at Harvard Medical School in 1966, as the person who drove the project of a unified neuroscience together, "surmount[ing] the conventional divisions that had separated the subdisciplines of the neurosciences, and succeed[ing] in establishing a unified discipline that represented the various subfields" (Cowan, Harter, and Kandel, 2000, 347). They argued that it was Kuffler's work that inspired the formation of similar departments in other universities and led to the formation of the Society for Neuroscience in 1969 "under the leadership of the psychologist Neil Miller, the biochemist Ralph Gerard, and the neurophysiologist Vernon Mountcastle," which in turn brought together researchers who had previously been associated with their own disciplinary societies—as physiologists, anatomists, biochemists, or psychologists (ibid.).[25] What was taking shape here was a rather different form of infrastructure from that developed in the NRP, linked to a different project of neuro-interdisciplinarity. Rioch certainly facilitated interdisciplinarity, as David Hubel later reflected: "In the neuropsychiatry division [of Walter Reed] David Rioch had assembled a broad and lively group of young neuroscientists, notably M.G.F. Fuortes and Robert Galambos in neurophysiology, Walle Nauta in neuroanatomy, Joseph Brady and Murray Sidman in experimental psychology and John Mason in chemistry. . . . As in Montreal, the focus was on the entire nervous system, not on a subdivision of biological subject matter based on methods."[26]

So indeed did Kuffler and his group at Johns Hopkins—a group particularly focused on the electrical physiology of the nervous system, and that included Vernon Mountcastle, David Hubel, and Torsten Wiesel. The scope of these collaborations widened when Kuffler moved to Harvard to set up a new Department of Neurobiology in 1966 (Katz 1982). As Timothy Harrison puts it:

> In the 1960's, many scientists in and outside Kuffler's orbit realized that their grasp of biology was too limited. A prime example of this limitation was that electrical recording techniques could give only electrical answers and information, whereas new questions constantly suggested themselves. What are optimal nerve membrane ionic and biochemical requirements for nerve impulse propagation? What are the brain's energy sources? Do they differ from those of peripheral nerves? To answer such questions, an army of scientists developed new areas of neurobiologic research. Molecular and sub-cellular physiologists were interested in regulation of protein synthesis. Other scientists, specifically neurochemists, immunologists, molecular geneticists and several varieties of biophysicists all focused on the biology of the nervous system. Many of these neurobiologists gathered in Kuffler's new department at Harvard. . . . Kuffler's department inspired many other scientists trying to organize their departments and research in neurobiology. In its way Harvard served as a microcosm of a greater and broader scientific world now available to the handful of interested scientists who started the Society of Neuroscience in 1969. (Harrison 2000, 177)

The achievement here was both conceptual—the need for some kind of integration of these distinct approached to the nervous system—and topographical—for, as the autobiographies by those associated with these developments show, it was the co-location in a single physical building, often along a single corridor, of researchers working on different questions, using different methods and approaches—that enabled ideas and techniques to travel and to hybridize (Purves 2010).[27]

As the researchers and postdocs moved from mentor to mentor, from the United States to London, to Australia, to Germany and back again, encountering enthusiasms and resistances to these new integrative approaches in equal measure, the linkages became stronger. By the end of the 1960s, then, neuroscience, as an interdisciplinary project with a neuromolecular gaze, was acquiring a spreading network of trained disciples of such charismatic people as Schmitt and Kuffler, although it took longer for other departments to form under the title of neurobiology.[28] By 1970, Schmitt was able to point to the "recent appearance of several journals and series of books with the words *brain* and *neuroscience* in their titles, and the formation of new multidisciplinary societies concerned with neuroscience. These include the Brain Research Association of the United Kingdom, the Society for Neuroscience of the United States, the European Brain and Behavior Organization and new specialty groups such as the International Neurochemical Society and the American Society for Neurochemistry" (Schmitt 1970, 1006). And Schmitt would have agreed with Kandel and his colleagues when they write that the existence of the new discipline of neuroscience was institutionalized in a new way when, a "decade later, in 1978, the first volume of the *Annual Review of Neuroscience* appeared . . . a clear recognition that this new synthesis of disciplines had matured and could stand on an equal footing with other fully mature basic science disciplines" (Cowan, Harter, and Kandel 2000, 348).

A Neuromolecular Style of Thought

In 1962, the International Brain Research Organization began its world survey of research facilities and manpower in the brain sciences, focusing on three broad areas: first, the basic sciences modified by the prefix *neuro-*: neuroanatomy, neurochemistry, neuroendocrinology, neuropathology, neuropharmacology, and neurophysiology; second, a group of disciplines consisting of neurocommunications and biophysics, including cybernetics and information theory, mathematical modeling, nerve cell and membrane biophysics, and neuromolecular biology; and third, the behavioral sciences, encompassing animal behavior, behavioral genetics, psycholinguistics, sensation and perception, and certain aspects of anthropology, psychology, psychiatry, and sociology. By 1968, using rather generous criteria, it could identify 880 different research groups and 4,245 investigators identified with the neurosciences in the United States alone (International Brain Research Organization 1968). And in the forty years that followed, it seemed self-evident that neuroscience was indeed a discipline, with all the implications of the term—institutional organization; authorities of truth; incorporation into the academy; textbooks and training; norms of experimentation, evidence, and argument; and much more.

But nonetheless, it is a discipline whose boundaries seem blurred and whose internal configuration is complex and shifting. Papers are published in hundreds of different specialist journals as well as those, such as *Nature*, intended for a more general scientific audience. The Neuroscience Citation Index (NSCI), which covers more than 350 journals, divides the neurosciences into eleven fields: Behavioral Neurology, Cardiovascular Disease and

Metabolism, Developmental Neuroscience, Electroencephalography, Epilepsy Research, Molecular Brain Research, Neural Networks, Neurogenetics, Neuroimaging, Neurosurgery, and Psychopharmacology.[29] Even in areas that might seem closely related to the outsider, interdisciplinary divisions into distinct organizations are evident, each with its own international conferences, recurrent themes, key reference points, and forms of argument and experimentation.[30]

Thus, despite the claims that neuroscience is now a discipline with a unified object of study and concern, what we have here is something different. But this heterogeneity and contestation is not surprising. For, as we have seen, the creation of a neuroscience community was not intended to eradicate the specific disciplinary allegiances of those who formed it, but rather to create a common space within which they could interact. In any event, we agree with those historians and philosophers of science who have pointed out that disunity, rather than unity, characterizes the sciences (Dupré 1993; Rosenberg 1994; Galison and Stump 1996; Hacking 1996; Galison 1997). There are a number of research fronts, with different sources of public and private research funding directed to different aspects and research strategies, and the multiplication of specialist institutes on everything from basic research on brain and behavior, to clinical and translational research. Efforts at unification, even when merely in the form of a consolidated textbook or introduction to neuroscience, can thus be seen as strategic, with political, cultural, epistemological, and ethical implications (Galison and Stump 1996).

Yet there is another way in which this appearance of diversity is misleading, because the event that we have traced to the 1960s is not really one of disciplinary formation. It is more an event in epistemology and ontology, in the nature of the object of the neurosciences and the forms of knowledge that can render it into thought. Is there any unity in the thought styles that characterize the diverse neuroscientific communities built around specific disciplines and research fronts? Is there something like an overarching "style of thought" (Fleck [1935] 1979)? Ian Hacking, drawing on A. C. Crombie more than Fleck, has identified six "styles of reasoning" in the sciences: (1) deducing from postulates, (2) experimental exploration and measurement of observational relations, (3) hypothetical construction of analogical models, (4) use of comparison and taxonomy to establish order, (5) statistical analysis of regularities in populations and the calculation of probabilities, and (6) analysis in terms of genesis or historical development (Hacking 1992b). John Forrester added another: (7) "thinking in cases" (Forrester 1996). Approached from this perspective, we would not find just single style of reasoning in neuroscience: in the same paper or the work of a single research group, we can often see thinking in terms of models, use of the laws of large numbers, analyses of the genesis and development of neuronal structures, concerns with traits or disorders, questions of taxonomy and classification, and even reasoning in cases.

Yet, approached in a different way, closer to the spirit of Ludwik Fleck, it is indeed possible, by 1970, to discern something like a style of thought what we term a "neuromolecular style of thought." At the risk of oversimplification, let us enumerate some of its key structuring principles:

1. The brain is an organ like any other (even if much more complex).
2. Like other organs, many basic neural processes and structures have been conserved by evolution, so that one can posit, and indeed identify, similarities not merely in other primates, but also in other vertebrate animals, and indeed, in some cases, even in other phyla. Thus one can use animal models to understand these features that humans share with other creatures.
3. Neural processes in the brain, and their common features across species, can and should be anatomized at a molecular level—that is to say, each event in the brain can and should in principle be traced to identifiable molecular events—thus investigations should proceed in a reductionist form, that is to say, by exploring these in the simplest possible systems.
4. At that level, the key processes are those of neurotransmission, that is to say, communication along and between neurons. Communication is a combination of chemical transmission (across the synaptic cleft) and electrical transmission (down the neuron itself), and neurons are of different types, in part depending on the neurotransmitters that they use—dopamine, serotonin, and so forth.
5. Neurotransmission also entails the function of multiple other entities: ion channels, transporters, receptors, enzymes that catalyze or metabolize neurotransmitters at different rates, and so forth. Variations in each of these elements have functional significance and can in principle account for processes at higher levels.
6. Different parts of the brain—both different structural areas and different types of neurons using different neurotransmitters—have different evolutionary histories, and have been evolved for different mental functions, and these networks of interconnected neurons can be identified anatomically and visualized by observing levels of activation in different brain regions when the organism is undertaking different tasks or is in different cognitive, emotional, or volitional states.
7. All mental processes reside *in the brain* (where else could they reside!), and each mental process will reflect, or be mediated by, or have something variously described as a correlate, an underpinning, or a basis, in brain events.
8. Thus any mental state or process (normal or abnormal), and the behavior associated with that state or process (normal or abnormal), will have a relation—exactly what relation is in dispute—with a potentially observable material process in the organic functioning of the neuromolecular processes in the brain.

Explicitly or implicitly, those who belonged to the invisible college of neuro-science shared this neuromolecular vision of the brain. However intellectu-ally challenging it was in practice, in principle the brain was to be understood as an organ like any other organ (Schmitt 1967). The mental functions of per-ception, cognition, emotion, volition, and so forth could all be accounted for, in principle, by processes operating at the molecular level that was their substrate, and that underpinned or subserved them.

Initially, the focus of this reductionist style of thought was on features that were common to all members of a species and especially those that were con-served across species by evolution. But what of variation? It was, of course, well understood that there were variations in all of the above, from individual to individual and across the life span, as well as across species. The belief that these variations were evolutionarily intelligible, meant, of course, that they were shaped by genetics, as did the fact that they could be affected by mutations, both naturally occurring and artificially induced though mutagenesis. This is not the place to trace the history of neurogenetics or the role played by such key figures as Seymour Benzer in identifying the genetic bases of behavioral characteris-tics in his favored model animal, the fruit fly *Drosophila melanogaster*.[31] In this style of thought, genes were conceived of as more or less individualized, distinct units of inherited information that determined the synthesis of proteins. They could occur in a small number of distinct forms, or alleles, which could affect the phenotype of the animal concerned—either its physical properties or in its behavior. These could be investigated in the laboratory by artificially inducing mutations in the genes of individual animals—that is to say by mutagenesis—and then by breeding from those animals and observing the effects in their offspring, and if necessary, in further generations. A whole range of technolo-gies were used to study these changes and the way in which they were carried in pedigrees or lineages, and to explore the relations between genes, brains, and behavior, not just in laboratory studies of animals, but also in field studies of the genetics of mental disorders in humans. While few questioned the advances made in behavioral genetics of animals, the use of these explanatory regimes in humans was highly socially and politically controversial, not least because of the historical association with eugenics: genetic accounts of mental illness ap-peared to naturalize mental distress, to make it a matter of pathological brains not of pathogenic social experiences, and to legitimate physical methods of in-tervention, notably through the use of drugs.[32]

In the aftermath of the sequencing of the human genome, however, a new style of thought took shape that may still speak of genes, but that focuses on variations at a different level—at the molecular level of the base pairs them-selves that code for proteins or that act to regulate gene expression. Variations between individuals were now thought to be related to variations at this mo-lecular level, where one base might be substituted for another, thus affecting the nature of the protein or the activity of the enzyme that was synthesized. Con-ceived in this way, molecular genomics came into alignment with molecular neurobiology. Research could now explore the genomic basis of variations in the molecular components of neural activity—of the neurotransmitters or the enzymes that metabolize them, the structures of receptor sites, ion channels, and so forth, and, indeed, of the factors that influence methylation and gene expression, or neurogenesis.

This molecular vision of genomic complexity thus mapped onto the vision of the neuromolecular brain. In this way of thinking, the properties and variations in the former underpinned all differences in human mental functioning, whether these be deemed normal variations or pathologies. Other molecular-level variations in DNA sequences were discovered, notably copy number variations (CNVs), where a relatively large region of sequence on a chromosome, sometimes within a coding sequence, sometimes involving several coding sequences, may be repeated several times. Taken together with research that was beginning to grasp the mysterious operations of the stretches of DNA previously thought of as 'junk' as well as the complex operations of RNA in gene transcription, by the first decade of the twenty-first century, genomics was a world away from notions of simple genetic determinisms of the 'gene for schizophrenia' variety. As we shall see in later chapters, this shift in thought was linked to a very significant shift in strategies of intervention and prevention.

Despite constant references to the social importance of the project of neuroscience by those who sought funding and recognition for this new discipline, and, indeed, the initial concerns of many of its key figures with problems of mental illness and practices of psychotherapy, until the closing decades of the twentieth century, most of the actual research undertaken in the institutions and laboratories of the new discipline was primarily focused on investigations of what were thought to be basic neural processes at the neuromolecular level. The reductionist approach that started to gain more prominence in neurobiology in the 1950s and 1960s aimed at studying primarily known and simple behaviors, usually in their normal, or physiological, state: "*normal* mentation: of perception, motor coordination, feeling, thought, and memory" (Kandel 1982, 299). It seemed that, in order to account for abnormal mental states, one first needed something like a map of the so-called normal nervous system: one had to proceed "one cell at a time" (Kandel 2006, 55). One needed to reduce the complex phenomenon to a simple or basic process. One needed to chart the normal neuronal circuitry involved in simple and normal behaviors; inducing abnormalities was indeed important experimentally, but only in order to assess their implications for the understanding of normal patterns and functions.

Nonetheless, the burgeoning neurosciences were full of hopes and also claims about their social relevance. Thus Schmitt, writing on the "promising trends in neuroscience" in 1970 suggested that it has "relevance to social needs" (Schmitt 1970, 1008): research in animals on hormonal reaction to psychosocial stress may have relevance to ghetto disturbances; basic and clinical research in neuroscience may help address the drug menace to society evidenced by "extensive use of various psychotomimetic substances by adolescents and young adults and the widespread addiction in suburbia to 'energizing' and tranquillizing drugs"; work with brain stimulation may help investigate "the neural, hormonal, biochemical and genetic mechanisms of aggressive and other aberrant behavior," and so forth. Such speculations, however, remained gestural; Schmitt concluded, "Perhaps still more important to man's progress is the possibility of a better understanding of *basic* mental processes which, in the end, is responsible for the advancement of science and society" (Schmitt 1970, 1008; emphasis added).

Nonetheless, the neuromolecular vision of the brain that was taking shape blurred two historically significant boundaries, and in so doing, paved the way for neuroscience to leave the lab for the world outside. *First*, what one might term the Cartesian boundary, so crucial to psychiatry since its beginnings in the mid-nineteenth century, between organic and functional disorders—the former being regarded as originating in identifiable lesions in the brain, the latter being considered to be a disturbance of mental functioning in which the symptoms have no known or detectable organic basis, and whose origin is thought to be in life history, biography, stress, or some other experience. In the neuromolecular style of thought, all mental disorders and disturbances must, in principle, and in some potentially identifiable way be related to anomalies within the brain. *Second*, the distinction between 'states' and 'traits,' which had grounded the division between psychiatry and psychology.[33] States were intermittent periods of illness—a person had been well, became depressed, and then, as a result of treatment or merely of time, returned to more or less his or her normal state of mind. Traits were pervasive features of mind or character, attributes one was born with, perhaps that one inherited, aspects of one's personality, such as, for example, that one was introverted or of a melancholy disposition. On this distinction between states and traits, a disciplinary stand-off between psychiatry and psychology was erected in the mid-twentieth century. Psychiatry would deal with abnormal states, for after all, these were illnesses, which could be treated and perhaps even cured, and psychiatry was a medical discipline. Psychology, on the other hand, would deal with traits, for these were aspects of character and personality, even if they were problematic or pathological: they could not be treated or cured, although they could be assessed and perhaps even managed better via forms of psychological therapy.[34] But if both states and traits essentially were variations of the same molecular mechanisms, that distinction blurred, and along with it the distinction between personality disorders and psychiatric illnesses—perhaps, even, the disciplinary divide between psychology and psychiatry when it comes to intervention.

And indeed, it was in and through practices of intervention, and research aimed at generating interventions, that the conceptual architecture of the neuromolecular brain was established. We have already remarked on the crucial role of psychopharmacological research in identifying processes of neurotransmission and in explicating its elements—as receptor sites, reuptake mechanisms, and the like—all of which were first invoked as hypothetical entities to explain differences in action of various drugs on the brain, and then, through sophisticated experiments in vitro and in vivo, accorded an anatomical reality. The researchers, and the pharmaceutical companies who were funding much of the research, became convinced, not only that this was the pathway for the development of powerful, specific, and effective psychiatric drugs, but also that one could divide up the psychiatric nosology according to the pharmaceuticals that would act on each condition, and then develop and market the drug for each.

It became increasingly common, and acceptable, to seek to modulate troublesome mental states—from the minor anxieties of family and work to the major disturbances of thoughts and feelings of schizophrenia and depression—by drugs that, according to their manufacturers and marketers, acted precisely on specific sites in the brain (Rose 2003b, 2004). Beyond the walls of any asylum, when troubles arose in the family, the school, the workplace, the prison, the military, and in the management of daily life, a plausible response—often the first response—was to seek a drug that would mitigate the problems by acting on the brain. The management of everyday life through neurobiology was no longer associated with science fiction. This neuromolecular vision of the brain was fundamental to the rise of 'psychopharmacological societies' over the closing decades of the twentieth century.

As the rising rates of consumption of minor and major tranquilizers, antidepressants, and antipsychotic medications showed, many were happy to accept these designations, or were persuaded or coerced to consume the products of these styles of thought.[35] But at the same time, a vocal movement began to take shape to contest them. The antipsychiatric movements of the 1960s and 1970s disputed the expansion of the categories of mental disorder to ways of thinking or acting that were at the least only unconventional, and at the most, were challenges to the existing social order. And they disputed the conviction that the use of psychiatric drugs should be understood as treatment of a disorder of the brain, rather than normalization of those who were socially challenging or disruptive. We return to these issues in chapter 4. However, whatever the fate of this sociopolitical challenge to the explanatory system and social practice of psychopharmacology, along with its founding mythologies of the neurobiological specificity of psychiatric classifications and the neurobiological specificities of the mode of action of drugs, the neuromolecular vision of the brain to which these were linked was to underpin research in the neurosciences for the next half century.

Enter Plasticity

It is no surprise to find that the research problems addressed by developmental neurobiologists were increasingly shaped by evidence and arguments from this growing neuromolecular vision of the brain. Those developments are illustrated by the pathway of research from the 1950s onward, which eventually led to the discovery of Nerve Growth Factor by Rita Levi-Montalcini and the identification of its molecular mechanisms in the 1970s (Levi-Montalcini 1982). Toward the end of the account of this journey that she published in 1982, Levi-Montalcini asked a key question: "Is the formation of neuronal circuits in the central nervous system and the establishment of specific connections between nerve fibers and peripheral end organs rigidly programmed and unmodifiable, or are nerve fibers endowed with sufficient plasticity to allow for deviation from predetermined routes, in response to chemical signals issued during neurogenesis and regeneration from neuronal and non-neuronal cells?" (ibid., 356). The growing attention to these questions of development, neurogenesis, and plasticity was to establish another key condition for the movement of neuroscience from the laboratory to the world. For it was linked to the growing belief that, when it comes to the *human* brain at least, neither structure nor function were fixed at birth, or inscribed in the genes. The neural architecture of the brain, it seemed, had to be located in the element of temporality—not just the time of development from fertilization to birth and into the early years of life, but also the time of the life-course, through adolescence, into young adulthood, and indeed across the decades.

By the close of the twentieth century, the brain had come to be envisaged as mutable across the whole of life, open to environmental influences, damaged by insults, and nourished and even reshaped by stimulation—in a word *plastic*. An ISI Web of Knowledge search reveals a steadily increasing rate of journal articles on this theme throughout the 1990s and 2000s. And in the first decade of the present century, more than a dozen semipopular books on this theme were published, mostly stories of hope invested in the power of the brain to change across a lifetime (for example, Doidge 2008; Begley 2009; Arden 2010). While the details of the history of the idea of neuroplasticity are disputed by specialists, a certain version is now entering common sense. The Wikipedia entry in December 2010 embodies this common sense when it asserts that neuroplasticity is "the ability of the human brain to change as a result of one's experience," that the brain is "plastic" and "malleable," and that the "discovery of this feature of the brain is rather modern; the previous belief amongst scientists was that the brain does not change after the critical period of infancy."[36]

The belief that up until about 1970, the human brain was viewed as immutable after the period of its development in utero and in early childhood has become something of a modern myth, though no less important for that. It is true that most researchers followed Ramón y Cajal's dogma from 1928 that: "[o]nce development has ended, the fonts of growth and regeneration of the axons and dendrites dried up irrevocably. In adult centers, the nerve paths are something fixed and immutable: everything may die, nothing may be regenerated" (Ramón y Cajal 1928, quoted from Rubin 2009). Other researchers demonstrated the importance of 'critical periods' in development of various

sensory abilities such as vision, hearing, language, and certain other capacities in humans. However, there was general acceptance that synaptic connections among neurons were constantly being created and pruned across the life
of the organism, as a result of neural activity arising from experience—even
Cajal, it seems, used the word *plasticity* in this context (Berlucchi and Buchtel
2009). This premise was most clearly stated in Hebb's famous postulate from
1949 that "any two cells or systems of cells that are repeatedly active at the
same time will tend to become 'associated,' so that activity in one facilitates
activity in the other" (Hebb 1949, 70)—usually glossed as 'what fires together,
wires together.'

A series of developments from the 1960s onward established neuroplasticity as a premise for those interested in the implications of neuroscience for
therapy and policy—and a premise entirely congruent with the hopeful ethos
that would imbue the life sciences over the closing decades of the twentieth
century. It had long been known from neurological studies of patients with
brain damage that the brain could somehow recover some of the functional-
ity lost as a result of stroke or injury that had caused a lesion in a key area,
suggesting that other areas of the adult brain were capable of taking over
those lost functions. But Paul Bach-y-Rita's famous work on the rehabilitation
of patients with brain injury would firmly establish this as the underpinning
of therapy in such cases (Bach-y-Rita 1967). Other research carried out on
animals, notably monkeys, strengthened this belief in plasticity.

In their popular account in *The Mind and the Brain: Neuroplasticity and
the Power of Mental Force*, Jeffrey Schwartz and Sharon Begley suggest that
one can trace the discovery of neuroplasticity to the Silver Spring monkey
experiments in the 1970s (Schwartz and Begley 2002). In these experiments,
carried out by at the Institute of Behavioral Research in Silver Spring, Maryland, Edward Taub cut the afferent nerves that carried sensations from the
limbs of the monkeys to their brains and then explored the ways in which
they might be trained to use those limbs despite the supposed absence of sensations. An undercover animal rights activist reported Taub's work to police,
the monkeys were seized by federal agents, and a series of trials ensued in
which the primary investigator, Edward Taub, was accused of animal cruelty
and initially convicted, although that verdict was overturned on appeal.[37] Following these trials, the monkeys were eventually killed, and on dissection it
appeared that significant cortical remapping had occurred; publications in
the 1990s took this as evidence of the plasticity of the somato-sensory cortex
in adult primates despite long-term deprivation of sensation (Jones and Pons
1998).

Taub eventually took a position at the University of Alabama and developed his findings into a program called constraint-induced movement therapy, or more simply Taub therapy, that "empowers people to improve the use
of their limbs, no matter how long ago their stroke or traumatic brain injury."[38] By the time of the publication of the research into the brains of these

monkeys, there was other evidence that showed that the mapping of sensory functions such as vision onto the cortex could be redrawn even into adulthood (Wall, Kaas, Sur, et al. 1986; Merzenich, Recanzone, Jenkins, et al. 1988; Jenkins, Merzenich, Ochs, et al. 1990). Like Taub, Merzenich was also to go on to found a number of companies seeking to apply such findings therapeutically to humans—to turn them into a technology. These include Scientific Learning, which uses his Fast for Word software, and Posit Science which seeks to develop behavioral therapies, including brain training software called Cortex and Insight.[39] Plasticity was to become one of the key dimensions of the matrix that linked the laboratory, the corporation, and the everyday world.

A further dimension was added by the work of Elizabeth Gould and her group, which seemed directly to contradict Cajal's 'dogma'.[40] Early claims to have discovered the formation of new neurons in the brains of birds and some other species had been discounted: leading authorities, notably Pasko Rakic, expressed the view that neurogenesis in the adult brain of primates could not occur, as the generation of new neurons would interfere with the neuronal storage of learning (Rakic 1985). But in the 1990s, using newly developed DNA labeling techniques, Gould and her colleagues were able to identify neurogenesis in adult mammals, first in adult rodents, where it was enhanced by training on learning tasks (Gould, Tanapat, Hastings, et al. 1999), and later in primates, where it was modulated by hormones and by stress (Gould and McEwen 1993; Cameron and Gould 1994; Cameron, McEwen, and Gould 1995; Gould, Reeves, Graziano, et al. 1999).

There remained many doubts about the functional implications of such newly generated neurons, let alone the extent to which such properties might be harnessed therapeutically. But by 2004, Gould herself was drawing social and policy implications, arguing that early adverse experience inhibits structural plasticity in responses to stress in adulthood, as did other factors such as social isolation (Mirescu, Peters, and Gould 2004; Stranahan, Khalil, and Gould 2006). By the end of the decade, she and her colleagues drew on evidence from neurogenesis to emphasize the need to understand the impacts of parenthood on the developing brain of the child to avoid potentially serious consequences for cognition and mental health (Leuner, Glasper, and Gould 2010). The idea of neurogenesis, and that external inputs might stimulate or inhibit it, was rapidly seized upon in sociopolitical debates. For some progressive thinkers, it supported what they had believed all along— that the environment—by which they seemed to mean that ill-defined social domain generated by political action—was a, perhaps *the*, crucial determinant of human abilities.[41] For others, for whom the environment was thought of, principally, as the family, it seemed to confirm that pathological consequences flowed from poor family environments, and hence that intensive family intervention was the key to addressing social problems (Allen and Duncan-Smith 2008).

These arguments for early intervention in the name of the brain drew support from another body of work that framed plasticity in terms of epigenetics. Michael Meaney and his group had been carrying out research in rodents from the 1980s on the effects of early experiences of maternal care on the developing brains of offspring. In the first decade of the twenty-first century, in a number of very widely cited papers, they came to understand these effects in terms of epigenetic programming, arguing that the mother's behavior toward her pup shapes the expression of genes in its brain through altering methylation, and that this shapes neuronal development, which in turn has consequences for that pup's later behavior toward its own offspring—hence providing a mechanism whereby these changes could be passed down the generations (Meaney and Stewart 1979; Meaney, Aitken, Bodnoff, et al. 1985; Champagne, Chretien, Stevenson, et al. 2004; Champagne and Meaney 2006; Szyf, McGowan, and Meaney 2008).

By 2009, Meaney and his colleagues were suggesting that these findings could be extended to humans, for example, suicide victims with a history of child abuse (McGowan, Sasaki, D'Alessio, et al. 2009; Meaney and Ferguson-Smith 2010). The brain now seemed open to environmental inputs, not just at the level of the synapse, at the level of cortical mapping, or at the level of the neuron and neurogenesis, but at the level of the molecular processes of the genome with consequences that might pass from parents to children and even on to *their* children. The developing brain of the child now seemed to be a key site through which a range of social problems could be understood; once mapped onto the brain, paradoxically, they became more, not less amenable to intervention—governing through, and in the name of the plastic brain.

A Neuromolecular and Plastic Brain

The processes and structures of the brain and central nervous system were now understandable as material processes of interaction among molecules in nerve fibers and the synapses between them, in terms of the chemical and electrical properties of their constituent parts. While the explanatory gap still remained, and the move from the molecular level to mental processes of volition, emotion, and cognition was recognized to be highly challenging, the dualism that had haunted philosophy and the sciences of mental life seemed increasingly anachronistic. In this profoundly reductionist approach, despite the recognition that there was much that could not yet be explained, there seemed nothing about the operations of the nervous system that could not potentially be explained in terms of the biophysical properties of its component parts.

But the growing acceptance of plasticity acted as something of a counterweight to such reductionism. Ideas about plasticity and the openness of brains to environmental influences, from initial evidence about nerve development, through the recognition that synaptic plasticity was the very basis of learning and memory, to evidence about the influence of environment on gene expression and the persistence throughout life of the capacity to make new neurons—all this made this neuromolecular brain seem exquisitely open to its milieu, with changes at this molecular level occurring throughout the course of a human life and thus shaping the growth, organization, and regeneration of neurons and neuronal circuits at time scales from the millisecond to the decade. This was an opportunity to explore the myriad ways in which the milieu got 'under the skin,' implying an openness of these molecular processes of the brain to biography, sociality, and culture, and hence perhaps even to history and politics. It could act as an antidote to suggestions of unidirectional causality. For if that openness was a condition of normal development, perhaps abnormal development might be understood in terms of such influences as well, and perhaps, by modulating those influences, one could indeed open the brain—and the person—up to calculated intervention at the neural level.

Hence the plastic brain itself became thinkable as open to government by experts.[42] It simultaneously became open to action by each individual themselves. Dozens of Internet sites were developed to instruct us how to harness brain plasticity for our personal growth—informed by books such as that by Norman Doidge—for "your brain constantly alters its structure according to your thoughts and actions has enormous implications for your personal growth, particularly in the areas of behavior change."[43] Correctly informed about plasticity, we could rewire our brains for love, rewire corporate brains, and so much more (Zohar 1997; Begley 2007; Lucas 2012). The plastic brain becomes a site of choice, prudence, and responsibility for each individual.

Chapter Two

· · · · · · · · · · · · · · · · ·

The Visible Invisible

The biological basis of mental illness is now demonstrable: no one can reasonably watch the frenzied, localized activity in the brain of a person driven by some obsession, or see the dull glow of a depressed brain, and still doubt that these are physical conditions rather than some ineffable sickness of the soul. Similarly, it is now possible to locate and observe the mechanics of rage, violence and misperception, and even to detect the physical signs of complex qualities of mind like kindness, humour, heartlessness, gregariousness, altruism, mother-love and self-awareness.

—**Rita Carter**, *Mapping the Mind*, 1998

Neuroscience, and the study of the activity of the brain, is beginning to bring its own illumination to our understanding of how art works, and what it is. I have come to see the delight in making connections—of which metaphor-making is one of the most intense—as perhaps the fundamental reason for art and its pleasures. Philip Davis, at Liverpool University, has been working with scientists on responses to Shakespeare's syntax, and has found that the connecting links between neurons stay "live"—lit up for longer—after responding to Shakespeare's words, especially his novel formations of verbs from nouns, than they do in the case of "ordinary" sentences.

—**A. S. Byatt**, "Novel Thoughts," 2007[1]

In mid-February 2007, a press release generated considerable interest in the popular media. Titled "The Brain Scan That Can Read People's Intentions," it reported the work of an international team led by John-Dylan Haynes at the Max Planck Institute for Human Cognitive and Brain Sciences in Germany. The team had conducted some laboratory experiments imaging the brains of subjects who were confronted with a simple task in which they had to decide between a small number of possible choices. The scans showed activity in particular brain regions *before* the action was undertaken; it seemed that one could predict the later choice from the earlier pattern of activation. On this basis, Haynes sketched out a mind-reading scenario in which such scans could reveal a person's intentions in advance of actions, and he called for an ethical debate about the implications. Neuroethicists pitched in, expressing their worries to the journalists, happy to take this prospect seriously. Of course, this scenario was a fantasy; the highly artificial and simplistic laboratory experiment told us nothing about the formation of actual complex human intentions to undertake meaningful actions in the messy real world. But it was premised on a belief that is coming to acquire the status of truth: that experiments scanning the brains of individuals in laboratory settings to identify patterns of activation while they undertake simple tasks or games can reveal the workings of the human mind in real time—and, moreover, that they can show exactly which regions of the brain are involved in exactly what specific mental states, not just emotion and cognition but also intentionality, volition, and more. It seems that these new technologies of visualization have finally and objectively revealed the physical basis of human mental life in patterns of activity in the living brain.

It would be misleading to claim that, by the end of the twentieth century, brain imagers believed that the age-old dilemmas concerning the relations of mind and brain had been resolved.[2] Most neuroscientists are careful to speak of 'the neural correlates of mental processes,' avoiding the language of causes, which suggests that the neural is prior to and determinative of such states, and the language of identity, which suggests that mental states are simply neural states.[3] Their press releases and media accounts are less cautious. And the images produced by brain scans are certainly central to the new ways in which this issue of mind and brain is being posed, not just in the laboratory, or in psychiatry, but in the popular media, in the arts, and in many other areas of human life. It is not simply that many now believe it to be possible to locate each faculty of the human mind—emotion, fear, self-control—in a specific region of the brain; it is also the belief that one might, in principle, use this new knowledge of the brain to target interventions into the mind on these faculties in practices from the therapeutic to the artistic and the commercial.

We have many good studies of the intertwined social, technological, and conceptual history of medical imaging in general (Kevles 1997; Burri and Dumit 2007; Saunders 2008), and of brain imaging in particular (Dumit 1997, 2003; Beaulieu 2000a, 2000b, 2002, 2004; Roepstorff 2001, 2002, 2007). While we draw upon these studies in this chapter, our focus is quite specific—it is on the diverse attempts to render 'mind' thinkable by means of images. Without pretense to a genealogy of our contemporary forms of visualization, we explore some of the technologies that have been invented and the kinds of information that they seem to provide. We examine the modes of interpretation and arguments that have been developed around those technologies and their products. And we sketch out the new ways of thinking about and articulating the relations of the corporeal, the mental, and the psychopathological that these entail and support. This visual imaginary has been one pathway along which neuroscience has been able to move out of the laboratory and into the territory of everyday life, and to play a role in the management of normal and problematic conduct.

The Clinical Gaze

When we speak of rendering the mind visible, we do not just mean the act of seeing. We mean the whole configuration by means of which a certain way of seeing becomes possible and can be articulated. The ways of seeing that characterize a scientific practice, or a mode of intervention, do not arise from an isolated act of looking, but from a conjunction of different elements and practices. One dimension is *spatial and temporal*. This distribution occurs along a number of planes: across the individual body (localization), across the collective body (social factors in causation and recovery), across time (the life-course and the course of illness), across generations (family pathology, heredity). A second dimension of this conjunction is *technical* or perhaps *technological*. It consists in the means, the apparatus and devices, that render that which is observed into marks, lines, colors, spaces and edges, patterns and patterning. As many recent studies from the social and cultural sciences have demonstrated, the images produced by the electroencephalographs, the CT scanners, the magnetic resonance imaging machines, and all the other technical devices of contemporary brain mapping are no more 'true to nature' than those produced by the other "engines of visualization," such as photographs, so tellingly analyzed by Patrick Maynard (Maynard 1997). But it is only through such artifacts that we have become able to see that which is not present to our un-augmented eyes.

A third dimension of this rendering visible, perhaps the most significant, consists in the practices within which acts of seeing are enmeshed. They are always bound up with specific ways of *practicing*, occurring within specific sites and by means of particular techniques of intervention upon those who are the subjects of that act of seeing. When it comes to seeing the brain, seeking to discover within its fleshy volume the traces of the pathological or normal mental processes that the brain might embody, it involves the designation of those who have the authority to see: doctors, neurologists, researchers, psychopharmacologists, geneticists, and now, of course, the imagers. It also involves the subjectification of those who are spoken about—subjectification in the sense that living creatures become subjects of these visualizing technologies only as a consequence of certain technical interventions, and subjectification in another sense, in the case of humans, whose sense of themselves may well be transformed as a result of the images of their brains with which they are presented. The act of seeing, or the practice of seeing, also involves a particular locale for the act of visualization—the designation of a case in the asylum, the 'demonstration' of the patient in the clinic, the dissection of the brain in the laboratory, the scanning of the subject in the imaging suite. A real space is required that is all too often absent when the image produced in the scanner is being interpreted. To render visible, that is to say, requires conditions of possibility within a larger networked, distributed, assembled field of intensities and powers—connecting up such diverse sites as the clinic, the lab, the pharmaceutical company, popular literature, and the mass media. Seeing, rendering visible, is thus part epistemology, part topography, part technology, part objectification and subjectification, part network of forces, even part ethics.

Our argument in this chapter is intended, in part, as a homage to Michel Foucault and to his 1963 book, *Naissance de la clinique*, translated into English under the title *The Birth of the Clinic: An Archaeology of Medical Perception* (Foucault 1973).[4] Alan Sheridan, the translator, remarks that the word *gaze* would have been a better choice in the subtitle than *perception*—the book is about the formation of a particular clinical gaze—a way of seeing, saying, and doing in relation to illness, the body, life itself. It charts a fundamental shift in the medical gaze, which occurred in Europe in the early nineteenth century. Previously medicine had focused on a two-dimensional space of tissues and symptoms, but from that time forth, the gaze of the doctor would plunge into the interior of the body. To diagnose an illness would be to interpret symptoms in terms of the internal organic malfunctions that were their cause. In life, the gaze of the doctor would now need to be augmented, initially by such devices as the stethoscope. In death, that which had been rendered into thought while life persisted would be confirmed—or disconfirmed—by the dissection of the corpse. This new way of seeing was inscribed into the anatomical textbooks and embodied in a multitude of anatomical models, educating the eye of the doctor by progressively revealing the layers of skin, muscles, blood vessels, nerves, and organs that made up the interiority of the living being.[5] It was the body itself that had become ill.[6] Through a knowledge of how that illness appeared in the sick body, it became possible to understand what the absence of illness might look like—to envisage the functioning of those interior processes of the body to produce the state we took to be health. To know normality, it was necessary to see pathology. But if this was the case for the body, what then for the afflictions of the mind?

Inscribed on the Body Itself

Sander Gilman is the preeminent historian of ways of "seeing the insane" (Gilman 1982).[7] Gilman tells us that, from the sixteenth century onward, to speak of madness was also to visualize it. Images of madness were central to the ways in which the mad person was turned into a proper object for rational knowledge and for the practice of a cure. Madness was represented iconically through the association of the mad person with images such as that of the divided stick or the pinwheel. It was represented in the complexion of the mad person, where the black bile that was at the root of the condition was manifested in the dark hue of complexion, and this also linked madness to the blackness of skin that was consequent upon the soul falling from grace. It embraced the posture, gesture, and movements of the figure. The melancholic was seated, eyes cast down, clothing disheveled, hands clenched or hidden to manifest a turning away from useful labor. The maniac raved in semi-nakedness, arms akimbo, and mouth agape. The epileptic and the ecstatic shared a posture: arms and legs flailing, body in the fixed curve known as the *arc de cercle* that would become, for Jean-Martin Charcot at the Salpêtrière, the emblem of the hysterical crisis.

From the mid-eighteenth century onward, from the essay on the passions in Diderot and d'Alembert's *Encyclopédie* through Johann Lavater's *Physiognomische Fragmente* (1774–78), madness became associated with a specific physiognomy, initially an emphasis on the fixed aspects of skull shape and facial appearance, which could be derived from particular techniques of inscription of the head on a homogeneous and regular plane of two dimensions. Some disputed the claim that there was a direct relation between psychopathological states and these fixed physical characteristics. They argued that other visible features—expression, gesture, posture—were the real external signs of the inner pathology. But in either case, few doubted that madness was inscribed on the surface of the individual, that, as Arthur Schopenhauer was to put it in his work on physiognomy published in the mid-nineteenth century, "the outer man is a graphic reproduction of the inner and the face the expression and revelation of his whole nature" (quoted in Gilman 1982, 164).

Every student of psychiatry is told the story of Philippe Pinel, acclaimed as the radical physician of the French Revolution who publicly struck the chains from the inmates at Bicêtre in Paris in October 1793, restoring the mad person to the status of an individual citizen with rights, and inaugurating 'moral treatment.'[8] The enlightened asylums of the early nineteenth century also flooded the inmates with a new light, a light that that made them individually visible in new ways. Pinel himself introduced illustrations of the insane in his *Traité médico-philosphique sur l'aliénation mentale, ou la manie* (Pinel 1801). These compared the physiognomy and skulls of "an idiot" and "a maniac" in a style familiar to Lavater and the craniologists, with a special emphasis on the proportionality of the parts and the relation of head size to body size: these drawings of particular individuals seemed to show that for each case, appearance was the key to diagnosis. In the ensuing years, Pinel's pupil Jean Esquirol documented the newly visible varieties of insanity in his *Dictionnaire des sciences médicales* (1812–22). He commissioned some two hundred drawings of the insane at the Salpêtrière, focusing both on facial expression and on bodily comportment, each of which was used to illustrate the varieties of insanity that he described. When he published his collected papers in 1830, he selected twenty-seven of these illustrations and appended them in the form of an *Atlas* (Esquirol 1838).

The use of the word *atlas* in such contexts is conventionally traced to Gerard Mercator, who adopted it "to honour the Titan, Atlas, King of Mauritania, a learned philosopher, mathematician, and astronomer." It appears first to have been used posthumously on the title page of a collection of his maps published by his son in 1595: the illustration of Atlas with the globe of the earth in his hands as the frontispiece was added to later editions (Hall and Brevoort 1878).[9] The atlas format was taken up by William Playfair, "who, in 1786 published a collection of forty-four diagrams and charts, the content of which was in the nature of political economy, with the title *The Commercial and Political Atlas*: he "founded a tradition . . . in which collections of charts and figures appertaining to disciplines like biology, medicine, physics as well as statistical overviews were also put into the category 'atlas'" (Krausse 1998, 11–12). The lines, curves, and shapes in the diagrams that Playfair produced, Krausse suggests, enable the eye to reckon without calculating—to commit information to memory in seconds that would take days to assimilate if it were in the form of words or numbers.

This was equally true of the lines and curves of Esquirol's drawings. They certainly recycled some of the old themes in the positioning of hands and body. They also followed the physiognomies in their ways of rendering the shape and proportions of the skull. But these plates did something more. They were part of a fundamental individualization that rendered madness into a possible object for a clinical medicine. The laborious procedures of observation and documentation that accompanied moral treatment visualized each insane person as an individual case, but one whose uniqueness was intelligible because it could be charted and measured in terms of the general norms of types of madness, their characteristics, etiology, and prognosis. Esquirol gave a detailed description of the visual appearance characteristic of each type of madness, accompanied by an account of a specific case—an individual diagnosed with mania, dementia, or lypemania—focusing not just on the course of that person's condition and its symptoms in speech, behavior, posture, and comportment, but also his or her visual appearance—the color of the hair, the shape of the eyebrows, the complexion of the skin, the look of the face, and much more. In this way, the image was fused to the biography in the form of the case, and inscribed at the heart of psychiatric epistemology and diagnostic practice. As Sander Gilman argues, such combinations of description, case study, and picture bridge the theoretical and the observable (Gilman 1982). The patient is rendered into thought in terms of the theory; the means of diagnosis have merged with the object of diagnosis itself. In these images of the mad person, the norms of psychiatric thought seem to have merged with the individual subject of psychiatric practice in the portrayal of each specific case.

Esquirol was not the first to use portrayals of the inmates of asylums to document the varieties of madness through appearance. As Gilman shows us, Charles Bell had done so at the very start of the nineteenth century in his vividly illustrated *Essays on the Anatomy of Expression in Painting* (Bell 1806) focusing in particular on the fear and terror of the mad, and Alexander Morrison, having visited Esquirol in Paris, published his own series of plates in his *Outline of Lectures on Mental Diseases* of 1825 (Albin 2002), which rapidly went into two further editions and led to his later *Physiognomy of Mental Diseases* (Sani, Jobe, Smith, et al. 2007). And the idea that madness could be captured in the image of the mad person was pervasive well beyond the asylum. From the drawings of William Blake and Thomas Rowlinson to those of Goya and Delacroix, whatever their stylistic differences, to see madness in the face, in the posture, in the comportment of the body, was to know it for what it was.

Clinical medicine in the nineteenth century mutated fundamentally when the gaze of the doctor focused beneath the skin, into the interior of the patient's body. But across the second half of the nineteenth century, the gaze of the mad doctors remained stubbornly on the surface; the images played the same epistemological role when they moved from drawings to still photographs and then to moving images. Whether in the photographs that Hugh Diamond presented to the Royal Society in 1856,[10] in William Hood's collection of Henry Hering's photographs of the inmates of the Bethlem Hospital of which he was the first superintendent (Gale and Howard 2003; Logan 2008), in Dietrich George Kieser's *Elemente der Psychiatrik* of 1855 (Kieser 1855), in Max Leidsdorf's *Lehrbuch der Psychischen Krankheiten* of 1865 (Leidesdorf 1865), in the nosological archives created in the San Clemente Hospital in Venice in 1873,[11] in the photographs taken at the Devon County Lunatic Asylum used by Bucknill and Tuke to illustrate their *Manual of Psychological Medicine* (Bucknill and Tuke 1874), or in Henri Dagonet's *Nouveau traité élémentaire et practique des maladies mentales* (Esman 1999), the photographs seemed to show, apparently without requiring the artifice of the portrait painter, that the invisible states of the human soul were visible in the proportions of the mouth, the hue of the complexion, the expression on the face, the posture of the body and the arrangement of the limbs (most of these examples are discussed in detail in Gilman [1982]).[12] Even Darwin was initially tempted to use similar photographs—supplied by James Crichton Browne, the medical director of the West Riding Asylum at Wakefield—to support his theory of the innate bases for the expression of the emotions. But Darwin recognized the circularity of the procedures by which the images were produced: the unfortunate individuals who appeared in these photographs were clothed and posed in terms of prevailing beliefs as to how emotions were manifested in visual expression—they could hardly serve as evidence for the inherited or universal basis of those expressions (Gilman 1982, 185).

In this continuous work of visual documentation of asylum inmates, we can learn a lot, not merely about the form of the gaze that the mad doctors directed to their patients, but also of the power of the visual in seeming to render the interior world of the human mind into thought—its passions, affections, thoughts—in the form of images. Such images were simultaneously to satisfy the requirements of knowledge (the image as both evidence and proof), of the clinic (the image as diagnostic), of pedagogy (for these pictures were teaching tools), of museumology (for the picture would enter the archive and the plethora of scientific collections gathered across the nineteenth century), of aesthetics (for there was a certain style of representing the interior world here that would go beyond the world of science), and of the popular imagination. Understanding, here, is played out, almost without the need for language, through the apparently empirical form of the image.[13] Despite its objectivity effect, a whole scientific and cultural imaginary underpins these images of mental states—this remains true today, despite the technological gulf between these images and those generated in the brain scanners of our own times.

"Behold the truth. . . . I am nothing more than a photographer; I inscribe what I see." With this quote from Jean-Martin Charcot, Georges Didi-Huberman opens the chapter on "Legends of Photography" in his provocative psychoanalytic study of *The Invention of Hysteria* (Didi-Huberman 2003, 29). Charcot, who worked in the Salpêtrière from the 1850s to 1893, represented the apotheosis of this tradition of visualization of the troubled or deranged mind through the appearance and comportment of the body and the expressions of the face. In 1877, he founded the *Iconographie photographique de la Salpêtrière*. It was published until 1880 and resumed publication under his editorship in 1888, as the *Nouvelle iconographie*. In this form it was to publish twenty-eight volumes until its demise in 1918. It illustrated not only the faces and postures of the varieties of madness, but introduces time into these pictorial representations, for example, providing a series of pictures demonstrating the sequences of epileptic attacks, hysteria, and the like. Vision was the key to grasping the nature and progress of the condition: the troubles of the mind were now intelligible to the extent that, before all else, they were visible. Charcot is, of course, now infamous for the way these demonstrations were carefully staged, and for how they relied on extensive coaching, suggestion, and simulation: yet the photographs became their inscribed and durable counterparts. Charcot's most famous pupil, Freud, referred to him in his obituary as a "visuel," one who sees (Freud [1893] 1962, 12). But Freud, and the psychoanalysis and dynamic psychiatries that followed, turned away from the image and paid scant attention to the visual appearance of the patient—*the voice* was the royal road to the unconscious in all of the 'talking cures' that were to follow.

The chapters on cretinism, dementia praecox, general paresis, manic depression, and idiocy in Kraepelin's *Textbook on Psychiatry* ([1899] 1902) all contained the carefully posed photographs familiar from earlier works, but these images were no longer accorded a particular diagnostic significance. In his *Lectures on Clinical Psychiatry* (translated into English in 1904) Kraepelin describes the patient to his audience: the clothes, the demeanor, the facial expression of emotion. But these are important only for what they reveal of the everyday life of the individual, they are not clues to the nature of his or her condition. The significance of the image of the mad person as the outward and visible sign of an internal constitutional defect or malady was beginning to fade (Kraepelin 1904). Kraepelinian diagnosis was based on the course of the condition and on etiology and prognosis as manifested in conduct, speech, emotion, thought. Illustrations still play a role, but this is, precisely, to illustrate—the pictures in Kraepelin's texts are presented neither as pedagogic nor as diagnostic. But nonetheless, in the closing decades of the nineteenth century, different kinds of image were beginning to become important—images of the brain itself.

Open Up a Few Brains

Of course, even nineteenth-century proponents of moral medicine like Esquirol thought madness was a disease of the brain. But it was in the work of Franz Joseph Gall that we can find the first modern articulation of the idea that the brain is a differentiated organ whose different regions are responsible for different aspects of human mental life.[14] While the physiognomists sought to correlate the form of the face and the proportions of the skull with mental characteristics, Gall's view was that it was the cerebral cortex that was the key (English and Annand 2010; Gall [1822] 1835). The cortex was composed of distinct "brain organs"—regions associated with different faculties—whose distribution could be represented visually to show exactly where under the external skull each organ was located. The size of each brain organ was linked both to its power in the individual and to the configuration of the part of the skull above it. Thus it could be detected externally by examining the form of the skull and the shape and size of the various regions, not just in death, but also in life. Condemned for his materialism in Austria, Gall traveled around Europe, giving demonstrations of his skill at brain dissection, and collecting material—brains and skulls—especially from prisons and asylums.

Gall's craniological work had many disciples; under the name of phrenology, it was promoted in Europe by his former assistant, Johann Spurzheim, and by George Combe.[15] It was institutionalized in phrenological societies and journals and was enthusiastically taken up by many in America. However, by the end of the 1830s, it had lost most of its credibility as a scientific analysis of mental functions, notably at the hands of Pierre Flourens, whose experimental work creating localized brain lesions in animals and documenting their consequences seemed to disprove the claim that there was a close relation between cerebral locales and specific behavioral capacities (Flourens 1824).[16] Nonetheless, phrenology retained considerable popular appeal, especially through the visual technologies developed by Spurzheim and Combe—

the familiar charts, diagrams, and model phrenological heads that showed the different mental faculties distributed across the brain and mapped to specific locations on the external surface of the skull.[17]

Gall's fate was to be derided by association with the role of phrenology in fortune telling and showmanship. But the concept of cerebral localization was to entrance European neurologists in the second half of the twentieth century. In 1861, Paul Broca seemed to have reestablished the truth of localization theory, when he described eight patients with loss of speech, each of whom had lesions in the third left frontal convolution (Broca 1861). He developed this thesis further, with arguments based on specific case studies, such as that of Leborgne (the famous "Tan"), claiming to have demonstrated the centrality of the frontal lobes in many higher functions, and he also developed the thesis of cerebral dominance, arguing for the distinct functions of the left and right cortex. Neurology seemed to have shown without doubt that the brain was an organ composed of regions with specific structural properties that were the basis of the various human mental functions.

In 1867, Wilhelm Griesinger, professor of psychiatry in Berlin from 1865 to 1868, founded the *Archive for Psychiatry and Nervous Diseases*, and this journal provided a focus and a forum for a succession of German psychiatrists and students of brain structure. It also stimulated the development of psychiatric clinics in Germany attached to universities, which helped provide the necessary infrastructure for this organic gaze. In the 1870s and 1880s, Theodore Meynert in Vienna pored over his microscope scrutinizing stained sections of the brains from psychiatric patients in a search for pathological lesions in the cerebral cortex, the frontal lobes, and the nerve fibers themselves. More work on cerebral localization was carried out by Carl Wernicke, also in Vienna, who in 1874 identified the specific part of the brain that, if damaged, led to an incapacity to understand the spoken word or to speak comprehensibly. Wernicke spent the rest of his days trying—and failing—to discover similar links between psychiatric symptoms and brain abnormalities.

Paul Flechsig worked in Leipzig from 1877 to 1922, seeking to localize the basis of many neurological disorders, and mapping the human brain into multiple areas according to the sequences in which they became myelinated. He used this as the basis for his speculations that particular areas in the cortex were most important for the exercise of intellectual functions, for the formation of mental images, the naming of objects, and so forth, and sought to confirm these by postmortem studies of the brains of patients with general paresis and other brain diseases. Flechsig had trained himself in psychiatry, visiting asylums in Germany and spending time with Charcot in Paris (Flechsig's work is usefully discussed in Finger 1994, 308–10). Daniel Paul Schreber, whose case was to be made famous by Sigmund Freud, spent time in his clinic, and Flechsig was the target of Schreber's allegations of "soul murder" (Schatzman 1973, 30–31). However limited his therapeutic achievements, Flechsig's cortical maps were highly influential visualizations of cerebral functions. He believed that what he termed "the posterior association center"—rather than the frontal lobes—was the part of the brain most important for intellectual functions, thus, like so many other neuroanatomists, linking the size of regions to intellectual capacity.

We should not underestimate the risky radicalism of many of the early researchers on cerebral localization: most held explicitly materialist philosophical positions, asserting that human mental capacities could be explained without reference to any metaphysics or religion, and without invoking an immaterial soul. We can see this clearly in the work of the Society for Mutual Autopsy studied by Jennifer Hecht (2003). This organization was established by a group of French anthropologists, psychologists, artists, and intellectuals in the 1870s, and was directly inspired by Paul Broca, who was one of the founding members, along with other neurologists and intellectuals. It was organized around specifically atheist values—notably the belief that there was no soul, only brain. Those who established the society were troubled by the fact that the only brains available for dissection were those of the abnormal, the damaged, the insane—not of the normal or the intellectuals. Membership was based on an agreement to examine each other's brains after death to see to what extent their eminence was inscribed within the brain itself, and what difference their lives may have made to the configuration of the cerebral fissures—a relationship that proved to be problematic when some of the more eminent intellects proved to have small brain volumes, and in some cases, the convolutions of their brains were not fine, as had been hoped, but somewhat coarse.

If the brain was the seat of mental life, it seemed self-evident that there must be a relation between the intellect and the size and configuration of the brain, if the calculations were done correctly. Given the ways in which racial science had long linked the hierarchy of races to the configuration of face and skull, it is not surprising that race was one focus of these investigations of the brain. Samuel Morton was one of the first systematically to measure brain size by calculating the volume of cranial capacity (initially filling the skull with mustard seed, then with lead shot) and comparing average skull volume by race (Morton 1844). But Broca was probably the most rigorous of all craniometrists (this work is discussed in detail by Stephen Jay Gould [1981, esp. chap. 3]). Gould points to Broca's careful refinement of the methods of craniometry and his undoubted respect for empirical and statistical methods. Nonetheless, Gould argues, Broca systematically distorted his data, perhaps unwittingly, and applied all manner of dubious corrections in order to make his results conform with his preconceptions concerning the differences in relative size of brain regions—for example frontal versus posterior cranial regions—in "inferior" versus "superior" races, and between men and women. A few neurologists disputed the basic premise of a relation between brain size and intelligence, but similar endeavors occupied many members of the anthropometric societies that flourished in many European countries and in the United States (Chimonas, Frosch, and Rothman 2011).[18]

Whatever the rationale for many of these early explorations of the shape, size, and anatomy of the brain, and despite the repeated failure of attempts to link intellect to brain size, a vision of the brain as an internally differentiated organ was taking shape, in which specific regions were associated with speech, language and comprehension, and other mental faculties. For obvious reasons, experiments to explore these relations between region and function were mostly undertaken on animals and involved creating lesions in specific brain areas and studying the consequences. The names of the researchers are known to all modern students of neuroanatomy. Fritsch and Hitzig's experiments on dogs in the 1870s seemed to identify a localized motor area in the cortex (republished as Fritsch and Hitzig 2009). David Ferrier's subsequent experiments on a range of animals, notably his work with monkeys, seemed to support the existence of distinct cortical areas for sensory and motor functions: the results were presented in the late 1870s and in the 1880s culminated in successive editions of *The Functions of the Brain*, replete with maps of brain areas associated with different function (Ferrier 1876, 1886). Others followed a similar route.

The architecture of the brain was also being opened to visualization in another way, thanks to a different technique—that of staining. The nerve cell was first described in the form we know it today by Otto Deiters in 1865, leading to much dispute as to whether such cells formed a continuous network or whether the cells were distinct and contiguous. Camillo Golgi, who invented his method of silver staining in the 1870s, was an ardent supporter of the first hypothesis; Santiago Ramón y Cajal used Golgi's method of staining to support the second.[19] Jean-Pierre Changeux quotes the account Cajal gave in 1909 of what he saw when he looked at a section of nervous tissue that had been left around for a few days in Müller's fluid and then immersed in silver nitrate: "Sections were made, dehydrated, illuminated, and observed. An unexpected spectacle! . . . Everything looked simple, clear, unconfused. Nothing remained but to interpret. One merely had to look and note" (Changeux 1997, 26).

Look and note: vision, once more, seems to reveal the truth without interpretation. Done correctly, everything looked clear. From 1903 to 1908, Korbinian Brodmann, working with Oscar Vogt in the Neurobiological Laboratory of Berlin,[20] used another staining technique, that developed by Franz Nissl, for his work on neuroanatomy. He published a series of papers in the *Journal für Psychologie und Neurologie*, setting out the results of his comparative studies of the mammalian cortex; these became the basis of the maps of the cerebral cortex of humans, monkeys, and other mammals contained in *Vergleichende Lokalisationslehre der Grosshirnrinde in ihren Prinzipien dargestellt auf Grund* (Rheinberger 2000). These maps delineated the so-called Brodmann areas of the cortex, which, in the words of Laurence Garey, "must be among the most commonly reproduced figures in neurobiological publishing" (in his introduction to J. Smith and Berlin 1999). The brain now appeared as a differentiated organ, with an anatomy that was consistent within species and comparable between species, and which could be localized and classified according to a standardized nomenclature agreed upon across laboratories and researchers.[21]

Ferrier's experiments had been carried out at the West Riding Lunatic Asylum, with the support of the asylum director, James Crichton-Browne; in 1878, Ferrier and Crichton-Browne together with John C. Bucknill—who had founded and edited the *Journal of Mental Science* in 1853—founded *Brain: A Journal of Neurology*. However, despite this close association of neuroanatomy with the asylum, and despite dissecting innumerable brains of deceased asylum inmates, nineteenth-century neurologists were unable to identify any visible abnormalities in the brain that correlated with abnormalities of thought, conduct, mood, or will. Charcot had carried out autopsies of his patients at the Salpêtrière, but was unable to find lesions that correlated with mental disturbances, except those that involved physical disabilities such as the scleroses. Freud had himself carried out neurological studies of the brain in the 1880s and was apparently a supporter of the theory of a continuous network of nerves. But while he remained committed in principle to the ultimate neurological basis of the processes that he hypothesized, these attempts to visualize psychiatric disorders in terms of lesions in the brains of dead patients, extracted, sliced, stained, and magnified, were derided by psychoanalysts in the early twentieth century. And it is true that they had little to offer psychiatrists. As Nissl concluded in 1908: "It was a bad mistake not to realize that the findings of brain anatomy bore no relationship to psychiatric findings, unless the relationships between brain anatomy and brain function were first clarified, and they certainly have not been up to the present" (quoted in Shorter 1997, 109).

The mapping of the functional architecture of the brain became ever more developed throughout the twentieth century (Frackowiak 2004). Yet even if anatomical variations between brains did have the functions attributed to them, they seemed to have no therapeutic implications. The study of the potential links between mental disorder and the anomalous brain did not seem able to generate any clinical interventions, and was only possible after the patient had died. How could one make this neuronal anatomy of the organ of the mind amenable to clinical thought and practice, let alone to those other practices concerned with the mental states and conduct of the living human being?

Seeing the Living Brain

Could one see the activity of the living human brain? Could the gaze of the neurologist actually plunge into the interior of that organ in life, rather than observing it dead, extracted, sliced, stained, and preserved, and thus inferring functions from accidentally produced lesions or experiments with animals? Central to the conquest of the living brain for knowledge in the twentieth century was the invention, assembly, refinement, and commercialization of a whole variety of devices that could "prosthetically augment . . . the range of the scientific observer's vision, allowing the perception of deep, minute, mobile and generally imperceptible structures" (L. Cartwright 1995, 83) within the skull of a living person. In the process, our very understanding of what it is to be human being—that is to say, an individual with a brain that makes mental life possible, has been irreversibly transformed.

The first conquest of the living brain by vision was to reveal its structure. At the turn of the nineteenth and twentieth centuries, X-rays transformed the way in which the interior of the body could be seen, and indeed transformed the very idea of what it was to have such a body. Bettyann Holzmann Kevles quotes an early enthusiast describing the experience as "the mind walk[ing] in among the tissues themselves" and from that time on, as she points out, much intellectual effort was spent visualizing what was once invisible: that which was once hidden was now amenable to a new kind of knowledge (Kevles 1997, 2). But when it came to the brain, while a great deal was known from brain dissections postmortem, the opacity of the skull made it impossible to produce X-ray pictures of the living brain. This was a challenge for surgeons seeking to understand symptoms that appeared to originate in damage to the brain from accidents, blood clots, or tumors.

What was needed, it seemed, was some kind of contrast agent that would show up the configuration of brain tissue in relation to the surrounding fluid. Some surgeons had already achieved results in a few cases where air had somehow entered the space occupied by the cerebrospinal fluid. These findings were drawn upon in 1918 by an American neurosurgeon, Walter Dandy, who developed a method that involved draining some of the cerebrospinal fluid via a lumbar puncture while simultaneously injecting an equivalent volume of air: the air found its way into the skull, filling the ventriculi, and subsequent X-rays revealed the position and size of tumors or blood clots. This rather painful technique, known as pneumoencephalography, enabled at least the gross features of the cerebral anatomy of living patients to be visualized, and hence surgeons were able to confirm the origins in the brain of symptoms such as extreme headaches or strange bodily sensations (Kevles 1997, 97–103).

Egas Moniz, who had returned to his original career in neurology after a successful spell in Portuguese politics, notably as ambassador to Madrid and minister of foreign affairs, is now famous or infamous for his development of the frontal lobotomy in the 1930s, for which he received the Nobel Prize in Physiology or Medicine in 1949. However, prior to that, he was already well known for inventing the technique of cerebral angiography, in which a contrast agent—initially sodium iodide and later thorium dioxide[22]—enabled the distribution of blood and other fluids in the brain to be captured on film when the head was X-rayed. This technique, developed in 1927, enabled both normal and abnormal blood vessels in and around the brain to be visualized:

X-rays of the skull now made visualization of intracranial tumors, vascular abnormalities, and aneurysms possible.

Of course, in some cases it proved necessary to actually expose the brain of an individual to the direct vision of the surgeon during an operation. No one made more of this opportunity than Wilder Penfield. While neurosurgeons had previously operated on the brain to remove tumors and scar tissue thought to be the focus of epileptic attacks, such operations often caused further scarring, with very undesirable consequences. Penfield, working in Montreal from the mid-1920s, developed a novel procedure involving the removal of considerably more tissue; initially controversial, this actually proved more effective (Penfield 1927). The procedure involved shaving the head of the patient, opening a "trapdoor" in the skull, and removing the protective layers of tissue to expose the surface of the cortex. The patient remained conscious at all times. The surgeon then stimulated different points on the cortex with a wire carrying a small electrical current, and the patient reported what he or she felt—a tingling in the left hand, or a flashing light, each understood, by Penfield, as indicating the part of the cortex where that organ or function was represented. The aim, in part, was to identify the region where the sensations that precede a fit were elicited. This is the so-called aura, which is different for each patient and sometimes involves a strong smell or flashing lights. Stimulation of different areas of the cortex would help identify the scar tissue or other lesion that was the focus of the seizure—a process aided initially by X-rays or ventriculograms, and later by the use of an EEG method developed by Hebert Jasper. Further, the mapping of vital functions enabled the surgeon to identify which regions should not be damaged in the operation if one was to preserve those functions. In the course of undertaking hundreds of these operations, Penfield was also able to produce his famous visual maps of the sensory and motor cortex, indicating the locations of different functions and sensory zones on its surface (Penfield and Erickson 1941; Penfield and Rasmussen 1950; Penfield, Jasper, and Macnaughton 1954).

In addition to mapping sensory and motor functions, this method of cortical stimulation produced some unexpected results. When some regions of the temporal lobes were stimulated in some patients, they recalled what seemed to be precise memories of long-forgotten events, sometimes mundane, sometimes frightening. In other cases, stimulation produced intense perceptual illusions or hallucinations, also usually seeming to recreate specific past experiences, such as an orchestra playing—the patient would hum along to the tune. The sensations, memories, and hallucinations would disappear when the probe was removed. For the first time, it seemed, Penfield had achieved the cerebral localization of conscious experiences and memories, although he doubted whether these were actually integrated in the cortex.[23]

But, starting in the 1950s, other methods would be found to enable the vision of the investigator to penetrate the skull without the need for surgery.

The first major development in imaging brain structure was computerized axial tomography—the CAT scan or CT scan.[24] Kevles recounts the long and winding road that finally led to the development of a scanner that would use computer algorithms to assemble the results of repeated X-ray 'slices' across the brain to produce an image of an anatomical slice, showing the internal structure to the extent that it was made up of tissues of different densities. Many relatively independent steps were involved—Bracewell's work in the 1950s developing mathematical techniques for constructing imagelike maps of astronomical bodies from multiple 'strips' of data; Olendorf's work in Los Angeles rigging up a device that sent collimated beams of high-energy particles through a plane in a model head; Cormack's work to develop a computer that could reconstruct images from X-ray scans.

But it was Godfrey Newbold Hounsfield who integrated all the elements into a functioning scanner that could image the internal structure of the brain in sufficient detail to be used for diagnostic and research purposes. Hounsfield developed and built the device for Electrical and Music Industries Limited (EMI), which financed this work with funds generated from the sale of records by bands such as the Beatles, together with support from the Department of Health and Social Security, which was interested in the possibility of early detection of brain tumors, and from the United Kingdom's Medical Research Council. CAT scanning—now simply CT scanning or CT—enabled three-dimensional images to be produced of any section of the body, including the brain. Indeed, the first CT scanner was designed for the head, and Hounsfield demonstrated it in 1971 at Atkinson Morley's Hospital in London. Cormack and Hounsfield won the 1979 Nobel Prize in Physiology or Medicine for this work, and such scanners—rapidly improving in speed and acuity—and scans—in pictorial rather than numerical form—rapidly became part of the everyday world of physicians treating brain injuries and neurological diseases in the clinic and hospital.

The development of the CT scanning technology was closely paralleled by another method for visualizing body structure, including brain structure—magnetic resonance imaging, or MRI (Kevles 1997). The principle of what was initially termed nuclear magnetic resonance (NMR) was that the protons in atoms would become magnetized and align themselves when exposed to a strong magnetic field; when that field was altered, the protons would also realign themselves. In 1946, Felix Bloch (Harvard) and Edward Purcell (Stanford) independently published papers that showed that the protons in a spinning nucleus would resonate when placed in an alternating magnetic field at a particular frequency: when the magnetic field was turned on, the protons would align themselves, and when it was turned off, they would 'relax.' The times taken between their original state and their relaxed state varied between substances, and could be detected by a receiver and turned into images.

Initially, the technique was used in chemistry to explore the structure of molecules, but in the 1950s, as the machines became more widespread, researchers began experimenting with slices of tissue—the images picked up variations in the water in tissues by focusing on spin in hydrogen atoms. In 1959, the first NMR scans were conducted to measure rates of blood flow in living mice by Jay Singer at Berkeley, and at the start of the 1970s, in an episode of apparently independent discovery plagued by controversy and competition, Raymond Damadian and Paul Lauterbur worked out the techniques for deriving images by reconstructing them from multiple scans across a body: Damadian was looking for a device that would map malignancies, and Lauterbur hoped to image fluids.[25] In 1973, Lauterbur used NMR to create cross-sectional images in much the same manner as CT and published the results in a paper in *Nature* (Lauterbur 1973). There was immediate interest not only because the technique was free of ionizing radiation but because it produced better body images than CT because of its sensitivity to soft issues.

Meanwhile, in the United Kingdom, Peter Mansfield was developing a related way to use NMR to image first solids, then liquids, in the human body; working with others, he developed a machine that would produce three-dimensional images, not by reconstructing them from successive slices, but by using complex algorithms to manipulate data. By the early 1980s, facilitated by the success of CT scanning and by the advantage that NMR did not use ionizing radiation, the machines that undertook what became known as MRI scanning entered clinical use. During the 1980s, many technical refinements of MR were made, and a number diagnostic MR applications were undertaken: MRI received FDA approval for clinical use in 1985. Of course problems remained, not least the great expense of the machines, the need to take special precautions around the very high magnetic fields that they produced (in supercooled magnets, cooled with liquid helium), and the disorienting effects on humans of the strong magnetic field itself, when the organ imaged is the brain. Nonetheless, given its capacity to reveal variations in soft tissue, the brain was one of the key targets of MRI scanning, which was used to detect injuries caused to babies by violent shaking, to reveal the demyelination of nerve fibers that was at the root of the previously mysterious symptoms of multiple sclerosis, and to identify the anomalies, lesions, and tumors that underlay many neurological disorders. Mansfield and Lauterbur were awarded the Nobel Prize in Physiology or Medicine in 2003 for their work in developing the technology.

But how to go from imaging *structure* to imaging *function*? CT and MRI scanning, however useful for detecting structural abnormalities, did not image what the living brain was *doing*. Of course, conclusions could be drawn from structure, however tentatively, where structural anomalies were thought to be linked to functional anomalies. For example, in the trial of John Hinckley for attempting to assassinate Ronald Reagan, CT scans were introduced by the defense, who argued that they showed abnormalities in the appearance of the brain—in its gyri and sulci—and that autopsies had shown these to be characteristic of people diagnosed with schizophrenia.[26] However, this route was indirect—was there a more direct route to render visible the invisible operations of the mind?

From the 1930s onward, some investigators had explored normal and abnormal brain activity using electroencephalography (EEG). Edgar Adrian had demonstrated that the Berger rhythm, discovered by Hans Berger in the 1920s, was genuine and not artifactual, reporting his findings in a famous paper of 1934 that argued that the rhythm was linked to attention (Adrian and Matthews 1934). But it was positron emission tomography, or PET, that first seemed to enable the gaze of the investigator to visualize mind in brain. PET appeared to permit direct imaging of very specific processes in the brain as it undertook certain tasks, by showing how certain radiolabeled molecules, such as glucose or drugs, were taken up into brain tissue over time. As Joseph Dumit puts it in his ethnographic study: "PET was the first noninvasive technology to permit direct quantitative assessment of regional physiological processes in the brain. . . . PET allowed scientists to make exacting measurements of blood flow, glucose uptake and dopamine receptor uptake" (Dumit 2003, 22).

There is a long and generally accepted history of the precursors of PET—those whose basic work established the various elements that would be brought together in the technology. But there is a short and rather disputed history of rivalry and competition when it comes to the last stages in developing PET itself, in particular between the groups around Michael Phelps, Michael Ter-Pogossian, and Henry Wagner (Dumit [2003] provides an excellent account of these rival versions of the history). PET assembled together three basic elements: the idea that a molecule could be labeled with a small amount of radioactivity and traced as it circulated through the body and was metabolized in particular cells; the idea of a device that could attach radioactive tracers attached to a range of molecules of interest; and idea of a scanner that could identify the location of the radioactively labeled molecules at multiple time points, and compile these data into an image.

All accounts credit Georg Van Hervsey with discovering the tracer principle in the early decades of the twentieth century, for which he received the Nobel Prize in Chemistry in 1943: he showed that "radioisotopes of elements participate in biochemical and physiological processes in the same way as the chemicals they have replaced"; that organisms absorb, circulate, metabolize, and excrete these labeled molecules; that their radioactivity decays in a predictable way; and that this process can be traced over time through the various organs and can be traced externally by a detector such as a Geiger counter (Kevles 1997, 202). Hervsey initially used naturally occurring radioisotopes. However, after the Curies had shown, in 1934, that artificial radioisotopes could be produced by bombarding elements with atomic particles, it became possible to create radioisotopes of a whole range of elements that were taken up in particular organs and tissues and in specific metabolic processes. At around the same time, Ernest Lawrence invented a machine, later termed a cyclotron, that used this process of bombarding elements with high-speed neutrons to produce a whole host of radioisotopes, and these began to be used in medical research, even though, at first, there was some concern that they might produce biological damage from radiation effects. In 1965, the first medical cyclotron was installed at the Hammersmith Hospital in London, followed by installations at Massachusetts General Hospital and Washington University's Mallinckrodt Institute of Radiology, also in 1965.

The next step was the development of the scanning technology. In the 1950s, David Kuhl worked out methods to create images by detecting and localizing the emissions of the radiolabeled ligands, and in particular of a radiolabeled form of deoxyglucose, termed FDG, which, like blood, would be taken up specifically in areas of cerebral activity. This work led to the development of the first scanner that used the method of imaging multiple slices through the brain and using algorithms to compile them together into a three-dimensional image—the process known as single photon emission computerized tomography, or SPECT. SPECT was first used to map brain function by tracking blood flow by Niels Lassen in 1972 (see Lassen 1978). It is the question of priorities in the research that led from SPECT to PET that is most disputed by the participants: the short version of this period is that Phelps, working in Ter-Pogossian's lab, used versions of the algorithms developed for CT scanning to generate similar images from the positron emissions—first dubbed PETT (Ter-Pogossian, Phelps, Hoffman, et al. 1975).

In the subsequent years, techniques were improved, leading to the commercialization of PET devices, although these remained costly and complex. This was partly because, in order to use them, a lab or clinic had to have a cyclotron close at hand: the radioisotopes that were required had to be produced in cyclotrons, but they decayed rapidly and had to be used in a short time; a considerable crew of technicians and others were required to deploy the procedures. Nonetheless, PET was a powerful research tool, one which, for example, enabled Henry Wagner, who became a powerful promoter of the virtues of nuclear medicine, to image the dopamine receptor (Wagner, Burns, Dannals, et al. 1983). PET was also used in clinical studies of cancer, heart disease, and many other areas, as well as in imaging studies of motor activity, attention, cognition, and various psychiatric disorders, even though its diagnostic capacities proved limited.

What scale was appropriate, and possible, for imaging neuronal activity? Some argued that such imaging should ideally be at the molecular level, and that nuclear medicine should become molecular imaging. The earliest studies to quantify neurotransmission and neurochemistry in humans were published in the 1980s (Wagner, Burns, Dannals, et al. 1983; Holman, Gibson, Hill, et al. 1985). Twenty years later, advocates of this approach argued for "the visual representation, characterization, and quantification of biological processes at the cellular and subcellular levels within intact living organisms"; they believed that "the change in emphasis from a non-specific to a specific approach represents a significant paradigm shift, the impact of which is that imaging can now provide the potential for understanding of integrative biology, earlier detection and characterization of disease, and evaluation of treatment" (Massoud and Gambhir 2003, 545). Shankar Vallabhajosula, the author of one of a flurry of textbooks on molecular imaging published in the first decade of the twenty-first century, claimed that it would be the future of clinical practice: "although molecular imaging is not necessarily new, what is new is 'molecular and anatomic correlation'" (Vallabhajosula 2009). The aim would be "to integrate patient-specific and disease-specific molecular information with traditional anatomical imaging readouts" with the eventual hope of achieving "non-invasive or minimally invasive molecular diagnostic capabilities, better clinical risk stratification, more optimal selection of disease therapy, and improved assessment of treatment efficacy" (ibid., 4).

From this perspective, the goal was to create a molecular and neuromolecular therapeutics. The Society for Molecular Imaging (SMI) was founded in 2000 (Provenzale and Mukundan 2005); the European Society for Molecular Imaging (ESMI) was created in 2006; and the journal *Molecular Imaging and Biology* was launched in 2005 as the official journal of the Academy of Molecular Imaging (AMI), the SMI, and the ESMI. According to Henry Wagner, molecular imaging can help create a more "cost-benefit knowledge-based health care system" by shifting the focus and the gaze to specific molecules that become the "nexus for understanding disease, developing both preventive measures and effective therapies and monitoring treatment efforts" (Wagner 2006). By 2009, when Wagner published a book titled *Brain Imaging: The Chemistry of Mental Activity*, it seemed that the technologies for rendering the mental visible in the material, in the chemistry of tissues, fluids, and membranes, had fused with the neuromolecular conception of the brain (Wagner 2009).[27]

It was another development, however, that was to underpin the proliferating belief that one could see the functional activity of the living brain—to see the mind in the brain. This was functional MRI or fMRI. MRI, as we have seen, focused on structure. How could one get from structure to function? Speed was one part of the answer to this question. As the MRI technology became faster, and the algorithms became more sophisticated, it became possible to take successive MRI images and link them together to capture up to 110 frames per second. The second requirement was to find a tissue that, on the one hand could be imaged, and on the other could be linked to functional activity in the brain. What could be more closely linked to levels of activity in a body than blood? And blood was one of the tissues that could be imaged by MRI techniques. It was via the medium of blood, and the long-standing observations that correlated increased brain activity with increased blood flow, that the functional activity of the living brain—mental states themselves— would seem to be captured in a scanner.

The link between brain activity and blood flow goes back to the 1870s, when Angelo Mosso, a prominent Italian physiologist, made the observation in humans that local blood flow to the brain is related to brain function: he published his results in 1880 (Mosso 1880). The actual physiological relation between brain function and blood flow was further explored in 1890 by Charles Roy and Charles Sherrington (Roy and Sherrington 1890). In 1948, Seymour Kety and Carl Schmidt developed the method of measuring cerebral blood flow by using the nitrous oxide method (Kety and Schmidt 1948), and in 1955, Kety and colleagues at the University of Pennsylvania and the NIH used radioisotopes to track brain blood flow in cats (Landau, Freygang, Roland, et al. 1955; Freygang and Sokoloff 1958; Kety 1960). This 1955 paper from Kety's group is often claimed to be the first quantification of blood flow in the brain related to brain function. Blood had another great advantage: its magnetic properties. Faraday had carried out studies of the magnetic properties of hemoglobin in 1845, and almost a century later, Linus Pauling and Charles Coryell found that oxygenated and deoxygenated hemoglobin have different magnetic susceptibilities (Pauling and Coryell 1936).

Seiji Ogawa, working at the AT&T Bell Laboratory in New Jersey, recognized that these findings, taken together, made functional imaging possible. He based his approach on the observation that brain cells, when activated, increase their uptake of energy from oxygenated blood, which then becomes deoxygenated (Kevles 1997, 198). Ogawa realized that the deoxygenation of the blood in the regions of the brain where activity was taking place could be imaged. As Kevles explains: "[V]enous blood, lacking a bound oxygen molecule, is paramagnetic, and when placed in a magnetic field changes the magnetism all around it. This distortion of the magnetic field, in turn, affects the magnetic resonance of nearby water protons. It amplifies their signal as much as 100,000 times. Ogawa called this effect BOLD (blood-oxygen-level-dependent) contrast imaging, and he published his first paper using this method in 1989 (Kevles 1997, 198). Other research groups were experimenting with the use of contrast agents with MRI to demonstrate changes in brain blood volume produced by physiological manipulations of brain blood flow; after some animal studies in the late 1980s, the first study using this approach with human volunteers to explore task activation brain mapping was published in 1991 (Belliveau, Kennedy, McKinstry, et al. 1991). However, the BOLD response, which required no administration of an external contrast agent, was to become the key to the widespread use of functional MRI (Ogawa, Lee, Kay, et al. 1990).

As functional brain imaging, whether by PET or by fMRI, came to supplement images of anatomical structures, we now seemed able to observe metabolic activity in living, functioning brains in four dimensions—three of space and one of time.[28] The very interior processes of the living and dynamic brain, its anatomical normalities and pathologies, its activity in delusions and normal perception, its secretion or blockade of neurotransmitters, could be rendered visible and correlated with a phenomenology of mental life. At last, it now seems that we can see the physical basis of mind in the activities of the living brain. To see is to know, and to know is to see. But what is it that is being seen, what is happening when, to quote Andreas Roepstorff, "knowing becomes seeing"? (Roepstorff 2007)

The Epidemiology of Visualization

In 2008, Nikos Logothetis reported that

> a recent database (ISI/Web of Science) query using the keywords "fMRI," or "functional MRI" or "functional magnetic resonance imaging" returned over 19,000 peer reviewed articles. . . . About 43% of papers explore functional localization and/or cognitive anatomy associated with some cognitive task or stimulus—constructing statistical parametric maps from changes in haemodynamic responses from every point in the brain. Another 22% are region of interest studies examining the physiological properties of different brain structures, analogous to single-unit recordings; 8% are on neuropsychology; 5% on the properties of the fMRI signal; and the rest is on a variety of other topics including plasticity, drug action, experimental designs and analysis methods. (Logothetis 2008, 869)

When we repeated this exercise in October 2010, the result was even more staggering—53,662 papers were identified, starting with one published in 1980 (Holland, Moore, and Hawkes in the *Journal of Computer Assisted Tomography*), 146 papers in 1990, 2,139 in 2000, 4,725 in 2005, and almost 7,000 in the one year of 2009 alone. Alongside the categories that Logothetis reports, we find papers relating changes in brain activity to responses to art in general and to the work of specific painters; to responses to music and to specific composers or performers; papers discussing the neurocognitive processes of religious leaders; studies of brain activation related to the use of language, metaphor, or responses to various novelists; papers that monitor brain activity in response to television commercials or to grief after the loss of a child; as well as hundreds of studies purporting to image love, hate, fear, and other emotions, all manner of cognitive processes, and volition and acts of will when individuals in scanners are given simple tasks to undertake. What is one to make of this industry of visualization?

One cannot fail to be tempted by the results of all this research—has not the activities of the living mind within living brain now come into view? Yet many of the researchers in this area, including many of those actually responsible for generating the statistical maps that make the images possible, feel profoundly uneasy about this translation of their quantitative data into pictorial form and the conclusions, both professional and popular, that are drawn from these studies.[29] As indeed do the social scientists who have explored these developments. For the sake of simplicity, we can group their concerns under four headings: localization, the lab, pixels to pictures, and evidence to interpretation.

Are Brain Functions Localized?

Are brain functions localized, and in what sense? And why should this be important? One of the most vociferous critics of the emphasis on localization in the practices of imaging has been Jerry Fodor who is a philosopher and cognitive scientist. Writing in the popular magazine the *London Review of Books*, he argued:

> I want, to begin with, to distinguish between the question whether mental functions are neurally localized in the brain, and the question where they are neurally localized in the brain. Though I find it hard to care about the second, the first clearly connects with deep issues about how the mind works; . . . [but] just what is the question about the mind-brain relation in general, or about language in particular, that turns on where the brain's linguistic capacities are? And if, as I suspect, none does, why are we spending so much time and money trying to find them? It isn't, after all, seriously in doubt that talking (or riding a bicycle, or building a bridge) depends on things that go on in the brain somewhere or other. If the mind happens in space at all, it happens somewhere north of the neck. What exactly turns on knowing how far north? (Fodor 1999)

This argument actually comes rather strangely from someone who is one of the strongest proponents of the mental ontology that seems quite consistent with localization, that is to say, a 'modular' theory of mind (Fodor 1983, 2001). Nonetheless, many others, including many neuroscientists, have also questioned a strict conception of localization, one that is based on the idea of hard-wired connections of neurons, clustered together, along with the belief that there can be a one-to-one mapping of specific mental processes to specific brain regions common to all 'normal' human beings' brains.

Consider this analysis from Russell Poldrack:

> One cannot infer that the activated regions are necessary or sufficient for the engagement of the mental process. Indeed, there are well-known examples of cases in which regions that are activated during a task are not necessary for the task. For example, the hippocampus is activated during delay classical conditioning, but lesions to the hippocampus do not impair this function. . . . Inferences from lesion studies are limited by the fact that the brain may often have multiple ways to perform a cognitive process. . . . Cognitive neuroscientists have generally adopted a strongly modular approach to structure–function relationships, perhaps driven by the facile leap to localizationist conclusions from le-

sion and neuroimaging results. Despite the longstanding appreciation for the importance of functional integration within the neuroimaging literature, the widespread use of functional and effective connectivity analyses has not yet come about. Given that many cognitive processes may be distinguished not by activity in specific regions but by patterns of activity across regions, there is reason for caution regarding many of the inferences that have been driven by highly modular approaches. (Poldrack 2008)

Neuroscientists critical of many of the conclusions drawn from brain imaging further point out the problem of scale. For despite all its increases in acuity, fMRI can only measure mass action. Nikos Logothetis has examined the three hundred top-cited fMRI studies and estimated that a typical voxel size (before filtering of data that is likely to increase this by a factor of two or three) contains 5.5 million neurons, between 2.2×10^{10} and 5.5×10^{10} synapses, 22 km of dendrites, and 220 km of axons. Leaving aside all the other complications that will shape the blood-oxygen-level variations within this cube of brain tissue that is somewhere between 9 and 16 mm square and around 7 mm in thickness— notably, the balance in every single neuron between excitation and inhibition— this is, to say the least, a large and heterogeneous population of cells.

In any event, how are we to know what scale is appropriate for examining mental processes? What proportion, in what configuration of the 10^{10} neurons and 10^{14} connections in the human cortex alone should be analyzed in relation to any task? Logothetis quotes neuroanatomist Valentino Braitenberg: "It makes no sense to read a newspaper with a microscope" (Logothetis 2008); one might add, as pointed out by Hauke Heekeren, of the Max Planck Institute for Human Development, that it makes no sense to read a book from a photograph of the bookshelf.[30] No wonder that Logothetis concludes that the "limitations of fMRI are not related to physics or poor engineering, and are unlikely to be resolved by increasing the sophistication and power of the scanners; they are instead due to the circuitry and functional organization of the brain, as well as to inappropriate experimental protocols that ignore this organization" (Logothetis 2008: 876–77).[31]

The Lab: Place or Non-Place?

Typically, reports of brain imaging experiments pay little attention to the material properties of the laboratory setting. This is unfortunate, for, as ethnographers such as Simon Cohn have pointed out, the scanning facility is not a 'non-place' for those who are being scanned, it is a particular and rather unusual arrangement of space, persons, machinery, sounds, and sights, not to mention the experience of being in the scanner itself. What are the subjects thinking and doing when they are in the scanner? The scanned brain is in a human body, the body of a human being who is being paid to lie in a machine to perform a task that the brain will never be confronted with in the world outside. These features, that cannot enter the modeling of the relation between the stimulus and response contained within the space of the scanner, cannot enter into the analysis that the scanning software uses to generate the images (Cohn 2008).

Yet as Cohn also points out, those social relations of the scanning facility, the "brief but intimate personal relations" between the researchers and their subjects, are central to the possibility of the scanning experiment. What impact do the social relations of the scanning facility have on the mental life of the scanned subject? How can the researchers set aside, for the purposes of the experiment, the arguments about the nature of 'the social brain' that neuroscientists themselves are generating—that the human brain is particularly evolved for interactions, including adapting itself in all sorts of ways to its attributions of wishes and intentions to others? How might this experience and response to sociality differ between subjects, for example, in the scanning of healthy volunteers and hospitalized patients diagnosed with depression? To what extent can there be generalization across subjects who are experiencing and understanding the scanning situation in different ways, let alone beyond the lab setting?

These are, indeed, key questions. What is involved here, as we will discuss in the next chapter in relation to animal models, is the creation of a certain kind of "setup" (Latour 1987) that brings together, within a particular institutional form and practice, experimenters, assorted technicians from various backgrounds, devices, concepts, theories, and subjects. Only through this setup is it possible to generate the data that are required to generate the images that can then be subject to interpretation at leisure and then deployed in presentations, articles, and books to make arguments. The brain of the individual is thus given the opportunity to produce something that it would not otherwise be able to produce. But that setup, with all its sociality, contingency, specificity, and embedded meaning, disappears in the interpretation of the measurements produced. This is not necessarily a problem in itself: it is an inescapable part of the incredibly difficult process by which data of this sort can be generated. But it does raise some rather important questions about what can be concluded from such data, or rather, from the images that the data are used to generate. What is the salience of the image, when it emerges from the secluded and simplified world of the laboratory, and makes claims for its relevance to the understanding of conduct in the wild and messy world of everyday life?

Pixels to Pictures

The artifactual character of the images produced in the various technologies of brain scanning has attracted much critical attention (Dumit 1997, 1999; Beaulieu 2000b, 2002; Roepstorff 2004). In the analysis and presentation of the scan data, computerized methods are used, involving choosing among a whole range of available algorithms to transform quantitative information into images of domains that can be explored spatially. This is a technically highly complex rendition, generated by mapping quantitative data on the emission of each positron in a scan of a particular subject (carefully chosen, screened, prepared) in four dimensions onto a morphological space usually drawn from a standardized MR atlas of slices of the typical or average brain. The process involves transforming, smoothing, warping, and stretching the data on each individual to fit the standard anatomical space; this is necessary to make the images comparable and cumulable. And the colors, so impressive in the proliferation of such images in the domains of nonspecialists, are similarly artifactual—there are no reasons apart from convention why areas seemingly highly active should be represented in reds and those less active in blues. Yet of course, those colors carry cultural resonances that have great significance for the meanings that are attributed to the images, at least when they are represented beyond the laboratory or the journal article.[32]

Joseph Dumit identifies four key stages in the amalgam of social and technical processes that generates the image (Dumit 2003). First, *devising the experimental design*: choosing participants, defining them (normal, schizophrenic, drug naive, etc.), controlling prior states, specifying the task unambiguously. Second, *managing the technical process* of measuring brain activity: conducting the scan, compiling the data, algorithmically reconstructing a three-dimensional map of activity. Third, *making the data comparable*: normalizing the brainsets, warping them onto a standard anatomical atlas of brain space, then combining brainsets by manipulating the datasets. Fourth, *making the data presentable*: coloring, contrasting, and production of the images. Much of this is handled by standard software packages, notably statistical parametric mapping (SPM). While such packages are regularly updated and incorporate input from the researchers in the field, like other such programs, many assumptions, some dating back many years, are 'black-boxed' within them—in this case literally, as the software is designed to work 'straight out of the box.'[33]

The sociologists of scientific knowledge have given us many ways to think of the process of black-boxing in scientific research. But here, in the image itself, as the process disappears into the product, the illusion of transparency is crucial to the effects of truth. To quote another ethnographer of brain imaging, this process does not so much "image" mind in brain, but constitutes a new object: "the mind-in-the-brain" (Beaulieu 2000b). Like the diagnostic space opened by Bucknill and Tuke's carefully posed pictorial representations of the varieties of madness, then, this new molecular vision opens onto an irreal space of thought and interventions. It represents not by the *simulacrum* but by the *simulation*—by generating a model of processes at the level of the living brain, a model that is, precisely, capable of modeling in apparently precise images and flows, the functioning and malfunctioning of those process under particular conditions.

It is undoubtedly true that such images provide powerful support for particular theories of brain functioning, by purporting to allow researchers to actually observe the mind/brain seeing, hearing, smelling, thinking, desiring, emoting, willing, or hallucinating. Yet the power of these images is located at their weakest point—the point where simulation claims not merely to be a way of isolating certain elements of a more complex system for analysis (as, for example, in a computer simulation of an earthquake), but to be the thought image of that system itself. Despite their apparent realism, claims to account for the phenomenology of mental states in terms of these simulated functions localized in the living brain require an act of faith in all those elements on which the image is premised. The visual imaginary that is the result is, in this sense, no different from that of the phrenologists of the nineteenth century.

Evidence and Interpretation

Of course, the experiments, and the images they produce, do not demonstrate *nothing*. But *what* they demonstrate is a matter of dispute. The 'objectivity effect' of the image seems to be clear: studies have shown that the presentation of a brain image can positively influence the reader's judgment about the quality of the reasoning of the article in which that image appears (McCabe and Castel 2008) and that of verbal descriptions of neuroimaging results can make weak arguments more persuasive—even when they are irrelevant to the argument (Weisberg, Keil, Goodstein, et al. 2008). We shall see in a later chapter, however, that this objectivity effect seems to be limited by the characteristics of the practice in which the image is deployed. For the image, despite suggestions to the contrary, and despite all the arguments black-boxed within it, does not speak for itself—it has to be spoken for. And in areas where those who speak in the name of the image are challenged, for example, in the criminal justice system, the image itself may get caught up in the dynamics of controversy.[34]

In other domains too, the interpretations necessary to give images their salience have come under significant criticism from imagers themselves. Let us take a single example here, a paper titled "Does Rejection Hurt? An fMRI Study of Social Exclusion" published in *Science* in 2003 (Eisenberger, Lieberman, and Williams 2003). Here is how the abstract describes the study:

> A neuroimaging study examined the neural correlates of social exclusion and tested the hypothesis that the brain bases of social pain are similar to those of physical pain. Participants were scanned while playing a virtual balltossing game in which they were ultimately excluded. Paralleling results from physical pain studies, the anterior cingulate cortex (ACC) was more active during exclusion than during inclusion and correlated positively with self-reported distress. Right ven-

tral prefrontal cortex (RVPFC) was active during exclusion and correlated negatively with self-reported distress. ACC changes mediated the RVPFC-distress correlation, suggesting that RVPFC regulates the distress of social exclusion by disrupting ACC activity.

This illustrates the problems of this style of reasoning rather clearly.[35] First, one creates a game that an individual can play in a scanner, that is taken to simulate a 'social' interaction (although the relation between virtual ball tossing and any salient social situation is somewhat tenuous). Second, one adjusts one's scanning parameters until one is able to observe increased brain activity—or rather an increased BOLD response—in a particular region of the brain—or rather, a region of the standard brain space onto which your data has been warped, in this case the anterior cingulate cortex, or ACC. Next, one refers to previous experiments claiming that this brain region shows an increase in activation in another condition: in this case, the ACC has been shown to be activated when the subject experiences physical pain. Then one concludes that, because the same region is activated in both conditions—here in both social exclusion and in physical pain processing—the two are causally linked: hence social exclusion hurts.

Among the many rather obvious flaws in this chain of reasoning is the fact that activation in the same brain region has been claimed in multiple tasks. For example, changes in amygdala activity have been found to occur in experiments seeking to generate or simulate a whole range of mental states: fear, reward, fairness, moral status, subjective reports of beauty, trustworthiness, and more. So to make such arguments work, either more and more baroque alignments must be made, or the experimenters must, as in this case, be exceptionally selective. The authors of the study cited above conclude: "This study suggests that social pain is analogous in its neurocognitive function to physical pain, alerting us when we have sustained injury to our social connections, allowing restorative measures to be taken. Understanding the underlying commonalities between physical and social pain unearths new perspectives on issues such as why physical and social pain are affected similarly by both social support and neurochemical interventions . . . and why it 'hurts' to lose someone we love" (Eisenberger, Lieberman, and Williams 2003, 292). On the basis of such extrapolations in hundreds of papers using fMRI to measure changes in brain activity when subjects undergo tasks said to simulate features of our form of life in the scanner, an empire of brain imaging research has been built.

The New Engines of Brain Visualization

What, then, are the consequences of this artful construction of the mind? Should we object to artfulness itself? This would be unwise, not least because it would tie us, even if tacitly, to the aspiration for an untenable realism.

Knowledge, as a whole tradition of thinkers about science have reminded us, always entails a break away from the immediate and the concrete to the visualization and manipulation of artificial objects moved by hypothetical forces in abstract space. It is not the irreality of this image of the mind that should concern us, but the attribution of false concreteness to that image by researchers, policymakers, and popular interpreters. The implications of these attributions of false concreteness are best discussed in relation to specific practices, and will be examined in our later chapters. But, more generally, can we say that we have traveled up a gradient of objectivity as we have moved from Esquirol's drawings through Diamond's posed photographs, Charcot's movies, Brodmann's maps, and Penfield's homunculi to the luminous images of fMRI investigations that stud the pages of our contemporary scientific journals?

It is hard to free ourselves from the photographic illusion that might support such a belief in progress toward the real. But it is nonetheless misleading. In saying this, we are not suggesting a fundamental rejection of the truth effects of brain imaging. To attend to the artifice of the image is not to critique it but to recognize the nature of the objectivity, the facticity that it produces. The image should be understood not as a picture to be judged via a criterion of realism, but as a tool, a move in an argument, an instrument in an intervention, to be judged by criteria of rationality, validity, or efficacy. Advances in clinical medicine from the nineteenth century onward went hand in hand with the penetration of the gaze of the doctor into depths of the body itself. There are now many examples of analogous advances linked to the structural imaging of the brain—in the detection of tumors, the identification of lesions, the mapping of the damage caused by injury or stroke. Thanks to such images, the mind of the neuroscientist, the neurologist, and the psychiatrist does now seem able to "walk among the tissues themselves." But however similar the images of brain *function* are to those of brain *structure*, they mislead if they seem to allow the mind of the neuroscientist to walk among thoughts, feelings, or desires. Technology alone, even where it appears to measure neural activity, cannot enable the gaze to bridge the gap between molecules and mental states.[36] We should not fall victim to the illusion that visualization itself has—or could—resolve the question of the relations between minds and brains. To recognize this is not to undermine the work of brain scanning, far from it. On the contrary, by enabling us critically to evaluate the different kinds of claims made in behalf of the diverse components of this avalanche of images, an analysis of these new engines of visualization can and should help improve the robustness of the truths they create.

Chapter Three

· · · · · · · · · · · · · · · · · · · ·

What's Wrong with Their Mice?

What a pitiful, what a sorry thing to have said that animals are machines bereft of understanding and feeling, which perform their operations always in the same way, which learn nothing, perfect nothing, etc.!

—**Voltaire**, *Dictionnaire philosophique portatif*, 1764[1]

Tread softly when approaching a mouse model of a human psychiatric disease.

—**Jacqueline Crawley**, *What's Wrong with My Mouse? Behavioral Phenotyping of Transgenic and Knockout Mice*, 2007

We take the title of this chapter from Jacqueline Crawley's *What's Wrong with My Mouse?* from which we have learned a great deal about the world of animal modeling of behavioral disorders (Crawley 2007).[2] Given the limitations that have long surrounded experimental intervention into the brains of living humans, animal research has been crucial to the project of neuroscience, from the molecular dissection of neurobiological processes such as neurotransmission, to research on plasticity and neurogenesis, and of course, to the development and testing of psychiatric drugs. But why frame the chapter in such terms—why might there be 'something wrong' with the mice—or for that matter the flies, rats, macaques, and the other creatures—that populate the labs, and the papers, of those who use animal models, and model animals, in the experimental investigation of the neurobiological underpinnings of human behavior? The question of *translation* poses this issue most clearly, for if there are persistent failures in using the findings in the laboratory to develop effective therapeutic interventions in practices that seek to treat, cure, or reform humans, then it might be sensible to ask whether there is 'something wrong' with the practices of animal modeling and experimentation themselves.

Sociologists, anthropologists, historians and philosophers have written extensively on models in science. There have been analyses of scientific modeling as a practice and as a "style of reasoning"—sometimes referred to as "the third dimension of science" (Morgan and Morrison 1999; Thagard, Magnani, and Nersessian 1999; de Chadarevian and Hopwood 2004). There has been a range of ethnographic and historical studies that have focused on the social relations within which models are shaped and used among communities of modelers (e.g., Star and Griesemer [1989] 1999; Löwy 1992; Kohler 1994; Gaudillière and Löwy 1998; Lemov 2005; Franklin 2007; Shostak 2007; Friese 2009).[3] But surprisingly, little has been written from the social and human sciences on animal models of *behavior*. And even less has been written on animal models *in psychiatry*—where the models are neither conceptual representations (although they embody these) nor three dimensional artifacts (although they are in three dimensions), but living, breathing, developing organisms (adding the dimension of vitality and temporality) that are of interest because of their *behavior* (adding a further dimension to the model), and of behavior deemed pathological or abnormal, which complicates things even further.

When we have discussed with colleagues from the social and human sciences the use of animals to explore issues relating to human cognition, emotion, volition, and their pathologies, the initial response has usually been skeptical. When we have described the methods of assessment used in such laboratory studies, for example, the use of the elevated plus maze in the study of anxiety, or the forced swim test in the study of depression, the reaction is often one of laughter. How, our naive colleagues ask, can anyone believe that there can be any similarity between mouse *behavior* in this strange apparatus in an artificial laboratory setting and the rich meaningful, culturally embedded, historically shaped, linguistically organized, situationally framed, *experience* of depression, anxiety, and whatever else in our everyday human world—especially, when we find it difficult enough to describe these human feelings in rigorous ways, to compare them between persons and societies, or to distinguish what is normal from what is not?

The immediate response of researchers who use such animal models in their work is to point to similarities in the genomes of the two species, in the structure of mouse and human brain, in patterns of brain activation, in neural mechanisms at the cellular and molecular level, in responses to drugs and so forth, perhaps with reference to evolution and the principle of conservation across species when it comes to the most basic aspects of living organisms, including their brains.[4] To which our colleagues in the human sciences might well respond—if they had read Wittgenstein—that this is all very well, but "[o]nly of a living human being and what resembles (behaves like) a living human being, can one say it has sensations; it sees, is blind; hears, is deaf; is conscious or unconscious" (Wittgenstein 1958, § 281)—or, to extend the argument, it is only of a person that one can say that he or she experiences anxiety or feels depressed.[5]

They might also suggest that the animal modelers were guilty of what Bennett and Hacker call the "mereological fallacy" (Bennett and Hacker 2003):[6]

that talk of similarities in genes or brains between animals and humans assigns properties to genes or brains that properly belong only to persons. And so, they might continue, showing similarities in genes or brains between animals and humans does not resolve the problem. For while the genes or the brain may be involved—certainly are involved—in anxiety and depression in humans, while there may be brain activity associated with these complex affects, they are distributed across whole embodied persons in specific social situations whose meaning is given by the whole complex of resources that we can briefly sum up as 'culture.' In short, the way in which humans are in the world, their emotions, actions, and desires, depends on the way in which they give meaning to themselves and to their world, their models of that world, their beliefs about themselves and that world, and the affects associated with those beliefs. While those may have conditions of possibility in brain states and gene states, they do not reflect them. Indeed such an idea is meaningless because, *as a result of their biology*, humans are the *only* creatures that live, act, feel, and will in a world of language, meanings, beliefs—sedimented and stabilized in artifacts, buildings, forms of life, human social practices, and much more.

Researchers who use animal models are familiar with such reactions and have several responses.[7] They claim that they are well aware of the problems of anthropomorphism and are taught from the outset not to speak of anxiety or depression in mice but of 'anxiety-like' or 'depression-related' states, because "[t]here is no way for a human investigator to know whether a mouse is feeling afraid, anxious, or depressed. These are subjective emotional experiences, existing in the mind and body of the individual" (Crawley 2007, 227). Thoughtful modelers such as Crawley insist that, at most, it is possible to develop a behavioral test for mice that has some validity for a symptom of the human condition, or an assay that is sensitive to therapeutic drug responses in humans (261). Nonetheless, those within the community of animal modelers in psychiatric research and behavioral neuroscience have pointed to at least six sets of problems that afflict such work:[8] (1) the problematic homologies and analogies between human and animal behaviors, especially in relation to psychiatry; (2) the precarious assumption of common mechanisms; (3) the difficulty of the phenotype—of the behavioral classification of mouse models of disorders generated by knockouts or other techniques, as these are seldom open to validation other than by means of pointing to apparent similarities in the observable behavior of the animal and the human;[9] (4) the difficulties of modeling human stressors in animals; (5) the weakness of current tests, such as forced swim test, the elevated plus maze, and so forth for simulating variations in *human* behavior; and (6) the fact that many trials are not properly blinded or randomized. This is not to mention the further difficulties arising from the great variety between strains, between labs, between particular practices for managing the animals, and so forth, all of which bedevil comparisons between results obtained by different research groups.

These technical difficulties are of great importance. However, we want to locate them in the context of some more fundamental issues. We can systematize these into four interconnected themes that will structure this chapter. The first three are rather general and conceptual, the fourth more specific and practical.

- First, the question of the *artificiality* of the laboratory situation within which animal experiments are conducted.[10]
- Second, the idea of a *model* in behavioral and psychiatric research.
- Third, the fundamental objection from the social and human sciences— the *specificity* of the human and the elision of history and human (as opposed to animal) sociality.
- Finally, the problem of *translation*. For if the criticisms from the social and human sciences are correct—of the artificiality of the laboratory experiment; of the idea of an animal model of human behavior; of the blindness of animal modeling to the realities of the social and cultural lives that shape human mental life and human actions, then the translational problems of behavioral and neuroscientific work on animals would be inevitable. And the question would then be, to what extent are any of these problems remediable.

Artificiality?

First Stabilization: The Experimenter

For those of us who have undergone training in biology, the cry "what's wrong with my mouse?"—or fruit fly or pigeon—is usually one of desperation from novitiates whose dratted creatures will not manifest the attributes or behavior that our textbooks, professors, and demonstrators assure us is there if only we knew how to look: how to mix the reagents, to manage the apparatus, to observe the experiment, to note the results, and so forth. This common experience, banal as it is, may serve to highlight one set of considerations when we are thinking about laboratory studies using animals as models of human behavior. The activity of modeling must be seen as a practice, a craft work, and one that involves training and inculcation of routines and habits. This is a process that involves a triple stabilization—of the experimenter, of the setup, and, of course, of the subject (the fly, mouse, rat, pigeon, etc.). No stabilization along these three dimensions—no laboratory, no result, no paper; and, for those who fail the test of stabilization, no more experimenter. So the stabilization of the researcher—or in most cases the research team—is the first task. We will not dwell on this process of creating the experimenter here, though there is much to be said about it, for we all know that this is what scientific apprenticeship is intended to achieve—to embed these ways of seeing and doing (and many other controls over conduct)—into the bodily and mental habits of the experimenter so that all the strange activities invoved in animal work become a kind of second nature.[11]

Second Stabilization: The Setup

Let us say more, however, about the idea of a setup, of which the experimenter is one element. We have taken the term *setup* from Bruno Latour (Latour 1987), who uses it to refer to the *dispositif* that brings together experimenters, colleagues from a range of disciplines, institutions, apparatuses, data, concepts, theories, subjects (in this case the animals themselves) and much more in order to produce what he terms "inscriptions." Inscriptions are the visual records of the experiment, turned into data that can be taken away from the lab, and worked upon long after the experiment itself is concluded and combined with many other inscriptions, in order to produce an argument. In that process, Latour suggests, the complex, heterogeneous, fragile, crafted assemblage of the setup retreats into the background, as its outcomes, in highly processed forms, appear in graphs, diagrams, formulas, and the like.

While some regard this account as a criticism, from our own point of view, to characterize the process through which data are produced is not to seek to undermine it. It is just to point out that this is one way of enabling the animal and its behavior to 'appear' in a form that enables one to do scientific work on it. One consequence, of course, is that the animals appear in a particular way. They do not appear as individual creatures, with their own peculiarities, as they are seen in the actual practice of the experiment, and as they are known by those who work in the animal labs—who often have a finely tuned knowledge of the specific 'character' of each animal, and an awareness of its 'emotional state' at the time of its entry to the experiment.[12] These individual differences among individual living organisms, which persist despite all the work of stabilization of the animals that we discuss later, disappear as the animal is transformed into a series of measurements that are aggregated with others and turned into the group data that can work in experimental analysis.

This is not the only way to work with animals, of course—one could observe them in their natural habitat—but this would require a different setup. Latour, who has done some interesting work in primatology with Shirley Strum, therefore suggests that it is not helpful to analyze this in terms of an opposition between naturalness and artificiality. We devise different setups to make visible, to elicit, features of the lives and potentialities of the animals that were previously invisible or hard to discern. We can make use of something that Latour points out in a debate with primatologists, drawing on the way in which Thelma Rowell speaks of her work on sheep—we "give the mouse the opportunity to show what it can do," we give it a chance to show us something, perhaps even give it a chance to enter into behavior in which it would not otherwise engage (Latour 2002). There are, of course, other ways to give the mouse, or the monkey, the chance to show what it can do—maybe better ways, maybe ways that have different consequences. But without some setup, producing some kinds of inscriptions that can be worked upon, the animal would not have a chance to show what it can do in a way that can enter scientific debate. So it is no criticism of animal models to say they produce their phenomena only in a certain setup. But this does exemplify one of the two translation problems that we are addressing—the translation from those features that are given permission to appear within the setup of the lab to the features that appear in the clinic and the world beyond.

A comment by Georges Canguilhem in his 1965 essay "The Living and Its Milieu" is relevant here. Discussing the work of Kurt Goldstein, he writes, "The situation of a living being commanded from the outside by the milieu is what Goldstein considers the archetype of a catastrophic situation. And that is the situation of the living in a laboratory. The relations between the living and the milieu as they are studied experimentally, objectively, are among all possible relations, those that make the least sense biologically; they are *pathological* relations" (Canguilhem [1965] 2008, 113; emphasis added). To live, to be an individual living being, Canguilhem argues, is to *organize* one's milieu—and to deprive the experimental animal of precisely that opportunity is to vitiate a crucial aspect of that mode of living that one wishes to analyze. And yet, as we shall see, this seems to be precisely the situation of the animal in behavioral neuroscience. Is this, then a catastrophic situation, and if so, with what consequences?

We need to situate this question in the more general context of the laboratory disciplines, which use a very particular sort of setup to produce truth. They make use of apparatus, instruments, and procedures in order to create phenomena in the laboratory that did not exist before: the truthfulness of their statements is then mapped onto the phenomena produced by that experimental apparatus (Shapin and Schaffer 1985; Hacking 2009, 43). Historians of science tell us that this laboratory form of life emerged across Europe in the seventeenth century (Salomon-Bayet 1978). The *OED* defines a laboratory as "a building set apart for conducting practical investigations in natural science, orig. and esp. in chemistry, and for the elaboration or manufacture of chemical, medicinal, and like products" and finds its first use in the early seventeenth century.[13] Laboratory researchers may bridle at the way in which some of these early usages think of the laboratory as place for alchemy, but undoubtedly, the emergence of laboratories led to a new way of making truths.

This way of truth-making was controversial at the outset, but is now so ordinary that its artifactual nature has become almost imperceptible. In the laboratory sciences, we no longer derive truth from speculation, nor from the observation of natural phenomena in the wild. From this time forward, scientific truths will have to answer to the phenomena that are produced in a novel, enclosed and most unnatural setting which is neither public nor is it exactly private: the phenomena and procedures are observed and recorded by reliable witnesses and only then communicated to others by means of words spoken or written. So the apparatus, the setting, the witnesses (which include, of course, the apparatus and the inscriptions it produces), the form of writing, the kind of fact, the means of making things true—all of these are specific to, constitutive of, the laboratory as a style of thought, and one that is embedded in the emerging form of scientific life. And while, in the early days of the laboratory, the emphasis was on the trustworthiness and reliability of the human witnesses to the experiment, today the crucial role of trust is retained, but displaced to other mechanisms such as peer review and the rhetoric of replication (Shapin 1994, 2008).[14] So, this is the second stabilization—of the apparatuses and the instruments.

If we are to understand the role of model animals and animal models in experimental investigations in neuroscience, then, it will not suffice to denounce them for their artificiality. This would require us to denounce most of what has come to be accepted as truth within the laboratory sciences, for those truths too have been created, manufactured, or at least exposed, intensified, given a chance, made visible, stabilized, within the peculiar world of the laboratory. But perhaps another line of criticism is more salient—that of self-vindication. Ian Hacking has argued that, in the case of those kinds of science that "study phenomena that seldom or never occur in a pure state before people have brought them under surveillance" (1992a, 34), the setups of the laboratory science are "self-vindicating":

> As a laboratory science matures, it develops a body of theory and types of apparatus and types of analysis that are mutually adjusted to each other. . . . They become . . . "a closed system" that is essentially unrefutable. . . . The theories of the laboratory sciences . . . persist because they are true to phenomena produced or even created by apparatus in the laboratory and are measured by instruments that we have engineered. . . . Our theories are at best true to the phenomena that were elicited by instrumentation in order to get a good mesh with theory. The process of modifying the workings of instruments—both materially (we fix them up) and intellectually (we redescribe what they do)—furnishes the glue that keeps our intellectual and material world together. It is what stabilizes science. (1992a, 30, 58)[15]

Animal research in behavioral neuroscience certainly studies phenomena that do not occur in nature but are produced in the laboratory setup. So is such research also self-vindicating? If so, the translation problem would inescapably follow.

This is a question raised within the community of animal modelers themselves. Thus Helene Richter, together with Joseph Garner and Hanno Würbel, argues that "mounting evidence indicates that even subtle differences in laboratory or test conditions can lead to conflicting test outcomes. Because experimental treatments may interact with environmental conditions, experiments conducted under highly standardized conditions may reveal local 'truths' with little external validity" (Richter, Garner, and Würbel 2009). Could one move from such local truths—the truths of a particular laboratory—to the truths of the lifeworld of persons, or even the lifeworld of mice? From our perspective, the key test should not merely be a one-off demonstration of a capacity to translate, as, for example, in the claim that some pharmaceuticals tested on animals seem effective when used to treat analogous conditions in humans. We need an integrated way of escaping self-vindication by establishing a continual relation of transaction between the world of the laboratory and that lifeworld world of humans that it seeks to elucidate. If the laboratory neuroscience of animal models cannot establish those relations, then their problem, perhaps, is not a technical matter, to be rectified by tinkering in various ways, but a conceptual matter about the very nature of the animal models themselves as they function in behavioral neuroscience. We will return to this point later. First, however, let us turn to the third stabilization, that of the subjects, the animals themselves.

Third Stabilization: The Animals

The stabilization of animals used in modeling has involved two aspects—the selection of species to use in modeling, and the stabilization of the creatures within that species to render them into 'model animals' appropriate for experimentation. How does one choose one's model animal? No doubt some animals are exceptionally convenient for laboratory research on behavior. Curt Richter, in his charmingly titled piece "Experiences of a Reluctant Rat Catcher," extols the Norway rat, which he studied for more than half a century, beginning in the 1920s: "If someone were to give me the power to create an animal most useful for all types of studies on problems concerned directly or indirectly with human welfare, I could not possibly improve on the Norway rat" (Richter 1968, 403). He lists the following advantages of his favored creature (377–78):

Norway Rat's Qualification for Behavior Studies

At this point I should like to call attention to the advantages that the rat offers for behavior studies as well as for scientific studies in general.

1. Its dietary needs are very nearly the same as man's. This explains how observations made on it can readily be extrapolated to man. It is a very stable and reliable animal.
2. It reproduces readily.
3. It is just the right size for all kinds of anatomical and physiological studies.
4. Its short lifespan of about three years makes it possible to study factors involved in growth, development and aging.

5. It is a very clean animal. Given the opportunity, it will keep itself perfectly clean.
6. It is well equipped with tools for manipulation of its environment: with very sharp and strong teeth, very dexterous paws and feet, and strong hind legs.
7. It is well equipped with sensory organs: for tasting, smelling, seeing, hearing, and tactile orienting (whiskers and tail).
8. It is available in large numbers in both the domesticated and wild forms.

When the *Scientist* produced its supplement on "Biology's Models" in 2003, its account of the criteria for the choice of model organism was articulated in pragmatic terms very similar to Richter's.[16] Introducing their collection of articles on the "motley collection of creatures" that "fly, swim, wiggle, scurry, or just blow in the wind" that "for the scientific community . . . has been elevated above all other species," the editors argue that "[r]esearchers selected this weird and wonderful assortment from tens of millions of possibilities because they have common attributes as well as unique characteristics. They're practical: A model must be cheap and plentiful; be inexpensive to house; be straightforward to propagate; have short gestation periods that produce large numbers of offspring; be easy to manipulate in the lab; and boast a fairly small and (relatively) uncomplicated genome. This type of tractability is a feature of all well-used models" (Bahls, Weitzman, and Gallagher 2003, 5).[17]

But how should one choose one's model organism? Is there one standard species that one should use to study everything, or are there some species that are particularly suitable for studying specific questions? For Hans Krebs, the answer was the latter: he articulated what he terms "the August Krogh principle," which was that "[f]or many problems there is an animal on which it can be most conveniently studied" (Krebs 1975). From this perspective, close attention to diversity is an important feature of biological research on animals, because that enables one to select the particular animal most appropriate for the purpose at hand.[18] Nonetheless, while a range of animals have been used as model organisms—the fly, the worm, the zebra fish, the mouse, the rat, the guinea pig, the macaque—the assumption has almost always been that what one is able to study most clearly in *each* is an exemplar of principles and processes that have generality across *all*—or many—species. One studies the fly or the mouse, not because one has a particular interest in the fly or the mouse, but because, one hopes that this will enable one to explore issues, processes, and functions that go beyond those specific animals. The particular species is thus not so much a model as an *example* of general principles, no doubt based on the idea that such general principles are preserved, or conserved, across species because of the conservatism of evolution: "Two premises underlie the use of simple organisms in medical research. First, most of the important biologic processes have remained essentially unchanged throughout evolution—

that is, they are conserved in humans and simpler organisms. Second, these processes are easier to unravel in simple organisms than in humans" (Hariharan and Haber 2003, 2457). But are these assumptions so unproblematic?

Cheryl Logan, who has examined the history of test animals in physiology, argues that this presupposition of uniformity across species diversity arises, at least in part, from the pragmatics of the modern research process. The use of a small number of 'standardized' animal species and strains for experimental research began around the start of the twentieth century. Until then, Logan points out, experimenters tended to stress the importance of exploring their hypotheses in a number of different species: "Experimental physiologists gave many reasons for the choice of test animals, some practical and others truly comparative. But, despite strong philosophical differences in the approaches they represented, the view that it was best to incorporate as many species as possible into research . . . was widespread. . . . Authors aimed for generality, but they treated it as a conclusion that would or would not follow from the examination of many species" (Logan 2002, 329).[19] However, there were numerous criticisms of the variations in results found between different laboratories, and Logan argues that this was partly responsible for an increasing emphasis on standardization, which was also linked to the growth and industrialization of scientific research, and the commercialization of animal breeding for research. This, she suggests, led to a restriction on the number of species used. "As animals were increasingly viewed as things that were assumed to be fundamentally similar, scientific generality became an *a priori* assumption rather than an empirical conclusion" (Logan 2002, 329). The standardized animals were thus regarded as laboratory materials, along with other materials that had to be combined in an experimental setup, and—like those other materials, measures, scales, and instruments—their uniformity was both required and, after a while, assumed. If they were obtained from reputable sources, they were, indeed, stable and stabilized.

The animals conventionally used as experimental materials have been raised through many generations of interbreeding of closely related individuals in order to create organisms that can be treated as interchangeable in repeated experimental situations. This has required much labor—often a labor of love—over many years and many generations—as the historical studies by Robert Kohler of the fly (Kohler 1993, 1994) and Karen Rader of the mouse (Rader, 1998, 2004) have shown. Thus Evelyn Fox Keller is uncharacteristically misleading when she suggests that "unlike mechanical and mathematical models (and this may be the crucial point), model organisms are exemplars or natural models—not artefactually constructed but selected from nature's very own workshop" (Keller 2002, 51). For while those species that are to become models may be selected from nature, the model animals are indeed artifactually constructed, and it takes a lot to sustain them as such— not only to breed, feed, and manage them, but crucially to standardize them and the experiments undertaken with them—it takes animal houses, trained persons, specially adapted feed cages, constant tending and familiarity, and much more. We thus agree with Canguilhem when he writes that "the study of such biological materials, whose elements are a given, is literally the study of an *artefact*" (Canguilhem [1965] 2008, 13). We have here another problem of translation, for even the translation between the stabilized model animal and its wild type contemporaries is assumed rather than demonstrated.[20] This brings us directly to the issue of models themselves.

Models¹, Models², Models³, Models⁴ (and Possibly Models⁵)

There are, of course, many types of models.[21] For our current purposes, we would like to distinguish four: conceptual models (*models¹*); mechanical or material models (*models²*); animal models of structures, functions, and diseases (*models³*); and animal models of behavior (*models⁴*). And perhaps there is even a fifth: animal models of human psychopathology (*models⁵*). Most epistemological work has been done on the first two of these—conceptual models and mechanical models. A conceptual model—*models¹* in our little classification—such as those in theoretical physics, is a kind of diagram of theoretical entities and the dynamic relations between them, which can be manipulated in thought—nowadays often in silica on-screen or in virtual reality—by varying values and the like in order to explore their consequences.[22] Nancy Cartwright suggests that, in such cases, "the model is the theory of the phenomenon" (Cartwright 1983, 159). Or, in a slightly different vein, that we should view such models as instruments that mediate between experiments and theory, not to be evaluated by their truth but by their ability to perform in particular experimental and practical contexts, "go-betweens" between the world of things and the world of theories (Cartwright, 1997; Morgan and Morrison 1999). Others have suggested their value is more practical—as Evelyn Fox Keller puts it, a model "suggests experiments, it enables one to predict the consequences of particular interventions, and it prompts the posing of new questions" (Keller 2000, S78).

The brain sciences have not been without their conceptual models that perform such functions. For example, the catecholamine hypothesis of affective disorders discussed in chapter 1, formulated by Schildkraut in 1965 explicitly as a heuristic device, brought together a range of empirical evidence and conceptualization into a diagram of the mode of action of the monoamines, and this functioned as a model to organize much research and explanation in this area for some four decades (Schildkraut 1965). But, as this example illustrates, a model of this sort, which abstracts and generalizes from the experimental moment and which then structures the path of future experiments and the interpretation of results, runs the risk of becoming self-vindicating. Indeed, the life span of the monoamine hypothesis of depression is a very good case in point. Simplified and abstracted, and represented in icons and simulations

(Stahl 1996), this theory structured clinical and experimental thought and practice and underpinned drug development and the interpretation of the effects of the drugs so developed (Kramer 1992).[23]

Most recent work in the history of science and technology has focused on mechanical or material models—which we think of as *models*[2]. Soraya de Chadarevian and Nick Hopwood have referred to these kinds of models as "the third dimension of science."[24] Their focus is on such things as wooden ships and plastic molecules, displays of stuffed animals, artfully constructed model bridges and other artifacts that "purported to bring the tiny, the huge, the past, or the future within reach, to make fruitful analogies, to demonstrate theories. . . . [A]ll are three-dimensional (3-D), and as such, their advocates claimed, displayed relations that could not easily be represented on paper" (de Chadarevian and Hopwood 2004, 1). Such models combine a certain sense of realism with the advantage of docility and manipulability. One can hold them in one's hands, examine them from different perspectives, and subject them to various stresses and other insults to see how they respond.

But the history of biology—more or less since the invention of the term at the start of the nineteenth century—has been coextensive with the use of a different kind of model—one that we term *models*[3]. These use certain animals, or parts of animals, to elucidate what are thought to be the basic features of the structure and function of organs, tissues, systems, cells, and other corporeal elements, which are believed to be generalizable to other living creatures, notably to humans.[25] This has certainly been the case in research on the nervous system. The frog was the animal of choice for the eighteenth- and nineteenth-century practice of neurophysiology. Frogs were used for the study of the concept of reflex and the discovery by Galvani of "animal electricity," thus displacing the long-standing belief in "animal spirits" (Meulders 2006; Harré 2009), and paving the way for the birth of a new avenue of inquiry: electrophysiology. In the 1950s, after his work with Eccles and Kuffler in Australia in the 1930s, Bernard Katz used the neuromuscular junction of the frog as a model system to explore synaptic transmission.[26]

The squid was also a model of this sort—or at least its giant axon was. Initially discovered by J. Z. Young in 1936 (Keynes 2005), the giant squid axon was subsequently used by several prominent scientists in the twentieth century, notably Alan Hodgkin and Andrew Huxley, who shared the Nobel Prize in Physiology or Medicine in 1963 for uncovering the ionic mechanisms in nervous cells. One chose the animal on the basis of its suitability for the purpose at hand—for instance, Kuffler, and a number of those who worked with him in the 1960s, chose the medicinal leech because of its relatively small number of neurons, whose size enabled them to be observed using an ordinary microscope and to be recorded by using intracellular microelectrodes. Not that the researcher was interested in the leech—or the crayfish, the lobster, the mudpuppy, or the sea slug, to enumerate some of the creatures popular with neurobiologists—but because each in some way provided a simple system for studying basic neurobiological mechanisms believed to be universal and hence generalizable. And, it was hoped, one would eventually be able to build up from such basic knowledge of isolated specific mechanisms to a knowledge of the whole—this was the promise, and the premise, of neuroreductionism.[27]

These brief examples serve to demonstrate that model organisms are rather different from the classes of models that we have termed *models¹* and *models²*. This is, in part, because they do not exist in just three dimensions—as in the model of a bridge, for example—but in four. The fourth dimension is time.[28] They have—indeed they exploit—the added feature of vitality—birth, development, change, and death. Model animals are certainly 'simplifications' in one sense: because they are thought to have eliminated much of the genetic variation that is found in the wild type and because of the apparent standardization of their rearing and their living environment.[29] But like other models, they are complex and heterogeneous products of persons, artifacts, theories, concepts, apparatus, and much craft work—to use the term coined by Carrie Friese, they are "model assemblages" wherein heterogeneous things with varying historical legacies are brought together in order to pursue certain goals (Friese 2009).[30]

Yet despite that labor of standardization, the animals used as *models³* remain individual living creatures. It may be hard to think this of the fly, the worm, or the zebra fish, but when one reaches rodents, it is hard to ignore the fact that each mouse is a unique life form struggling to survive in its own way. Animals, unlike most other model entities, are also dynamic, homeostatic, self-regenerating, and self-compensating, continually adapting in various ways to internal and external signals. As Canguilhem put it in his essay titled "Machine and Organism," animals are polyvalent: "An organism has greater latitude of function than a machine. . . . Life . . . is experience, that is to say, improvisation, the utilization of occurrences" (Canguilhem [1965] 2008, 90). Indeed, it is often this internal systemic vitality that is precisely the issue for investigation. Perhaps this is why, to paraphrase Canguilhem again, it is only of living creatures—and not of machines—that one can distinguish the normal and the pathological (ibid., 88–90).

This vitality, this purposiveness, this 'holism,' this teleology of each individual organism has implications for the particular problems of animal models: the practice of treating one organism as a model for another. Every biology student has heard of François Jacob's confident assertion that what is true for *Escherichia coli* is also true for the elephant (Jacob 1988).[31] Canguilhem takes a different view. In his classic paper "Experimentation in Animal Biology," he emphasizes the specificity of the animal that is used in each experimental demonstration:

> It must be said that, in biology, the specificity of an object of observation or experiment limits unpredictably any logical generalization. . . .
> [I]t would often be both prudent and honest to add to the title of a

chapter of physiology that it concerns the physiology of a certain animal, so that the laws of phenomena, which almost always bear the
name of the person who formulated them, would also bear the name of
the animal used for the experiment: the dog for conditioned reflexes,
the pigeon for equilibration, the hydra for regeneration, the rat for vi-
tamins and maternal comportment, the frog (that "Job of biology") for
reflexes, the sea urchin for the fertilization and segmentation of the egg,
the drosophila for heredity, the horse for the circulation of blood. (Canguilhem [1965] 2008, 11)

In the experimental approach of biology, he argues, we need to accept "that no
experimentally acquired fact (whether it deals with structures, functions, or
comportments) can be generalized either from one variety to another within
a single species, or from one species to another, or from animal to man without express reservations" (Canguilhem [1965] 2008, 12). When it comes to the
action of caffeine on striated muscles, for example, what is true for the green
frog is not for the red frog; when it comes to the reflex, the cat differs from
the rabbit but resembles the newt, and when it comes to the repair of bone
fractures, the normal development of the callus in humans differs from that
in dogs. That is to say, there can be no *assumption* of common mechanisms
between different species or strains of animals: even rather basic mechanisms
are *not* conserved by evolution but differ, even between closely related varieties, let alone between animals and humans.

If Canguilhem may be regarded as a rather obsolete source, more recent
studies have similar findings.[32] Thus Andrea Gawrylewski, in her article "The
Trouble with Animal Models," says: "A recent study at the Massachusetts Institute of Technology shows distinct differences between gene regulation in
humans and mouse liver, particularly how the master regulatory proteins
function. In a comparison of 4,000 genes in humans and mice, the researchers expected to see identical behavior, that is, the binding of transcription
factors to the same sites in most pairs of homologous genes. However, they
found that transcription factor binding sites differed between the species in
41% to 89% of the cases" (Gawrylewski 2007, 44).

This issue becomes more important when one begins directly to approach
issues of disease. *Models*[3] have long been used to study human disease, either
by making use of an apparently analogous pathology that occurs spontane-
ously in the animal, by breeding the animal to manifest the disease, or otherwise inducing it experimentally. Mice, for example, have served as models
for the study of cancer since the elucidation of the nature of leukaemias in the
bone marrow of rats in 1933 by the French oncologist and hematologist Jean
Bernard (Degos 2006; Irino 2006). However, translation is not as straightforward as this might imply. As Francesco Marincola put it in the opening editorial of his new *Journal of Translational Medicine*, animal models "do not represent the basic essence of human diseases. . . . Prestigious journals, however,

appear more fascinated with the modern mythology of transgenic and knock-out mice than the humble reality of human disease" (Marincola 2003, 1, 3). Or, in the words of Thomas Insel, director of the National Institute of Mental Health (NIMH), in a paper titled "From Model Animals to Animal Models":

> [M]any biological psychiatrists seem to assume that a mouse is a small rat, a rat is a small monkey, a monkey is a small human, and that all of these are "models" for studying abnormal behavior or abnormal brain function in humans. The failure to attend to species differences not only ignores the opportunity for understanding mechanisms of diversity, it will doom anyone who wants to make facile comparisons. The broad use of laboratory mice and rats for clinical drug development has demonstrated repeatedly that not only are these species inconsistent predictors of clinical response, they are poor predictors of toxicity in humans. (Insel 2007, 1337)[33]

Insel provides a convenient link to what we have termed *models*[4]: animal models of *behavior*. The use of such models dates at least back to the nineteenth century. David Ferrier's studies in the 1870s, carried out while he was the director of the laboratory of experimental neurology at the West Riding Lunatic Asylum, involved ablating cortical regions in dogs and monkeys in the 1870s to cast light on the issue of the localization of functions in the human brain.[34] Better known are the experiments in Pavlov's laboratory in 1912, notably the work of Yerofeeva on behavioral disturbances in dogs caused by conditioning techniques, which were followed by Pavlov's own experiments from the early 1920s on "experimental neurosis" (Broadhurst 1960; Abramson and Seligman 1977). That type of neurosis was very different from the human neuroses as they were conceived by Freud or the other dynamic psychologies of the time—the Pavlovian neurosis was either "excitatory" or "inhibitory" as opposed to the "ideal normal type in which both opposing nervous processes exist in equilibrium" (Pavlov 1941, 73).

Related work was carried out in the United States in the early 1940s, where a number of research groups created experimental neuroses and other conditioned reflexes primarily in sheep. Models, here, were experimental mediums through which a behavior, like salivation or other physiological phenomena like respiration and heart rate could be measured and tested. From this point onward, we begin to see the use of animal models for therapeutic investigation. For example, in the 1930s, phenobarbital, a sedative drug, was investigated by Liddell and his group, who tested its effect on the "excitatory neurosis" in sheep (Liddell, Anderson, Kotyuka, et al. 1935). Pavlov also used bromides, another sedative drug, in the treatment of the experimental neurosis of his dogs (Pavlov 1941). Indeed, experimental neurosis was investigated in all kinds of species—sheep, pigs, rats, cats, and monkeys—with a view to the psychiatric relevance of the findings: this was an "experimental approach to psychiatry" (Gantt 1936, 1942, 1953; Liddell 1938, 1947). This brings us to

Harry Harlow, whose early work was on the effects of brain lesions and radiation on learning in monkeys, but who is now best known for his studies of infant macaque monkeys and their wire and cloth mother surrogates, as models for the importance of mother love for normal human development—or, in the words of the title of his presidential address to the sixty-sixth annual convention of the American Psychological Association—on "the nature of love" (Harlow 1958).[35]

Neurosis, anxiety, depression, love—how is one to characterize the *human states* that the animal models model? Let us take one example—aggression. Jacqueline Crawley opens her chapter "Emotional Behaviors: Animal Models of Psychiatric Diseases" with the firm injunction "Step one: *Don't anthropomorphize!*" (Crawley 2007, 227).[36] In her discussion of animal models of emotional behavior, for example, in anxiety research, she expresses great caution about the dangers of anthropomorphization; we must always talk not of anxiety in a mouse but of "anxiety-like behavior." But surprisingly, in her discussion of "social behaviors" she does not write of "aggression-like behavior" but of "aggressive behavior." Perhaps we can learn something from this apparent contradiction.

It seems that, for Crawley and many others working in this field, aggressive behavior can be identified without reference to internal states. The behavior shown by mice in the lab is self-evidently aggressive and reflects that which is present in aspects of the "natural mouse repertoire": namely, attacks that "serve the functions of defending progeny, mates, and self from attack by a predator, defending the resources of the home territory, gaining access to a sexual partner, and achieving dominant status within the social group" (Crawley 2007, 209).[37] It is on this basis that strains of mice can be developed that are more or less aggressive, and knockouts can be created, entailing null mutants in many of the genes affecting aspects of the neurotransmitter system—subtypes of serotonin receptors, monoamine oxidase, the serotonin transporter, nitric oxide, adrenergic receptors, adenosine receptors, Substance P, vasopressin, and many more. On the basis of the behavioral definition of aggression in mice, these variations certainly increase aggressive behavior—in mice. But what relevance does this have for the kinds of actions in humans that we think of as aggression? In what sense are these artificially produced, engineered mice, developed in close association with the setups, apparatus, and testing procedures that measure their behavior, a mouse model of human aggression?

And, while "don't anthropomorphize" might be a maxim ritually invoked, how about the following breathless translation between species when it comes to aggression:[38]

Psychopathological violence in criminals and intense aggression in fruit flies and rodents are studied with novel behavioral, neurobiological, and genetic approaches that characterize the escalation from adaptive aggression to violence. . . . One goal is to delineate the type of aggressive behavior and its escalation with greater precision; second, the prefrontal cortex (PFC) and brainstem structures emerge as pivotal nodes in the limbic circuitry mediating escalated aggressive behavior. . . . By manipulating either the fruitless or transformer genes in the brains of male or female flies, patterns of aggression can be switched with males using female patterns and vice versa. . . . New data from feral rats point to the regulatory influences on mesocortical serotonin circuits in highly aggressive animals via feedback to autoreceptors and via GABAergic and glutamatergic inputs. Imaging data lead to the hypothesis that antisocial, violent, and psychopathic behavior may in part be attributable to impairments in some of the brain structures (dorsal and ventral PFC, amygdala, and angular gyrus) subserving moral cognition and emotion. (Miczek, de Almeida, Kravitz, et al. 2007, 11,803)

While few make the leap from flies to humans in quite such a startling manner,[39] there is a long history of ethological studies of aggression, with many arguing for its universality across species and drawing analogies between humans and animals. We see this, for example, in the work of John Paul Scott (Scott 1958), but the most emblematic of authors here is Konrad Lorenz (Lorenz 1964; see also Ardrey 1966; Tinbergen 1968). J. Z. Young notes, in his classic *Introduction to the Study of Man*, that the approach of these authors was undoubtedly shaped by the doleful history of human conflict across the twentieth century and indeed by the threat of nuclear war in the 1960s (Young 1974). Even among ethologists, arguments that human conflict manifests internal states of aggression, even in animals, were highly contentious and were criticized by eminent ethologists such as Robert Hinde (1967, 1970).

Of course, most of those from the social sciences were skeptical to say the least. And with good reason. Many circumstances commonly thought of as human aggression—fighting in wars, violence between gangs of youth, knife crime, for example—are not necessarily accompanied by an internal mental state of aggression.[40] Soldiers fight for their comrades, rarely because they feel aggressive toward their enemies; indeed, an aggressive soldier is often a liability. Gang members fight for social recognition or esteem, and for many other reasons related more to personal anxieties and social values than to aggression. Knife crime is as much a matter of fear as it is of aggression, and it of course depends on fashion, availability, town planning, alcohol, and much more. Suicide bombers kill as an act of faith or resistance and rarely out of misanthropy. Genocide, especially where groups mobilized by nationalist or other rhetoric brutally murder others with whom they have lived side by side for years, is another example that indicates that even apparently highly aggressive actions by human beings are shaped and often instigated by language and the mobilization of meanings, ideologies, and memories.

Are these actions based on the same evolutionary mechanisms that generated aggressive behavior in mice? Doubtful. Do drugs that reduce aggressive behavior in mice do the same in humans? Alcohol (when one can get a mouse to ingest it) does not seem to increase aggression in mice, while it certainly facilitates aggressive actions in some humans—but only in some circumstances (in other circumstances it facilitates singing, sex, or sleep). Is aggression part of "the natural human repertoire"? Given the enormous historical, geographical, cultural, and religious variations in what we consider human aggression, and the varieties of types of behavior that are often grouped under that one name, it seems problematic to place all this under one common heading and to imagine common neural underpinnings, let along to seek correlates of superficially analogous behavior in other species and then to grant those correlates a causal status.[41] And even if such behavior did have neural correlates, at what point and in what contexts would it become pathological? And, more relevant to the present discussion, what are the analogies between the 'natural' repertoires placed under this description in animals and those called aggression in humans? For example, according to Crawley, observations of aggression in mice entail such behaviors as general body sniffs, anogenital sniffs, tail rattling, rump biting, and allogrooming (Crawley 2007, 213–17): behaviors hardly thought of as components of aggression in humans, although some would undoubtedly elicit a punch on the (human) nose if displayed on the streets of London or New York on a Friday night.

This brief overview of some of the more obvious problems in using animal models to explore the genetic or neural correlates of human aggression brings us to *models*[5]—animal models of human psychiatric disorders. It is commonly thought that chlorpromazine, the first of the modern generation of psychiatric drugs, was developed by serendipity while the French pharmaceutical company Rhône-Poulenc was seeking potential molecules that might be marketed as antihistamines. However, even at this early point, the properties of novel molecules were first explored through experiments on animals.[42] The first experimental setups that used the elevated plus maze to explore anxiety in animals began in the 1950s, but it was not until the 1980s that this apparatus was validated as a *measure* of anxiety and then used to assess the efficacy of anxiolytic drugs (Pellow, Chopin, File, et al. 1985; Pellow and File 1986).[43] The role of the third edition of the *Diagnostic and Statistical Manual of Mental Disorders* (*DSM-III*) published in 1980 by the American Psychiatric Association was crucial here (American Psychiatric Association 1980). As we discuss in detail in chapter 4, *DSM-III* and its successors provided checklists of apparently objectively observable phenotypic features that could then be reproduced in animals. That is to say, one could try to produce a breeding line of mice or other animals that would exhibit these overt and visible behavioral characteristics, initially by conventional breeding techniques or by the use of drugs to modify animal behavior and later via transgenic and knockout technologies.

This procedure produces a paradoxical looping effect, in which the testing regime plays a key role. Suppose one is trying to develop a psychiatric drug for a given condition included in the *DSM*. One first uses selective breeding or genetic modification to produce a strain of mice that show analogous behaviors to those listed in the *DSM* as criteria for diagnosis of the disorder in humans. Whether or not they do have those analogous characteristics is assessed, in large part, by their performance on certain tests. For instance, the elevated plus maze is used to assess anxiety by measuring how much time a mouse or other rodent spends in the exposed arms of the maze, as opposed to the enclosed parts—presumably on the basis of an simple analogy that an 'anxious' mouse will 'prefer' to be in an enclosed space, while a less 'anxious' mouse will be more ready to explore. The forced swim test and the tail suspension test are thought to assess depression, based on the idea that depression in humans is characterized by 'learned helplessness'—that is to say, the tendency for the depressed person to give up rapidly in the face of obstacles. Hence, it is suggested, one can assess the level of 'depression' in a mouse by measuring the length of time that it will continue to swim and make attempts to escape when immersed in a bath of cold water before switching to mere floating, or the length of time that it will struggle when suspended by its tail before merely hanging passively: each is taken as a measure of learned helplessness and hence of depression.[44]

One then takes individuals from the strain of mice that seem to exhibit high levels of anxiety or depression, administers the drug to them, and sees whether the anxious or depressed behavior is mitigated. Once more, one uses the same tests—elevated plus maze, forced swim test, tail suspension test, and so forth—to assess the behavioral change. If the results are positive, then (subject to all the necessary protocols and phases of drug regulation) one trials the drugs on humans who are selected for the trial on the basis of the *DSM* criteria. One then interprets the results of these trials as evidence that the drug is indeed treating the condition that one has modeled in the animal. One then explores the gene expression pattern, the patterns of brain activation, or the structural features of the brains of the animals, in an attempt to learn about the molecular mechanisms of the condition in humans. The *DSM* diagnosis is thus the fulcrum of such a setup. The self-vindicating loop is designed to find the neural correlates of the *DSM* category that shaped the research design in the first place. And, given that such experiments seek to reduce, eliminate, or ignore all factors other than those modeled in the individual animal—that is to say, those that can be thought of as internal to its genome and brain—the translation to the humans entails an almost inescapable reduction of human action and human affect to expressions of such somatic determinants.

Crawley titles her excellent book on animal models in behavioral neuroscience *What's Wrong with My Mouse?* because this is the question repeatedly addressed to her by researchers who are trying to work out what they have actually modeled when they have genetically modified an animal (Crawley 2007). Have they produced a mouse (and hence a potential mouse line) that is depressed, anxious, aggressive, schizophrenic, or what? The process of characterizing an animal phenotype of a human disorder created by genetic manipulation is no easy matter, even in the hands of an expert such as Crawley:

> Evaluating the behavioral phenotype of a new line of transgenic or knockout mice involves a multi-tiered approach that begins with the systematic assessment of general health, home cage behaviors, neurological reflexes, sensory abilities, and motor functions. Intact performance on these initial measures of procedural abilities allows investigators to proceed with specific complex tests in the behavioral domain of interest (e.g. anxiety-like tests, feeding, learning and memory tasks). If preliminary analyzes indicate deficits in procedural abilities, then investigators may be able to select or modify complex behavioral tests to avoid false-positive results and artifactual interpretations of genotype differences. . . . In addition, careful attention to the baseline behaviors of the background strain(s), and to the possibility that flanking genes and environmental factors may affect phenotypic effects, will maximise the correct attribution of interesting phenotypic anomalies to the targeted gene mutation. (Bailey, Rustay, and Crawley 2006, 129)

But this technical sophistication does not remove the fundamental problem. Even if one goes through all these onerous steps, what is one to say of the relation between the phenotype that accompanies the knockout and the human actions and emotions that it is supposed to model? How is that translation accomplished?

Thomas Insel has some cautionary words on such an approach. Taking Fragile X syndrome as an example,[45] he argues that the animal model approach

> might have created a cognitive deficit in a mouse or rat to resemble Fragile X . . . and then hunted for altered molecular and cellular function that could be targets for drug development. This approach to model Fragile X might work if cognitive deficits were simple lesions with a few predictable causes, as, for instance, in a model of ischemic stroke. This approach might work if we knew that the deficit created in the mouse was identical to the cognitive deficit in the human syndrome, as, for instance, in a lesion model of Parkinson's syndrome. This approach might work if we knew much more about the pathophysiology of the human syndrome, as, for instance, in diabetes. The problem for autism, anorexia nervosa, schizophrenia, and mood and anxiety disorders is that we know so little about the biology of these disorders; there is no credible way to validate (or to falsify) an "animal model" to learn more about the pathophysiology of the human syndrome. (Insel 2007, 1338)

The problem, in Insel's view, is that these animal models are analogies rather than homologies—they produce a phenotype that superficially resembles the condition in humans, but the condition in humans is complex, with multiple causes; in the absence of a clear understanding of the pathophysiology, the model animal is merely a simulacrum.[46] Insel himself has some suggestions as to how to overcome such simulation-based experimental practices, and hence, he believes, to have more success in translation from the laboratory to the clinic and from animals to humans. But before turning to these, let us consider the primary objection from the human sciences: the specificity of the human.

The Specificity of the Human

A belief in the specificity of human beings has been foundational for the social and human sciences. Humans are unique among living beings, it is argued, in having language, culture, and history. Human action differs from animal behavior as a wink differs from a blink: the one has meaning, the other, though it involves the same set of physiological pathways, does not.[47] Even behavior in humans that resembles that in animals bears only a superficial similarity, as human action embodies intention, and intentions are linguistically and culturally shaped. Human affects may seem similar to those of animals, but that is misleading. Although, as Darwin argued in the late nineteenth century, there are similarities in "the expression of the emotions in animals and man" (Darwin [1872] 2009), there are also very significant historical and cross-cultural variations in the recognition, experience, naming, and consequences of human emotions (Lutz and White 1986; Reddy 2001; Oatley 2004). Even the expression of pain in humans, is a communicative act, as Erving Goffman points out in his wonderful little essay on response cries. The utterance "ouch" on hitting one's thumb with a hammer is a call for attention, a request for sympathy, a complaint about having been required to undertake a task unwillingly ("now see what you've made me do!"); it is not an immediate physiological and instinctive response to the sensation of the blow (Goffman 1978).

   The role of historical, social, and cultural processes is even more evident in human psychopathologies, which are shaped, organized, differentiated, and framed in terms of language and meaning. The categories of the diagnostic manuals certainly do not 'cut nature at its joints,' and psychiatric phenotypes from schizophrenia to anxiety are notoriously heterogeneous.[48] The thresholds of disorder—where one might mark the divisions of normality and pathology—are difficult to define and vary across history and culture, as is shown in the difficulties in delimiting attention deficit hyperactivity disorder or antisocial personality disorder, or in differentiating clinical depression from normal sadness. Features often considered to be unique symptoms of psychiatric disorder, for example hearing voices, often turn out to be quite widespread in the population, and only contingent factors lead to them being treated as diagnostic indicators (Romme and Escher 1993). Cognitive psychiatrists, whatever one may think of their general claims, are surely correct in the argument that beliefs are of crucial importance in the development and experience of psychiatric symptoms, and there is some plausibility to their argument that "disruption or dysfunction in psychological processes is a final common pathway in the development of mental disorder [and that] biological and social factors, together with a person's individual experiences, lead to mental disorder through their conjoint effects on those psychological processes" (Kinderman 2005, 206).

Thus, until about a decade ago, the standard social science position would be to deny, and decry, all endeavors that sought to account for human thoughts, beliefs, actions, and psychopathologies by relating them to supposedly analogous matters in nonhuman animals. Anthropomorphism was a cardinal error, as was its reciprocal, the biologization of the human. Such biologization, or animalization, was the province of those deemed reactionary, part of a lineage linked historically with racism and eugenics, and more recently with the reductionist simplicities of evolutionary psychology and sociobiology, with their beliefs in fixed human capacities and predilections, laid down over millennia, and merely played out in human social life and relations.

But perhaps things are changing. This is not merely because animal rights activists have repeatedly challenged the privilege that humans accord to themselves over other animals and ecologists have located humans squarely within a natural ecosphere. It is also because a theoretical avant-garde is beginning to question the dichotomy between animals and humans, which they now see as perhaps the final unchallenged legacy of the Enlightenment. Donna Haraway has pointed to the ways in which humans have evolved along with their companion species and has explored the anthropomorphism that seems endemic to the human condition itself—which may have provided some benefits for humans, though probably not for animals (Haraway 2003, 2007). But further, today, the most radical theoretical move of all is to question the ontological divide between humans and animals. For is it not the final rebuff to the narcissism of the human, to recognize that we are, after all, just animals—that the divide between humans and all other beings needs to be overcome, along with all the other humanist dichotomies of self and other, reason and passion, mind and body, society and nature, organism and machine (Wolfe 2009)?[49]

For present purposes, we do not need to follow that line of argument to its conclusion. We merely need to refuse the simplifications that see human animals as the only living creatures who exhibit the features of vitality that we have pointed to earlier—polyvalence, individuality, improvisation, and constant dynamic transactions with their milieu. What would it mean, then, for those who conduct this work with animals in psychiatry to 'think with the animal' in this sense—to locate their experimental results, and, even more important, their experimental procedures, in the mental life of the animal, and indeed in the form of life of the animal? Perhaps this is impossible. We might recall Wittgenstein once more, and ask whether, if a lion could talk, we would understand him (Wittgenstein 1958, 223). But we should not satisfy ourselves by repeating the nostrum that only humans have language or that animals' communication is radically different, a matter of signals eliciting 'fixed action patterns' rather than of signs embedded in systems of meaning. A host of evidence argues to the contrary, from Gregory Bateson's studies of play among mammals in his essay "A Theory of Play and Fantasy" (Bateson 1972) through discussions of linguistic competence in great apes (Maturana and Varela 1987) to studies of the domestication and training of animals, such as Paul Patton's intriguing use of his knowledge of French philosophy to explore his communication with his horse in the practice of dressage (Patton 2003).[50]

Anthropomorphism is almost inescapable among those who work with animals, whether in or outside the lab. But perhaps we can learn a better way of attributing mental life to animals, a better way of thinking with the animal. If we were to couple this kind of attitude in our animal experimentation with a genuine zoomorphism, then, this would challenge the standard social science critique of modeling humans, in whole or in part, in animal experimentation. Not that humans and animals would be rendered equivalent, but that the capacities, pathologies, and behaviors of animals, like those of humans, would have to be located in their form of life and their constant dynamic interchanges with the specificities of their milieu.[51] The price of breaching of this imagined boundary is, of course, to trouble the ethical sensibilities of those who undertake such experiments—though such concern with the welfare of laboratory animals is somewhat hypocritical in a culture where so many living creatures are bred for slaughter to serve human desires.[52] But even if one can get beyond the old oppositions between anthropomorphism and anthropodenial, the assumptions underlying animal models and model animals remain problematic and need to be opened to empirical examination. And that brings us back to the place where that examination of the distinctions between humans and other animals has consequences for medical and psychiatric practice—the question of translation.

Translation

Many leading biological, biomedical, and neuroscientific researchers have criticized the conception of translational medicine as a pathway "from bench to bedside" and suggested that their criticism might in turn challenge the usage of animal models. Thus Sydney Brenner has argued, "I think that the focus should be the other way around. We should have "from bedside to bench" research by bringing more basic science directly into the clinic. We should also stop working exclusively on model organisms like mice and shift the emphasis to humans, viewing each individual patient as an experimental model" (Brenner 2008, 9).[53]

Leading neuropsychiatric researchers such as Steven Hyman, former director of the NIMH, and Thomas Insel, its current director, share Brenner's critical assessment of much work with animal models, and his view that the failures of results from animal research to translate to effective understanding of the molecular underpinnings of human disorders, let alone to result in the radical improvement of pharmaceutical or other treatments that was promised, should lead us to question some of the fundamental premises on which this work has been based. Thomas Insel, in the paper discussed earlier, suggests that a more effective approach would be what he terms "reverse translation," starting not from the creation of model organisms whose behavior bears a phenotypical resemblance to a human psychopathology, but from the identification of the molecular pathways in humans with mental disorders, and then modeling those pathways in animals: "The good news is that studies of mental disorders might finally be able to exploit reverse translation, moving from studies in humans with a mental disorder to understand how risk genes or abnormal proteins alter brain development and function. . . . The next decade should see a shift from creating abnormal animal behaviors that phenotypically resemble aspects of mental disorders (i.e., animal models) to increasingly understanding the mechanisms of disease by developing animals with the molecular and cellular abnormalities found in mental disorders (i.e., model organisms)" (Insel 2007, 1339).[54]

At first sight this seems a very significant conceptual shift, embodied in a distinct idea of the model, and a distinct setup. One identifies molecular anomalies in humans diagnosed with a condition such as schizophrenia; one seeks to find their genomic basis, perhaps using genome-wide association studies (GWAS); and one then tries to produce animals with a similar genomic anomaly and investigates to see if they show the same disrupted molecular pathways and also display the same behavioral anomaly—if so, one has grounds for arguing that one has modeled one element of the pathway for the disorder.[55] Once more, however, this runs into the problem of replicating—perhaps self-vindicating—a particular *DSM*-derived model of a psychiatric disorder as an amalgam of discrete elements, each of which has a specific neurobiological basis. Insel himself seems to hint at that problem:

> Conditioned fear in a rat is not an anxiety disorder, and social defeat in a mouse is not major depressive disorder. Yes, we have made great progress in "models" of individual features of psychiatric syndromes. For example, in common laboratory animals, conditioned fear has been highly informative of human fear. But how well does this paradigm serve as a model of any human syndrome? At best, we can focus on singular features of disorders, such as vulnerability to conditioned fear,

assuming that human syndromes are composites of isolated abnormal behavior, cognition, or emotion. This assumption remains unproven. (Insel 2007, 1337)

But even so, this argument retains the idea of identifying singular features and of isolating, characterizing, and modeling what are taken to be the distinct components of a mental disorder. Perhaps we should remind ourselves of a caution expressed by Claude Bernard over 150 years ago in his *Introduction to the Study of Experimental Medicine*: "Physiologists and physicians must never forget that a living being is an organism with its own individuality. We really must learn, then, that if we break up a living organism by isolating its different parts, it is only for the sake of ease in experimental analysis and by no means in order to conceive them separately. Indeed when we wish to ascribe to a physiological quality its value and true significance, we must always refer it to this whole and draw our final conclusion only in relation to its effects in the whole" (Bernard [1865] 1957, 88–89). For Bernard, then, when it comes to a living creature, the whole is not merely the sum of its parts. Even if we had an effective animal homology, based on a specific neurobiological pathway, for one aspect of the particular behavior in question, can we isolate one feature from the individuality of the organism as a whole? How can we avoid the self-vindication of *DSM*-type neuropsychiatry?

Perhaps this problem is best illustrated by an example. In July 2007, a team from Johns Hopkins University published a paper that was referred to as "the first mouse model of schizophrenia" in the press release that accompanied it: "Johns Hopkins researchers have genetically engineered the first mouse that models both the anatomical and behavioral defects of schizophrenia, a complex and debilitating brain disorder that affects over 2 million Americans. In contrast to current animal studies that rely on drugs that can only mimic the manifestations of schizophrenia, such as delusions, mood changes and paranoia, this new mouse is based on a genetic change relevant to the disease. Thus, this mouse should greatly help with understanding disease progression and developing new therapies."[56]

The work, by Akira Sawa's lab, was published in a paper titled "Dominant-Negative DISC1 Transgenic Mice Display Schizophrenia-Associated Phenotypes Detected by Measures Translatable to Humans" (Hikida, Jaaro-Peled, Seshadri, et al. 2007). The researchers generated mice that make an incomplete, shortened form of a protein thought to play an important role in neural development—called DISC1 (short for "Disrupted-in-Schizophrenia-1")—in addition to the regular type. The short form of the protein attaches to the full-length one, disrupting its normal duties. As these mice matured, they became more agitated when placed in an open field, had trouble finding hidden food, and did not swim as long as regular mice; such behaviors allegedly parallel the hyperactivity, smell defects, and apathy observed in schizophrenia patients. The mice were also subjected to MRI and showed certain brain defects, notably enlarged lateral ventricles. The press release reports Sawa as noting that the defects in these mice were not as severe as those typically seen in people with schizophrenia, because more than one gene is required to trigger the clinical disease. "However, this mouse model will help us fill many gaps in schizophrenia research," he said. "We can use them to explore how external factors like stress or viruses may worsen symptoms. The animals can also be bred with other strains of genetically engineered mice to try to pinpoint additional schizophrenia genes."

Here we see the idea of reverse translation—and its inescapable link to *DSM* diagnoses. The belief that there is a relation between DISC1 and schizophrenia is based on a number of family studies, starting in 2000, that show statistical associations between mutations in this coding sequence and a psychiatric diagnosis. However, the original study was in a Scottish family with a very wide range of psychiatric symptoms—including bipolar affective disorder, unipolar affective disorder, and conduct disorder: the authors themselves remark that this may be somewhat "atypical" (Millar, Wilson-Annan, Anderson, et al. 2000). The significance of this gene in the original Scottish family is still controversial, as is the specificity of the link to a particular psychiatric diagnosis of schizophrenia, and the mechanisms of action in the brain are also disputed (Ishizuka, Paek, Kamiya, et al. 2006). The functional properties of the protein expressed by the DISC1 gene seem to relate to some very basic pathways and processes in the brain at key points in development, and it appears to interact with a wide variety of proteins linked to neural growth, signal transduction pathways, and proteins linked to neocortical development (Morris, Kandpal, Ma, et al. 2003). Thus one might expect disruption of this gene, and hence of these various pathways, to be nonspecific to this particular diagnosis, and this is indeed the case: "genetic analysis has implicated DISC1 in schizophrenia, schizoaffective disorder, bipolar affective disorder, and major depression" (Hennah, Thomson, Peltonen, et al. 2006, 409).[57]

Nonetheless, the mouse knockout made on the basis of the debatable claim concerning DISC1 is deemed a mouse model of basic pathways in schizophrenia, and then attempts are made to see if behaviors in the mouse parallel those associated with the diagnosis in humans. This relation is established by a rather simplistic set of analogies of just the sort that troubled Insel in his criticism of earlier animal model research: agitation of the mouse when placed in an open field reminds us of hyperactivity, trouble finding hidden food reminds us of smell defects apparently shown by persons diagnosed with schizophrenia, and poor performance in the forced swim test reminds us of apathy—or even "anhedonia" (Hikida, Jaaro-Peled, Seshadri, et al. 2007, 14,503)—observed in those diagnosed with schizophrenia. For those apprenticed in animal models in psychiatry (though not perhaps to others) the effect is, to quote John Medina "evocative of some of the information processing anomalies observed in patients with mood disorders and psychoses" (Medina 2008, 26). But would we be surprised—given the basic pathways being disrupted here—that mouse behavior was somewhat abnormal in a whole variety of ways? No. Do we know how widely distributed are disruptions in DISC1 in the general population of those without a psychiatric diagnosis, and in different countries or regions? No. Do we know whether DISC1 has the same developmental pathways and functions in mouse and human? No.[58] In what sense, then, do we have "a mouse model of schizophrenia"?

What indeed, about schizophrenia itself—a diagnosis given its name a century ago, whose symptomatology and prognosis have varied greatly across that period (changing dramatically, for example, when those living under that diagnosis were no longer subject to long-term confinement in psychiatric institutions) and where recent failures in GWAS confirm what has long been argued by critics—that this is a problematic label for an array of heterogeneous anomalies in human mental life and social skills.[59] If we are not very careful with our thinking, reverse translation from humans into animal models will indeed get caught in the spirals of self-vindication of the diagnosis through these multiple translations between the level of *DSM*, the level of the genome, and the level of the rodent performance in those setups we have discussed earlier.[60]

Life as Creation

Animal models have been crucial in the experimental practices of the brain sciences and central to the development of neuroscience over the past fifty years. Yet we have argued in this chapter that they have some quite fundamental flaws when used as the basis of claims concerning human mental life and psychopathology. Many of our criticisms, though perhaps not our conclusions, are shared by the most careful researchers who work in this field and, as we have seen, some are trying to rethink the practice and conceptualization of animal models and model animals.

However, it seems we are left with more questions than answers. How can we avoid self-vindication in behavioral neuroscience? Can one resolve some of these problems by a move toward 'ecological validity' in animal experimentation? Can one develop forms of assessment of animal behavior that do not rely on crude analogies with human stressors, such as the elevated plus maze or the forced swim test? Can we avoid self-vindication by moving beyond the checklist approach to mental health problems, with all the presumptions embodied in the *DSM*? Can we overcome the epistemological and ontological divide between animal work on human mental states and all we know of humans—and of animals? Or are model organisms in psychiatry doomed to disappear, as some suggest (Debru 2006)? Will they be displaced by the power of novel biotechnologies to create stem cell lines from tissue derived from those diagnosed with disorders and to use these for the conduct of experiments and the trialing of therapies,[61] or by simulation in computer models in silica,[62] or by virtual experimental situations,[63] or by a return to the human as its own model (Rheinberger 2006; Brenner 2008)?

Perhaps there is something to be learned from our earlier insistence on the animal experiment as craft work—for given that modeling is craft work, is there a way in which we can setup the animals differently?[64] Can we reimagine the epistemological status of animal models? And perhaps, just perhaps, a genuine interdisciplinary, transdisciplinary dialogue between animal modelers in neuroscience and those from the social and human sciences can contribute to this. So let us conclude by returning once more to the reflections of Georges Canguilhem in his essay on "Experimentation in Animal Biology," written almost half a century ago: "The use of concepts and intellectual tools [of biology] is thus at once both inevitable and artificial. We will not conclude from this that experimentation in biology is useless or impossible. Instead, keeping in mind Bernard's formulation that 'life is creation,' we will say that the knowledge of life must take place through unpredictable conversions, as it strives to grasp a becoming whose meaning is never so clearly revealed to our understanding as when it disconcerts it" (Canguilhem [1965] 2008, 22).

Chapter Four

· · · · · · · · · · · · · · · · ·

All in the Brain?

Nothing can be more slightly defined that the line of demarcation between sanity and insanity. Physicians and lawyers have vexed themselves with attempts at definition in a case where definition is impossible. There has never yet been given to the world anything in the shape of a formula upon this subject which may not be torn to shreds in five minutes by any ordinary logician. Make the definition too narrow, it becomes meaningless; make it too wide, the whole human race are involved in the drag-net. In strictness, we are all mad as often we give way to passion, to prejudice, to vice, to vanity; but if all the passionate, prejudiced, vicious, and vain people in this world are to be locked up as lunatics, who is to keep the key of the asylum?

—*The Times* (London), July 22, 1854[1]

Psychiatry is a biomedical science. This is not a trivial statement. Nor could psychiatry have been called a science twenty years ago. . . . [T]oday a remarkable body of new knowledge relates neurobiology to psychiatric disorders.

—**Philip Berger and Harlow Brodie**, preface to *Biological Psychiatry*, 1986[2]

Here's how doctors decide which mental or neurological disorder their troubled patients suffer from: they ask questions like, "Are you hearing voices?" and "Do you feel like people are out to get you?" . . . Not all that different from how they used to do it about 100 years ago. New techniques are set to radically change that approach . . . relying more on changes in physiology than in behavior. . . . Scientists are increasingly turning to biomarkers— such as genes or proteins in tissues, blood and body fluids—to distinguish between symptomatically indistinct illnesses. . . . Using techniques that have only become available in the past few years, scientists are looking at the brain, serum and spinal fluid and asking which combinations of genes or proteins might be expressed differently in these disorders. The field is still far from ready for the clinic, but holds great promise: markers could allow for earlier treatment and a better prognosis, ultimately enabling scientists to intervene even before the full-blown disease strikes. . . . The most challenging aspect might yet turn out to be redefining a 100-year-old classification system.

—**Amanda Haag**, "Biomarkers Trump Behavior in Mental Illness Diagnosis," 2007

In 1953, Dr. Henry Yellowlees, chief medical officer at the United Kingdom's Department of Health, published a small volume described as "commonsense psychiatry for lay people"; its title was *To Define True Madness*.[3] This phrase is, of course, a quote from Shakespeare's *Hamlet*, published in 1602. Polonius, in act 2, scene 2, having told the King that his noble son is mad, then pauses for a second: "Mad call I it; for to define true madness, What is't but to be nothing else but mad? But let that go." Psychiatry, since its invention in the nineteenth century, has found it less easy to let that question go—the question of what it actually is to be 'mad,' of the boundaries of madness and sanity, and of the differentiation between the various forms that madness—whether termed lunacy, mental illness, or mental disorder—can take. Indeed, it has been haunted by this problem, framing it as the problem of diagnosis, with theorists, practitioners, and critics returning to it again and again (cf. Porter 2002, 1). Some feel that we are, at last, on the verge of putting these troubles behind us, that neurobiology has enabled psychiatry to become a science, and as psychiatry becomes neuropsychiatry it will be able to answer, finally, the question of who is 'a suitable case for treatment.'[4] Will an appeal to the brain, or to specific genetic sequences associated with particular neurobiological anomalies, or to the new visibility apparently conferred by neuroimaging, finally enable clinicians to delineate the boundaries of normality, and to differentiate between disorders? Will neuroscience at last provide psychiatry with the objectivity that it seeks?

While the term *neuropsychiatry* had been in use both in English and French since the early twentieth century and probably since the late nineteenth century,[5] it was not until after the Second World War that it gained popularity in the European and American literature.[6] While it was used in a variety of senses, by the start of the twenty-first century the term was being used to argue that the future of psychiatry lay in the integration of insights from genetics and neurobiology into clinical practice (Healy 1997; J. Martin 2002; Sachdev 2002, 2005; Yudofsky and Hales 2002; Lee, Ng, and Lee 2008). There is a weak version and a strong version of this argument. Some advocate the formation of neuropsychiatry as a particular professional specialty, appropriate for those disorders whose understanding and treatment requires both neurology and psychiatry. This, for example, is the argument of Yudofsky and Hales in an article titled "Neuropsychiatry and the Future of Psychiatry and Neurology":

> [There are] many problems inherent in separating psychiatry and neurology into discrete medical specialties. First, the separation perpetuates mind-body dualism that is a perennial source of the public's confusion about whether or not mental illnesses have biological bases and whether psychiatry is a credible medical specialty. This confusion is also a source of the stigmatization of the mentally ill and leads to a lack of parity in the reimbursement for psychiatric treatments with those for other medical conditions. Second, this separation of the specialties promotes the condensation of complex brain disorders into simplistic categorizations that are based on single symptoms. . . . [N]europsychiatry . . . bridges conventional boundaries imposed between mind and matter as well as between intention and function [and] vaults the guild-based theoretical and clinical schisms between psychiatry and neurology. (2002, 1262)

Others argue more expansively and either propose, or assume, that all psychiatric disorders are ultimately grounded in anomalies in the brain and nervous system—that is to say, the disorders that psychiatry deals with are neurological disorders. This, for instance, is the U.S. Surgeon General writing in 1999: "Mental disorders are characterized by abnormalities in cognition, emotion or mood, or the highest integrative aspects of behavior, such as social interactions or planning of future activities. These mental functions are all *mediated* by the brain. It is, in fact, a core tenet of modern science that behavior and our subjective mental lives *reflect* the overall workings of the brain. Thus, symptoms related to behavior or our mental lives *clearly reflect* variations or abnormalities in brain function" (United States, Center for Mental Health Services 1999, 39; emphases added).

We will not dwell on the slippery terms used here—*mediated, reflect, clearly reflect*—we should not expect the Surgeon General to resolve the age-old problem of mind and brain. However, he is just one of the many who invest their hopes in a fusion of psychiatry and neuroscience. It will, some hope, overcome the legacies of the Cartesian division between mind and body that are still enshrined in the distinction of organic and functional disorders in psychiatry, proving conclusively that neurobiological anomalies underpin mental disorders. It could help challenge stigma, they believe, by demonstrating that psychiatric disorders are genuine conditions, not just 'all in the mind' and hence not problems for which the sufferers themselves should be held responsible. And, as we have said, some expressed the hope that such developments could overcome the multiple problems that have surrounded psychiatric diagnosis, generating new methods that rely on measurable and objective markers (Hyman 2003; Haag 2007; G. Miller 2010). Many made the linked argument that this would also enable early, presymptomatic identification of incipient disorders, allowing early intervention and preventive treatment. In this chapter we focus upon this question of diagnosis and explore some of the hopes for the relations between neuroscience and psychiatry from this perspective. For, as we have already suggested, this question of diagnosis goes to the heart of the key question, not just for neuropsychiatry, but for psychiatry in general—the question of what is it to be mad, what kind of a thing is a 'mental disorder'?

To Define True Madness

Some of us are old enough to have lived through—indeed participated in—the contentious disputes of the decades of the 1960s and 1970s that troubled psychiatry to its core. At that time madness, as mental illness, came to be regarded as an 'essentially contested category.' These contests were not restricted to academic debate. They entered popular imagination, in films like *One Flew over the Cuckoo's Nest* (1975) or *Morgan: A Suitable Case for Treatment* (1966), and penetrated general cultural awareness in the array of critiques that became known as antipsychiatry. David Rosenhan, in a classic and widely known experiment published in 1973, used this instability of diagnosis to ask a key question: "if sanity and insanity exist, how shall we know them?" Pseudopatients (students with no existing psychiatric problems) presented themselves at several psychiatric hospitals, claiming to hear voices that corresponded with prevailing existential philosophies—"thud," "empty," "hollow"—and each was admitted to a psychiatric ward. Their histories in their case records were reworked by their psychiatrists to fit prevailing conceptions of the etiology of schizophrenia. None were unmasked by psychiatric professionals, and all were eventually discharged with the diagnosis of "schizophrenia in remission." For Rosenhan, the message was stark: normality is hard to diagnose.

These criticisms were linked with others, familiar from the sociology of deviance then in its heyday—that a psychiatric diagnosis was actually the outcome of a social process of labeling; that mental illness was a residual category, applied to those whose violation of cultural norms could not be explained in any other way; that diagnosis is the outcome of a social process and retrospectively gathers together all manner of previous episodes in a life history and reframes them as evidence of developing mental breakdown; and that the label, once applied, is sticky and hard to dislodge, leading to stigma, exclusion, and isolation. In labeling deviance as mental illness, psychiatry, it was claimed, was actually creating the forms of damaged and disordered selfhood that it claimed to explain. If these concerns had been restricted to critical social scientists, they would have been of minor significance. But, as we shall see, others were also troubled by psychiatric diagnosis, including psychiatrists themselves.

What is diagnosis? A layperson might say that a medical diagnosis is intended to identify the nature of the disease or condition that ails an individual, in order to clarify its causes and suggest an appropriate treatment. Its Greek root διαγνωστικός (*diagnostikos*) means "capable of discerning." The *OED* gives a similar definition, dating it to Willis in 1681: "Determination of the nature of a diseased condition; identification of a disease by careful investigation of its symptoms and history; also, the opinion (formally stated) resulting from such investigation."[7]

Robert Kendell, who in 1975 was a professor of psychiatry at the University of Edinburgh, published *The Role of Diagnosis in Psychiatry* in the throes of these debates over the objective status of psychiatric diagnoses. He opens the book with a quote from Daniel Hack Tuke, the great-grandson of William Tuke, founder of the York Retreat: "Eighty years ago Hack Tuke observed that 'The wit of man has rarely been more exercised than in the attempt to classify the morbid mental phenomena covered by the term insanity,' and went on to add that the result had been disappointing. The remark remains as apposite now as it was then" (Kendell 1975, vii). Kendell argues that "the failure of both psychiatrists and psychologists to develop a satisfactory classification of their subject matter, or even to agree on the principles on which that classification should be based, is a most serious barrier to fruitful research into the aetiology of mental illness and even into the efficacy of treatment regimes" (ibid.).

Others took the fact that psychiatrists could not agree on the application of their own categories to indicate that those categories themselves were unstable subjective judgments. The decision by the American Psychiatric Association (APA) in 1973 to remove homosexuality as a category of disorder, the regular ridiculing of psychiatric evidence in legal cases, and the association of psychiatry with confinement for political reasons in the Soviet Union also led many to wonder about the capacity of psychiatrists to legitimately categorize an individual as suffering from a particular mental disorder, or even a mental disorder at all. We shall return to these debates presently. But if psychiatry ran into so much trouble over its diagnoses, one might be forgiven for thinking that this occurred for clinical reasons—that diagnoses were failing to identify an underlying disease from a set of symptoms or to clarify etiology and enable treatment and prognosis. However, a brief glance at the history of psychiatric classification is enough to show us that such categorization has not been a purely clinical matter.

Many have written histories of attempts to classify diseases, have described the work of pioneers in medical nosology, and have examined the debates over the status of such classifications—whether they were, for example, 'natural' or 'artificial' (see, e.g., Roy Porter's chapter in L. Conrad et al. 1995). Critical sociologists, especially in the United States, have focused on the development of the *Diagnostic and Statistical Manuals* produced by the APA and have tended to regard the whole enterprise as suspect, as part of an entrepreneurial strategy to expand the empire of psychiatry and enhance its power and legiti-

macy by pathologizing everyday life, hand in glove with a pharmaceutical industry looking to increase the market for its wares (Kutchins and Kirk 1997).
Such overly general critiques usually fail to consider the work of classification
in psychiatry in other national contexts, and greatly underestimate the level
of awareness of those undertaking the exercises about the problematic epistemological and ontological issues involved in classification. It is not only sociologists who understand the performative character of categorization, as the
debates among those responsible for creating and using the manuals themselves testify. It is, therefore, worth sketching the development of psychiatric
classification and examining some of the debates within psychiatry itself.

Although we can find many earlier attempts at classifying forms of madness, notably Robert Burton's *The Anatomy of Melancholy*, published in 1621,
or Thomas Sydenham's *Epistolary Dissertation on the Hysterical Affections* of
1682, the project of classification in modern psychiatry can be traced to the
visibility cast on inmates by the reformed asylum of the nineteenth century.
Indeed, the asylum provided key conditions for the birth of psychiatry, because it first assembled together the diverse characters who were to form its
object. It was only the asylum that seemed to confer some homogeneity upon
those disparate individuals—displaced, desolate, despairing, deranged—who
were collected within its walls. What first linked these troubled and troublesome people was not a medical gaze, but a civilizing project: the asylum
was the place for all those who offended the norms of self-conduct and self-
control that were beginning to grid the emerging urban, liberal social orders
in the mid-nineteenth century (Foucault [1961] 2005; Rothman 1971; Castel
1976; Scull 1979; Goldstein 2002).

As doctors conquered the space of asylum over the second half of the
nineteenth century, guarding its points of entry and of exit, and managing its
interior world, the assorted figures who were confined within became psychiatry's subjects. The doctors of the mad began to search for a single classificatory schema that would embrace them all, that would grasp them all as
sufferers from varieties of a common condition—madness, mental illness, or
mental disorder. These schemes of classification drew upon older themes, it
is true, but attempted to give them clinical utility, linking each category to a
description intended to educate the gaze of the doctor, and often adding some
case histories and recommendations as to appropriate treatment (Esquirol
1838, 1845, [1845] 1965; Bucknill and Tuke 1874).[8]

In the United States, this clinical desire to classify was supplemented by
another, a *statistical* project driven by the national census, and its demand
that every person within the national territory should be enumerated (on the
history of the U.S. Census, see Alonso and Starr 1987; Anderson 1988). The
1830 U.S. census had counted the deaf, dumb, and blind, and the 1840 U.S.
census added a count of the insane and idiots, distinguished by race and by
mode of institutional or other support. When the results of this census were
published in 1841, the total number of those reported as insane or feeble-

minded was more than seventeen thousand, of whom nearly three thousand were black. The rate of insanity among free blacks was eleven times higher than that of slaves and six times higher than that of the white population.

For those who opposed abolition, like U.S. Vice President John C. Calhoun, these census figures proved that blacks were congenitally unfit for freedom: "Here is proof of the necessity of slavery. The African is incapable of self-care and sinks into lunacy under the burden of freedom. It is a mercy to him to give him the guardianship and protection from mental death" (Calhoun, 1843; quoted in Deutsch 1944, 473). Abolition, he believed, far from improving the condition of "the African," worsened it: where "the ancient relations" between the races had been retained, the condition of the slaves had improved "in every respect—in number, comfort, intelligence and morals" (quoted in Gilman 1985, 137). A political dispute about madness and slavery followed, waged in the language of numbers (Deutsch 1944).

By the time of the 1880 U.S. census, seven categories were used to count the institutionalized population of the mad—mania, melancholia, monomania, paresis, dementia, dipsomania, and epilepsy. It was in order to provide the U.S. census with a better basis for its classification of those confined that the first official U.S. nosology of mental illnesses, the *Statistical Manual for the Use of Institutions for the Insane*, was prepared in 1918 by the American-Medico Psychological Association (the successor of the Association of Medical Superintendents of American Institutions for the Insane, which was to become the American Psychiatric Association in 1921) and the National Committee for Mental Hygiene (Grob 1991). That manual was the forerunner of the *Diagnostic and Statistical Manuals* that would follow from the 1950s onward.

The history of these manuals has often been told (Spitzer, Williams, and Skodol 1980; Bayer and Spitzer 1985; Grob 1991; M. Wilson 1993; Spitzer and Williams 1994; A. Young 1995, 89–117; Spitzer 2001; Cooper 2004; Mayes and Horwitz 2005). The first editions, as we have said, focused on classification of those who were confined in the asylums, with their growing populations of inmates. This remained their main concern until the 1940s. However, by then a different kind of psychiatry was beginning to flourish outside the asylum and was difficult to ignore. From the spas and water treatments of the nineteenth century, through Mesmerism and the work of the nerve doctors like Breuer and Freud, and through the pathologies of children; the problems of the maladjusted worker or soldier; the work of psychiatric social workers in courts, schools, and elsewhere; and in the growth of individual and group psychotherapeutics, a much wider territory for psychiatry had taken shape, not defined by the walls of the asylum. This was the territory of anxieties and minor mental troubles, of problems of the adjustment of the individual to the demands of family and workplace, of delinquency and criminality—in short, the territory of the neuroses and of problems of conduct (Rose 1989).

Hence it is not surprising that the tenth edition of the *Statistical Manual* included classifications for the psychoneuroses (hysteria, compulsive states, neurasthenia, hypochondriasis, reactive depression, anxiety state, anorexia nervosa, and mixed psychoneuroses) and primary behavior disorders (adult maladjustment and primary behavior disorders in children), although it nonetheless conceived of these states, like the other conditions it named, primarily as somatic conditions. It is also not surprising that the term *neuropsychiatry* was used at the time to refer to the study and management of those *neuroses* precipitated by industrialization and transformations at the workplace. Thus, for example, in an article published in 1921 under the title "The Service of Neuropsychiatry to Industrial Medicine," Dr. Harold Wright defines neuropsychiatry as a "synthetic" specialty and the neuropsychiatrist as "a synthesist" who "considers the personality as a whole with all its physical and psychological faults of adjustment. And if the psychiatrist or general medical man is to do effective work in an industrial plant, he must consider things outside of the plant as well as those influences which are present within it and within the patient's body" (Wright 1921, 465).

In the Second World War, in the United States, as in the United Kingdom, psychiatrists were widely deployed in the military apparatus, in branches ranging from psychological warfare and propaganda, through personnel selection and allocation, to the diagnosis and rehabilitation of those invalided out of the forces with mental troubles apparently brought on by the conflict (Rose 1989; Shephard 2003). A whole array of minor neuropathologies were now being diagnosed and treated by psychiatrists, and at the same time, psychiatry began to argue that these mild mental pathologies were at the root of problems in many social institutions and practices; they thus argued that there was a positive role for their expertise in the promotion of mental health. As Grob puts it (1991, 427): "In the postwar era, the traditional preoccupation with the severely mentally ill in public mental hospitals slowly gave way to a concern with the psychological problems of a far larger and more diverse population as well as social problems generally. Persuaded that there was a continuum from mental health to mental illness, psychiatrists increasingly shifted their activities away from the psychoses toward the other end of the spectrum in the hope that early treatment of functional but troubled individuals would ultimately diminish the incidence of the more serious mental illnesses."

The American psychiatrist William C. Menninger, head of the newly created Neuropsychiatric Division of the U.S. Army in 1943, perhaps captured this approach best. In treating war neurosis, he wrote, one should take into account the "whole environment," that is, "everything outside ourselves, the thing to which we have to adjust—our mates and our in-laws, the boss and the work, friends and enemies, bacteria and bullets, ease and hardship" (W. Menninger [1947] 1994, 352); in the postwar years, he was to play an important role in the revision of psychiatric nomenclature (Houts 2000), However, it was his brother, Karl Menninger, who made the most uncompromising statements of this idea that there was an essential unity of all mental troubles:

they were failures to adapt to pressures of the environment, and hence conditions should be explained and symptoms interpreted in psychosocial terms (K. Menninger 1965; cf. M. Wilson 1993). For Karl Menninger, "most people have some degree of mental illness at some time, and many of them have a degree of mental illness most of the time" (K. Menninger 1965, 33). Hence, as William Menninger put it in 1947: "Psychiatry should place high priority on its efforts to provide the 'average' person with psychiatric information he can apply to his own problems. . . . The public wants this education. . . . Very possibly it may increase the number of patients who seek help from a psychiatrist, just as a campaign about cancer or tuberculosis increases the number of patients who go to doctors about these problems" (W. Menninger [1947] 1994, 79).

It was in the context of this radically expanded field that the *Diagnostic and Statistical Manual*, published in 1952, was prepared by the APA's Committee on Nomenclature and Statistics (American Psychiatric Association 1952). This manual construed mental disorders as reactions of the personality to psychological, social, and biological factors. Following the lesson of wartime, both minor and major mental troubles were included in the classificatory framework—although distinguished as neuroses and psychoses. Psychiatry, it was argued, now had a key role outside the hospital, in addressing all those minor troubles, and in prevention. Accurate diagnosis was the key to competent prognosis and appropriate treatment, especially in the name of preventive intervention. And diagnosis required interpretation: symptoms needed to be understood as meaningful responses to events in the life of the patient.

It was not until sixteen years later, in 1968, that the first revision of the *DSM* was published. The revision arose, according to its foreword, from the need for American psychiatrists to collaborate with those who were preparing and using the new revision of the International Classification of Diseases (ICD-8), which had been approved by the WHO and was to come into use in 1968 (American Psychiatric Association 1968, vii). *DSM-II*, as it was known, was developed by a Committee on Nomenclature and Statistics of the APA, with advice from a team of three consultants—Bernard Glueck, Morton Kramer (who had worked on *DSM-I*), and Robert Spitzer (then director of the Evaluation Unit of Biometrics Research of the New York Psychiatric Institute). For Kramer, the earlier versions of ICD were unsuitable "for use in the United States for compiling statistics on the diagnostic characteristics of patients with mental disorders or for indexing medical records in psychiatric treatment facilities" (ibid., xi). The APA committee worked in consultation with the WHO team that was revising the ICD, though "the Committee had to make adjustments" to conform to U.S. usage. *DSM-II* was 134 pages long and had 180 disease classifications organized into ten categories. The aim was "to avoid terms which carry with them *implications* regarding either the nature of a disorder or its causes" (ibid., vii). The foreword, by Ernest Gruenberg, who was chairman of the APA's Committee on Nomenclature and Statistics, drew particular attention to the renaming of "schizophrenic reaction" to "schizophrenia" to eliminate such causal assumptions, adding, somewhat poignantly, "Even if it had tried, the Committee could not establish agreement about what this disorder is; it could only agree on what to name it" (ibid., ix).

On the occasion of the publication of *DSM-II*, Robert Spitzer, together with Paul Wilson, who had done much of the editing work on the *Manual*, published an account of the new diagnostic nomenclature and the ways in which it differed both from *DSM-I* and from ICD-8. They claimed that the new classification system was not a "return to a Kraeplinian way of thinking, which views mental disorders as fixed disease entities" but an attempt to avoid terms that carry implications about the causes of a disorder, especially where there is controversy over its nature or cause (Spitzer and Wilson 1968, 1621). While *DSM-II* reduced the previous emphasis on disorders as reactions to precipitating life experiences, it did not eliminate this suggestion entirely, and it also retained, and indeed extended, the categories of neurosis. It also retained the categorical distinctions between mental retardation, organic brain syndromes, neuroses, and personality disorders (American Psychiatric Association 1968). Spitzer and Wilson concluded their presentation thus: "With the adoption of *DSM-II*, American psychiatrists for the first time in history will be using diagnostic categories that are part of an international classification of diseases. While this is an important first step, it is only an agreement to use the same sets of categories for classifying disorders" (Spitzer and Wilson 1968, 1629).

Despite these high hopes, by the time of its publication in 1968, the expansive project of psychosocial psychiatry that *DSM-II* sought to underpin was already under attack (cf. M. Wilson 1993). As we have seen, over the 1960s, antipsychiatry, in its various manifestations, cast doubt on the conceptual underpinning and sociopolitical consequences of this broadening of the remit of psychiatry. The debate in the early 1970s over the inclusion or exclusion of homosexuality as a disease category seemed to some to mark a turning point (Bayer 1981). The text of *DSM* was altered for every printing of *DSM-II*, and in the seventh printing homosexuality was removed as a disease category, seemingly confirming the view of critics that its categories were just sociopolitical labeling of behavior considered to be socially undesirable by the authorities at any given time.

Within psychiatry itself, concern began to mount about the lack of standardized criteria for diagnosis and the poor reliability of diagnosis in practice—studies such as that by Rosenhan exacerbated this anxiety. Further, as funding for psychiatry was restricted in the United States in the 1970s, government and private health insurers became concerned about the lack of specificity of classification, the confusions of terminology, and the difficulty of establishing clear boundaries between those who were eligible for treatment—and hence for insurance reimbursement—because they were suffering from an illness and those who were not, even if they were receiving some kind of psychological therapy. The response by many within the profession was to emphasize the classical medical model of disease and to argue that this should form the basis and the limits of psychiatry's vocation, which should not seek an extension of its ministrations to the ills of society as a whole (Klerman 1978).

DSM-III, published in 1980, is often seen as a response to this crisis in legitimacy of psychiatry (American Psychiatric Association 1980; Mayes and Horwitz 2005). It had begun with the perceived imperative to narrow the definition of mental disorder, to raise the threshold for diagnosis of illness, and to set out a classificatory scheme based on observable symptoms. This was no simple "return to Kraepelin" (Decker 2007),[9] but it owed a great deal to the so-called Feighner criteria, associated with the Department of Psychiatry at Washington University School of Medicine in St. Louis (Kendler, Munoz, and Murphy 2010). These were most influentially set out in a 1972 article by John Feighner and his colleagues as diagnostic criteria to be used in psychiatric research (Feighner, Robins, Guze, et al. 1972).[10] In contrast to *DSM-II*, say the authors, "in which the diagnostic classification is based upon the 'best clinical judgment and experience' of a committee and its consultants, this communication will present a diagnostic classification validated primarily by follow up and family studies" (ibid., 57). The diagnostic criteria that they presented for fourteen psychiatric illnesses each contained a general clinical description together with a checklist of observable symptoms, with the stipulation that a certain number of these must be present to warrant a diagnosis, and a criterion concerning the time scale of the condition and the implications of preexisting psychiatric conditions. The authors referred to the desirability of "laboratory studies"—chemical, physiological, radiological, and anatomical, which they considered to be "generally more reliable, precise and reproducible than are clinical descriptions" but remarked that "[u]nfortunately, consistent and reliable laboratory findings have not yet been demonstrated in the more common psychiatric disorders" (ibid.). Nonetheless, they concluded that what they presented was "a synthesis [of existing information] based on data rather than opinion or tradition" (ibid., 62).

This paper was receiving some two hundred citations a year by the mid-1970s (Garfield 1989; Kendler, Munoz, and Murphy 2010), and by 1974, two of the authors consolidated this way of thinking into a textbook on psychiatric diagnosis (Woodruff, Goodwin, and Guze 1974). But the view that this was in some way a return to Kraepelin was hard to shake off, despite the protestations of the authors. Thus, writing in 1978, while head of the Alcohol, Drug Abuse and Mental Health Administration, Gerald Klerman referred to them as "neo-Kraeplinian," and portrayed them as engaged in a long struggle against what they felt to be an entrenched Meyerian domination of the field in the NIMH in the United States and the MRC in the United Kingdom (Klerman 1978).

Klerman, like Spitzer, had been working since the early 1970s with psychiatric research groups developing "research diagnostic criteria" based on the

Feighner approach. He argued that this classificatory ethos does not necessarily imply a commitment to a biological etiology for psychiatric problems,
let alone a biological approach to treatment—indeed Klerman himself was
a key contributor to the development of interpersonal psychotherapy. However, Spitzer and his team *were* committed to biological approaches from the
outset; they were working on the basis of a commission from the NIMH as
part of their Collaborative Program on the Psychobiology of the Depressive
Disorders: "biological and clinical collaborative programs . . . initiated in the
early 1970s [spanned] research on nosology, genetics, neurochemistry, neuroendocrinology, and psychosocial factors . . . [arising from the] excitement
created by the introduction of the new antidepressant drugs and the controversy surrounding some inspired hypotheses about the potential genetic
and biochemical causes of the major depressive psychoses" (Katz, Secunda,
Hirschfeld, et al. 1979, 765).

By 1975, writing with Eli Robins, a biological psychiatrist who was one
of the authors of the Feighner paper, Spitzer was commending the Feighner
approach to diagnostic classification for the forthcoming *DSM-III*, in contrast to that taken in *DSM-II* (Spitzer, Endicott, and Robins 1975, 1187–88).
Urging the use of what they now termed "specified diagnostic criteria" in
DSM-III, Spitzer and his colleagues argued that these would not eliminate
clinical judgment or displace validity in the quest for reliability, but would
improve training of psychiatrists and improve communication; in conclusion
they contended that "[t]he use of such criteria also forces clinicians to share
their definitions of clinical terms and concepts, overcoming the tendency to
be idiosyncratic. . . . We believe that the major potential beneficial effect of
including specified diagnostic criteria in *DSM-III* would be to improve the
reliability and validity of routine psychiatric diagnosis. This in turn should
improve the value of the standard nomenclature for all of its many uses, clinical, research and administrative" (ibid., 1191).

DSM-III was, indeed, reshaped along these lines. In particular, much to
the irritation of psychodynamic psychiatrists (Frances and Cooper 1981),
it eliminated the use of the term *neurosis*, which was thought inescapably
tainted by psychodynamic theory, although in a moment of appeasement to
pressures from psychoanalytic practitioners, the team preparing the revision
agreed to put some reference to this term in parentheses, next to the disorders
that had been termed neuroses in *DSM-II*. Thus 300.40 "Dysthymia" was followed by "(or Depressive Neurosis)," and the section headed "Anxiety Disorders" followed this title with "(or Anxiety and Phobic Neuroses)": "the term
'neurotic disorder' is used in DSM-III without any implication of a special
[and invisible] etiological process" (American Psychiatric Association 1980,
9–10; quoted from A. Young 1995, 100).

Although much criticized over the years of its development, its publication
marked a crucial moment in the development of American psychiatry. *DSM-
III* and its subsequent versions were sanctioned by the NIMH, and diagno-

ses using its classification became required elements in research protocols, in articles submitted for publication in psychiatric journals, in the management of clinical trials and recruitment of subjects, in reimbursement criteria of health management organizations (HMOs), and in training programs for psychiatrists. *DSM-III* stressed that diagnosis must be based on a pattern of symptoms that was not merely an expectable response to an event, but a manifestation of a dysfunction in the person: "Neither deviant behavior . . . nor conflicts that are primarily between the individual and society are mental disorders unless . . . a symptom of a dysfunction in the person" (American Psychiatric Association 1980). Its 494 pages classified 265 distinct diagnoses; the revised version of 1987 was 567 pages and classified 292 diagnoses (American Psychiatric Association 1987). Each was to be diagnosed in terms of a set of "observable" criteria, and each of these categories was construed as a distinct disorder, with a unique etiology and prognosis, amenable to a specific kind of treatment.

But despite the fact that the demand for revision of *DSM-II* grew, in part, out of disquiet with the expanding scope of psychiatry, *DSM-III* did not restrict the field of action of the experts of mental disorder. Its rejection of the fundamental division of neuroses and psychoses, together with its claims to provide neutral and theory-free empirical descriptions of symptom patterns did quite the reverse. It actually reinforced the claim that psychiatrists could diagnose as mental disorders conditions ranging from troublesome conduct in children to men's problems in getting an erection. All the travails of everyday life now seemed to come within its purview. And, indeed, by the 1980s this expanded territory of psychiatry had become a reality. At the risk of overgeneralization, we could say that up until the 1960s, there was a reasonably clear distinction—professional, clinical, conceptual—between what one might term 'asylum psychiatry,' based in the mental hospitals and often associated with the use of compulsory powers of confinement, and the psychiatry of everyday life, of the neuroses, of minor mental troubles of childhood, of psychotherapy and psychoanalysis. The former were the focus of the disquiet of the critics and were associated with the use of harsh physical treatments, drugs, and electroshock; the latter were associated with talking cures and increasingly with the use of minor tranquilizers such as Valium and the other benzodiazepines. But by the 1980s, in almost all developed countries a reconfiguration and reunification of the territory of psychiatry was under way (Castel, Castel, and Lovell 1982). The antipsychiatric critiques of the custodial and segregative project of the asylum were welcomed by many psychiatrists, who were frustrated by the fact that they were confined to the mental hospitals along with their patients, shut off from the world of therapeutic optimism outside the asylum, and from the general hospitals and general medicine where the real progress appeared to be being made.

The movement for deinstitutionalizing the mental hospitals was much more than a shift of the incarcerated population from one locale to another.

A new configuration of sites for the practice of psychiatry was taking shape, together with their populations—halfway houses; residential homes; specialized units for children, alcoholics, anorexics, or drug users; and clinics of many different sorts. Children, delinquents, criminals, alcoholics, vagrants, the work shy, unhappy sexual partners, troubled husbands and wives all became possible objects for this extended practice of psychiatry. The unrivaled dominance of the doctor gave way to teams of multiple professionals with diverse claims to expertise. That monologue of reason about madness, of which Foucault wrote so eloquently, was supplemented by an array of sites where the incitement of the patient or client to speech had become obligatory.

The voice of the patient, expressed in therapies, group sessions, confessional literature, and the like was initially encouraged by professionals for reasons of therapy: few regarded it as the pathway to a transformation of the relations of power and authority within the psychiatric system. However, it presaged the emergence of psychiatric users into the mainstream of debates about psychiatry and mental health, with narratives of illness and recovery, pressure groups, and support networks playing an increasing role in psychiatric politics. By the 1980s, psychiatry had become much more than a specialty for the management of a small minority of persons unable to live in the world of work, family, and civility—it had become a widespread "discipline of mental health" whose rationale was not so much cure as "coping"—helping troubled individuals manage themselves in their everyday lives (Rose 1986). And, by the closing decade of the twentieth century, this focus had widened to a positive vocation for the maximization of individual potential, the minimization of sadness and anxiety, the promotion of well-being, even happiness (Rose 1989, 1999).

The diagnostic manuals of our own age are, in a certain respect, analogous to those that arose from within the nineteenth century asylums. Like those 'atlases' that we discussed in chapter 2, they suggest that there are fundamental connections between all those who come under the gaze of the psychiatrist, a gaze now not confined to the asylum, or the closed world of the hospital, but directed across the open space of the community. Now, however, these connections among the inhabitants of this territory are established through making connections between the 'molar' and the 'molecular.'[11] That is to say, the diversity of problems in managing a life in the everyday world is nonetheless unified—at least in aspiration—at the level of the brain. For it is in the brain, in its patterns of activity, in the structures of its regions and networks, in the functioning of its systems of neurotransmission, in its capacities to adapt to stimuli, and in its molecular and cellular processes that these heterogeneous problems find their substrate. But we are anticipating our argument. Let us return to the *DSM*.

From the 1980s onward, the expanded yet differentiated idea of mental disorder that we have described became indelibly inscribed in the style of thought of modern psychiatry.[12] *DSM-IV*, published in 1994, runs to 886 pages and continues the trend of identifying even more distinct conditions. It classifies some 350 distinct disorders, from acute stress disorder to voyeurism (American Psychiatric Association 1994). It cautions that individuals within any diagnostic group are heterogeneous: its categories are only intended as aids to clinical judgment. But it promotes an idea of specificity in diagnosis that is linked to a conception of specificity in underlying pathology. The broad categories of the start of the twentieth century—depression, schizophrenia, neurosis—no longer appeared adequate: it seems that pathologies of mood, cognition, will, or affect must be dissected at a finer level of discrimination.

In the course of these developments, key distinctions began to wither away. The distinction between inside and outside the asylum began to blur in the 1950s. The distinction between the neuroses and the psychoses began to blur in the 1960s. By the 1980s, the distinction between mental and physical disorders itself began to blur. As *DSM-IV* put it: "although this volume is titled the *Diagnostic and Statistical Manual of Mental Disorders*, the term *mental disorder* unfortunately implies a distinction between 'mental' disorders and 'physical' disorders that is a reductionist anachronism of mind/body dualism" (1994, xxi). Over the 1980s, another key distinction, which had been the basis of a division of professional labor between psychiatrists and psychologists, also began to blur—the distinction between states and traits. As discussed in chapter 1, this distinction between a relatively enduring feature of personality (a trait) and a particular episode of illness (a state) were the province of distinct disciplines—psychology and psychiatry—distinct explanatory regimes, and distinct modes of intervention, as well as grounding legal distinctions between those who were, or were not 'treatable' for their condition. But by the 1980s, it was being argued that there was no such ontological distinction between states of illness and traits of personality: for example, episodes both of illness, such as depression, and variations in traits, such as shyness or hostility, could be altered by psychiatric drugs (Knutson, Wolkowitz, Cole, et al. 1998). In the new neuromolecular gaze, variations in both states and traits could be analyzed in terms of the same molecular mechanisms, differing in many things, of course, but not in their underlying nature.

DSM-IV included, for the first time, a "clinical significance" criterion, which requires that, to be classified as a mental disorder, symptoms must cause "clinically significant distress or impairment in social, occupational, or other important areas of functioning" (1994, 7). In the words of Robert Spitzer and Jerome Wakefield, "In response to concerns that the DSM criteria are overly inclusive, the clinical significance criterion attempts to minimize false positive diagnoses in situations in which the symptom criteria do not necessarily indicate pathology" (Spitzer and Wakefield 1999, 1856). But Spitzer and Wakefield suggest that this delimitation, though helpful, is not in itself sufficient to eliminate the problem of false positives, because it still does not sufficiently consider questions of social context, and discriminate between normal and abnormal reactions to "psychosocial stress—it does not restrict the diagnosis of a disorder to situations where there is what they term 'a dysfunction'" (ibid., 1863). Their cautions do not seem to have influenced the ways in which *DSM–IV* has been used to estimate the prevalence of psychiatric disorder. It seemed that, according to its criteria, the diagnosable psychiatric population was no longer a small minority; quite the reverse. Psychiatric morbidity was now common, widespread, almost ubiquitous. And, for many, this 'burden of mental disorder' had to be placed firmly on the agenda of policymakers.

The Burden of Mental Disorder

How many people suffer from psychiatric illness? This deceptively simple question turns out to be rather difficult to answer. One indicator is the high and increasing rate of use of psychiatric drugs (Rose 2006a, 2006b). But this is not the only one. In 2001, the WHO published a report titled *Mental Health: New Understanding, New Hope.* In the report, based on its own epidemiological work and that of others, it estimated that: "mental and neurological conditions account for 30.8% of all years lived with disability (YLDs). . . . Six neuropsychiatric conditions figured in the top twenty causes of disability (YLDs) in the world, these being unipolar depressive disorders, alcohol use disorders, schizophrenia, bipolar affective disorder, Alzheimer's and other dementias, and migraine" (World Health Organization 2001, 26). And it concluded: "By the year 2020, if current trends for demographic and epidemiological transition continue, the burden of depression will increase to 5.7% of the total burden of disease, becoming the second leading cause of DALYs (disability adjusted life years) lost. Worldwide it will be second only to ischemic heart disease for DALYs lost for both sexes. In the developed regions, depression will then be the highest ranking cause of burden of disease" (ibid., 30).

Epidemiological studies of the prevalence of psychiatric disorders in the general population seem to give empirical support to this picture. Both in Europe and in the United States, such research claimed very high rates of prevalence of untreated psychiatric illness, as diagnosed by checklist interviews based on the descriptions of disorders in *DSM-IV*. The classic studies were undertaken by a team led by Ronald Kessler, originally trained as a sociologist and later a professor of health policy at Harvard Medical School. By 2003, the first of these surveys, published in 1994, had received more citations than any other paper published in the field of psychiatry and psychology (Kessler, McGonagle, Zhao, et al. 1994). In an interview at that time, Kessler noted that his survey, carried out by a household interview survey, was fundamentally dependent on the "operational precision" of the *DSM* diagnostic criteria and the structured diagnostic interview developed by Lee Robins.

Kessler believed that the first of its findings—that as many as half of all Americans have met criteria for some mental disorder at some time in their lives—was so significant because "it addressed the issue of stigma that has for so long interfered with rational thinking about mental illness. The mentally ill are not some distinct set of 'them' out there who are distinct from 'us' sane people. Instead, the vast majority of us have been touched by some form of mental illness at some time in our lives either through personal experience or through the illness of a close loved one. In many cases these illnesses are either mild or transient or both, but they certainly should not be considered in any way foreign."[13] Kessler and his colleagues repeated the survey in 2005 with rather similar results: in any one year, 26.2% of adult Americans reported having symptoms that would qualify them for a *DSM-IV* diagnosis of mental disorder, with the anxiety disorders leading at 18.1%, followed by mood disorders, notably depression, at 9.5% and impulse control disorders at 8.9%—"a total of over 57 million Americans diagnosable with mental disorder in any one year" (Kessler, Chiu, Demler, et al. 2005). The authors concluded: "About half of Americans will meet the criteria for a *DSM-IV* disorder sometime in their life, with first onset usually in childhood or adolescence" (Kessler, Berglund, Demler, et al. 2005).[14]

We are not speaking here of the kinds of conditions that might previously been thought of as mild neuroses. Take, for example, personality disorder. While we often think of personality disorder as a relatively rare condition, the 2001–2 National Epidemiologic Survey on Alcohol and Related Conditions (NESARC), which administered a standardized diagnostic instrument to 43,093 adults over eighteen years old in the United States, reported that almost 15% of Americans, or 30.8 million adults, met *DSM-IV* diagnostic criteria for at least one personality disorder (Grant, Hasin, Stinson, et al. 2004). The most common types of this disorder diagnosed were obsessive-compulsive personality disorder (16.4 million people), followed by paranoid personality disorder (9.2 million), and antisocial personality disorder (7.6 million). Other personality disorders affecting substantial numbers of Americans were schizoid (6.5 million), avoidant (4.9 million), histrionic (3.8 million), and dependent (1 million) (Bender 2004). This survey excluded borderline, schizotypal, and narcissistic disorders; however, NIMH estimates that borderline personality disorder alone affects a further 5.5 million Americans (National Institute for Mental Health 2001). All in all, then, these figures suggest that some 20% of Americans have diagnosable personality disorders. Which, of course, does make one wonder about the very idea of a normal personality.

These estimates have been much criticized, with many suggesting that they are the consequence of the *DSM* itself (Kutchins and Kirk 1997). But they are not confined to the United States, nor to the use of the *DSM*. In 2005, a task force of the European College of Neuropsychopharmacology published the results of a Europe-wide survey of the size and burden of mental disorders in Europe, which claimed that some 27%, or 82.7 million of the adult EU population (eighteen to sixty-five years of age) "is or has been affected by at least one mental disorder in the past 12 months. . . . [T]he most frequent disorders are anxiety disorders, depressive, somatoform and substance dependency disorders. . . . Only 26% of cases had any consultation with professional health care services, a finding suggesting a considerable degree of unmet need" (Wittchen and Jacobi 2005, 357). It further estimated that over their lifetimes 50% of the EU population would be affected by at least one diagnosable psychiatric disorder, if lifetime risk was considered, and that mental disorders were associated with costs of more than €290 billion, the majority of which were not health-care costs (ibid., 355–56).

The matter was put plainly by the European Brain Council (EBC): "There are an estimated 127 million Europeans currently living with a brain disorder out of a population of 466 million. The total annual cost of brain disorders in Europe was estimated to €386 billion in 2004" (European Brain Council 2005, x).[15] The diagnostic and epidemiological data used in these estimates were not solely based on the *DSM*, but included the diagnostic criteria more often used in Europe, such as the ICD-9 or -10, or other diagnostic assessment scales. For the EBC, not just neurodegenerative disorders and major mental illnesses are considered brain disorders, but so are addiction, depression, anxiety and panic disorder—what else could they be? How can we understand this simultaneous extension and cerebralization of the idea of mental abnormality?

Let us return to the United States and the *DSM*. It is, of course, something of a paradox that while a key objective of those developing *DSM* had been the *delimitation* of the territory of psychiatry, the argument now is reversed. Kessler, speaking in 2003, said that the initial skepticism about his findings had been based on the error that "the term 'mentally ill' is being taken too seriously. . . . [W]e invest the term 'mentally ill' with excess meaning. A number of common mental illnesses, like adjustment disorders and brief episodes of depression, are usually mild and self-limiting."[16] In his view, not everything classified in the *DSM*—and hence included in his estimates—required treatment. However, this was a little disingenuous as, in the same year of this interview, he and his colleagues argued that it was important to include mild disorders in the forthcoming *DSM-5* because mild disorders may progress to major ones, and diagnosis at an early stage provides an opportunity for preventive intervention (Kessler, Merikangas, Berglund, et al. 2003). Indeed, most of those who use these data to urge policymakers to focus funding and energy on mental health argue along the same lines: the burden of untreated mental disorder shown by such studies, they argue, shows that we need a step change in our capacity to understand and to treat these disorders, and in our capacity to screen our populations for them, to pick up presymptomatic susceptibilities and prevent the development of disorder.

Thus, Steven Hyman opens his paper "A Glimmer of Light for Neuropsychiatric Disorders" by referring explicitly to estimates of the prevalence of psychiatric disorders—or what he terms "neuropsychiatric disorders"—and suggesting that neuropsychiatry may soon be able to meet the challenge they pose:

Neuropsychiatric disorders severely compromise the well-being of those affected, with their negative effects on general health and on the ability of children to learn and of adults to work. These disorders have a relatively high prevalence[,] . . . can have an early onset (for example, autism in childhood and schizophrenia in young adulthood) or a relapsing–remitting course (as in mood and anxiety disorders and obsessive–compulsive disorder), and often have disabling symptoms. For these reasons, they exert, in aggregate, a devastating impact on the families involved and on the human capital of a society. . . . Despite the disease burden attributable to neuropsychiatric disorders, and despite significant research, their mechanisms of pathogenesis and precise genetic and non-genetic risk factors have remained stubbornly out of reach. Although sustained and clever exploitation of clinical observations has produced a useful pharmacopoeia, almost all recently introduced psychotherapeutic drugs are based on reverse engineering the mechanisms of existing drugs. Arguably, no new drug targets or therapeutic mechanisms of real significance have been identified for more than four decades. . . . This parlous state of affairs is finally beginning to improve, in part through the application of new genomic technologies coupled to advances in neuroscience. (Hyman 2008, 890)

Two questions thus seem relevant. First, how much credibility can we give to these estimates? And second, why should we look to neuropsychiatry for the answer?

Let us begin with the first question. The debate about the utility of these epidemiological studies has not abated. Some have suggested that the high prevalence rates reported arose because researchers failed to address such issues as symptom threshold, severity and level of impairment, and duration of symptoms (Regier, Kaelber, Rae, et al. 1998; Regier 2000).[17] Insel and Fenton believe that this was indeed a problem for earlier epidemiological research: "Many critical issues were not addressed by earlier studies. While the overall 12-month prevalence of any mental illness was reported to be in the range of 30%, significant questions about the disability associated with these syndromes remain. How severe are the disorders reported to be present in 30% of the population? What is the economic and public health impact of these conditions? How long is the delay between onset and diagnosis?" (Insel and Fenton 2005, 590).

But their comment is in a commentary on more recent studies by Kessler and his colleagues that replicated the earlier research and sought to refine it; they claimed that the majority of those diagnosed by such methods are rated as having a serious or moderate disorder, with substantial disruption to their ability to carry out their normal activities and, further, that retrospective studies such as theirs actually *underestimate* prevalence because respondents underreport the symptoms that they have had in the past, perhaps because of poor recall (Kessler, Berglund, Demler, et al. 2005a; Kessler, Chiu, Demler, et al. 2005; Wang, Berglund, Olfson, et al. 2005; Wang, Lane, Olfson, et al. 2005). Thus, in their commentary, Insel and Fenton concur with those who believe that such surveys do clearly demonstrate the high level of unmet need for mental health care. If this was not enough, researchers using a different, 'prospective' approach, based on their work with the Dunedin cohort study in New Zealand, argue that this method confirms the underestimation of prevalence in existing research and suggest that the actual lifetime prevalence rate of *DSM*-diagnosed mental disorders is about twice that shown in the retrospective studies such as those by Kessler and his team (Moffitt, Caspi, Taylor, et al. 2009).

Others, however, continue to question the high prevalence rates and the methods used to generate them.[18] Those from the team developing *DSM-5* suggested technical changes, for example, requiring more evidence of severity (Narrow, Rae, Robins, et al. 2002; Narrow, Kuhl, and Regier 2009). Others warned those developing *DSM-5* of the pitfalls they might face. Thus Allen Frances, who was chair of the *DSM-IV* task force, argued that the problems in delimiting the diagnostic criteria in *DSM-IV* had produced the consequences of contributing to "the false epidemic of autism and attention-deficit disorder and a forensic disaster that has led to the inappropriate psychiatric commitment of sexually violent offenders": he warned those responsible for producing *DSM-5* of the danger of exacerbating these problems if they produce a set of criteria that "capture" prodromal and subthreshold conditions (Frances 2010, 2).[19]

Perhaps the most through interrogation of the methods used to create the prevalence data on the burden of brain disorders has been undertaken by Allan Horwitz and Jerome Wakefield (Horwitz and Wakefield 2007). Their principal argument is that psychiatric diagnoses using *DSM* criteria generate many false positives because they fail adequately to discriminate between 'normal' responses to experiential factors—that is to say, normal anxiety, normal sadness, or normal grief in the face of the difficulties that life presents us with—and pathological conditions requiring specialized medical intervention. The presenting features of these normal responses overlap with those designated as symptoms of disorder in *DSM*, and because *DSM* fails to recognize the importance and take adequate account of context, it thus fails to differentiate between normal reactions that have a clear cause (for example, loss of job, ending of a relationship, reaction to a life-threatening medical condition, etc.) and those that, because of their severity, duration, or other features, that is, those without a cause, reveal that something has gone wrong that requires specialized psychiatric intervention.

When framed as 'symptoms,' these normal reactions may well appear to psychiatrists, or those conducting community epidemiological surveys, as indicating a high level of undiagnosed psychiatric disorder in the general population. While clinicians may be able to make good and appropriate use of

DSM-type diagnostic procedures in clinical settings, they are not appropriate for use in the community. Even the clinical significance criterion introduced into *DSM-IV* is not very helpful here,[20] they argue, because normal reactions can indeed reach the threshold of significant distress or impairment in functioning in one's allocated social roles without indicating some underlying harmful dysfunction.[21] The apparent epidemic that such community studies seem to reveal, and that mental health policy is urged to address, is, in their view, merely a consequence of the inappropriate decontextualized conceptual framework used in the research—it is an epidemic of false positives.

It is not surprising, then, that in the light of these continued questions over the validity of diagnostic methods, some argue that there is really only one way to settle this matter—to identify some biological abnormality, lesion, malfunction, or marker that will demonstrate objectively, once and for all, who is, and who is not, suffering from a mental disorder.[22] Will neuroscience enable us, at last, 'to define true madness'?

All in the Brain?

Our chapter opened with a quote from Amanda Haag, but we do not have to look to science journalists for the hope that neuroscience will resolve the diagnostic problems of psychiatry by demarcating disorders in terms of their underlying neurobiological bases. Let us quote Steven Hyman again. In 2003, writing while director of the NIMH for a popular audience in *Scientific American*, he put it thus: "By combining neuroimaging with genetic studies, physicians may eventually be able to move psychiatric diagnoses out of the realm of symptom checklists and into the domain of objective medical tests. Genetic testing of patients could reveal who is at high risk for developing a disorder such as schizophrenia or depression. Doctors could then use neuroimaging on the high-risk patients to determine whether the disorder has actually set in. I do not want to sound too optimistic—the task is daunting. But the current pace of technological development augurs well for progress."

The hope at that time was threefold. First, in pinning psychiatric classification and categorization to a neurobiological basis, using identifiable markers in genetics and in patterns of brain activation, it would be possible not just to make accurate distinctions between normality and pathology, but also to distinguish subtypes of pathology according to their organic and functional pathways. Psychiatric classification would move from the surface to the interior, from symptom checklists to the brain itself. Second, in doing so, one would gain understandings of etiology, which would not only enable accurate diagnosis and prognosis, but also would make it possible to screen presymptomatic individuals thought to be at risk and intervene preventively before the disease developed. Third, on this basis, one might be able to develop effective treatments that target the neurobiological basis of mental disorders.

One organizing focus for these hopes of reform was, once again, the revision process for the *DSM*.[23] The process of revision of the fourth edition of that manual began in 1999, with the *DSM-5* scheduled for publication in 2013. In 2002, David Kupfer, Michael First, and Darrel Regier published *A Research Agenda for DSM-V*, which among other things, argued for greater neurobiological research on nosological issues—that is to say, questions of classification of disorders (Kupfer, First, and Regier 2002). The volume collected five "white papers" from different work groups, the first substantive one of which was titled "Neuroscience Research Agenda to Guide Development of a Pathophysiologically Based Classification System." The authors—Dennis Charney, David Barlow, Kelly Botteron, Jonathan Cohen, David Goldman, Raquel Gur, Keh-Ming Lin, Juan López, James Meador-Woodruff, Steven Moldin, Eric Nestler, Stanley Watson, and Steven Zalcman stated the problem boldly, if pessimistically, early in their discussion: "At the risk of making an overly broad statement of the status of neurobiological investigations of the major psychiatric disorders noted above, it can be concluded that the field of psychiatry has thus far failed to identify a single neurobiological phenotypic marker or gene that is useful in making a diagnosis of a major psychiatric disorder or for predicting response to psychopharmacologic treatment" (Charney, Barlow, Botteron, et al. 2002, 33).

They argued that it would be years—and possibly decades— before a fully explicated etiologically and pathophysiologically based classification system for psychiatry would exist. Nonetheless, at the close of their review of a whole variety of methods—genetic research, brain imaging, postmortem studies, and work with animal models—relevant to elucidating the pathophysiology of mental disorders, they allowed themselves to imagine:

> We speculate that single genes will be discovered that map onto specific cognitive, emotional, and behavioral disturbances but will not correspond neatly to currently defined diagnostic entities. Rather, it will be discovered that specific combinations of genes will relate to constellations of abnormalities in many brain-based functions—including but not limited to the regulation of mood, anxiety, perception, learning, memory, aggression, eating, sleeping, and sexual function—that will coalesce to form disease states heretofore unrecognized. On the other hand, genes that confer resilience and protection will also be identified, and their interaction with disease-related genes will be clarified. The impact of environmental factors on gene expression and phenotype expression will be defined. The ability to discover intermediate phenotypes will be improved with advances in techniques such as neuroimaging. This will all lead to novel therapeutic targets of greater efficacy and specificity to disease states. Prediction of therapeutic response will be possible through genetic analysis and phenotype analysis. Disease

prevention will become a realistic goal. (Charney, Barlow, Botteron et al., 2002: 70)[24]

And they even suggested a hypothetical radical revision of the *DSM* axes. Instead of the current multiaxial system (introduced in *DSM-III*), which provides information for clinical disorders (Axis I), personality disorders and mental retardation (Axis II), any general medical condition (Axis III), psychosocial and environmental problems (Axis IV), and global functioning (Axis V), they proposed the following: Axis I for *genotype*, identifying symptom- or disease-related genes, resiliency genes, and genes related to therapeutic responses and side effects to specific psychotropic drugs; Axis II for *neurobiological phenotype*, such as identifying intermediate phenotypes discovered by neuroimaging, cognitive evaluation, and neurophysiological testing that could aid in selecting targeted pharmacotherapies and psychotherapies and monitoring the neurobiological response; Axis III for *behavioral phenotype*, severity and frequency of specific cognitive, emotional, and behavioral disturbances; Axis IV for *environmental modifiers or precipitants* to be evaluated in the context of genotype; and Axis V for *therapeutics*.

Seven years later, the prospects for a radical revision of diagnostic criteria along neurobiological lines looked, if anything, even more remote than they did in 2002. When Regier and his colleagues from the team overseeing the revision discussed the conceptual basis of *DSM-5*, their diagnosis of the problems of *DSM-IV* was damning, and it seemed that the prospects for a neurobiological solution were remote (Regier, Narrow, Kuhl, et al. 2009). They were particularly concerned by the lack of clear separation between *DSM*-defined syndromes, as there seemed to be high levels of "comorbidity" between mood, somatic, and anxiety syndromes. The question, for them, was how to "update our classification to recognize the most prominent syndromes that are actually present in nature, rather than in the heuristic and anachronistic pure types of previous scientific eras" (ibid., 646).

The hope that psychopharmaceuticals would 'cut nature at its joints'—that one might differentiate disorders by identifying which conditions responded to which drugs—now seemed forlorn. The SSRIs had proved not to be specific for depression but were now prescribed for a range of anxiety, mood, and eating disorders;[25] the 'atypical antipsychotics' were now licensed not only for schizophrenia but also bipolar disorder and for treatment-resistant major depression. More troubling, given so much investment in genomic research, the *DSM-5* revision team was skeptical about the prospects for large-scale genetic studies using GWAS: they thought it would be "unlikely to find single gene underpinnings for most mental disorders, which are more likely to have polygenetic vulnerabilities interacting with epigenetic factors (that switch genes on and off) and environmental exposures to produce disorders" (Regier, Narrow, Kuhl, et al. 2009, 646). But hope does not die easily, and they remained optimistic that, in some way mental disorder syndromes could be redefined so that they were useful diagnostically and related to objective neurobiological features—that one could recognize the "dimensional" nature of disorders, yet still set clear thresholds between pathology and normality.

Indeed by the end of the decade, the prevailing attitude among neuropsychiatric researchers seemed to be a combination of disappointment in the present with hope for the future. Take, for example, Chou and Chouard, in their introduction to a *Nature* "Insights" supplement published in October 2008:

> Despite decades of . . . research, the prevalence of neuropsychiatric diseases has not decreased. Our understanding of the biological mechanisms of diseases such as mood disorders, schizophrenia and autism is frustratingly limited. And, although it has long been clear that most such diseases have a strong genetic component, the identities of the genes involved have proved elusive. There is also a lack of reliable biological markers for characterizing these diseases and, perhaps unsurprisingly, treatment options are far from optimal in terms of efficacy and specificity. There is, however, some cause for optimism. Recent advances in genomic technology and large-scale studies are helping to identify genetic variants associated with diseases. In addition, new animal models of disorders such as depression and autism are providing ways to test hypotheses about the underlying neuropathology—at the molecular, neural-circuit and behavioral levels. . . . The hope is that developments such as these will lead to integrative approaches for designing better therapeutic strategies. (Chou and Chouard 2008, 889)

Others writing in the same issue were less optimistic and doubted the very premises on which neuropsychiatric hope was based. Take, for example, Steven Hyman, who, as we have seen, was an optimist in 2003 and even quite upbeat when writing for a popular audience five years later. Yet in the same year, writing for his peers, he expressed a rather different opinion: that we have made little progress even in those disorders where conventional measures of heritability based on family studies indicate a strong genetic component, such as autism, schizophrenia, and bipolar disorder:

> Efforts to identify risk-conferring alleles . . . have been largely unrewarding. . . . The underlying genetics of common neuropsychiatric disorders has proved highly complex. . . . [T]here is much evidence that similar neuropsychiatric symptoms can result from different combinations of genetic risk factors [and] that the same genetic variant may be associated with multiple DSM-IV diagnoses. . . . The identification of common risk-conferring variants has . . . proved extremely challenging in most cases, because of their relatively small contribution to the disease phenotype . . . because of the diverse genetic, environmental and random factors that lead to these common disease. (Hyman 2008, 891)

Hyman, concurring with several generations of sociological critics, now concluded that "what we think of as a single 'disease' is not in a strict sense a homogeneous entity for which there is a 'Platonic' ideal phenotype. Common diseases are more likely to represent families of diseases that share major pathophysiological and symptomatic features, but can differ in important characteristics such as age of onset, severity of symptoms, rate of progression, and response to treatment" (2008, 891). He argued that this problem of the phenotype is one of the reasons why GWAS fail in psychiatry, but that they also fail because they do not take account of the fact that genes for psychiatric disorders are pleiotropic in effects (i.e., a single gene can produce many effects), and crucially that the pathways from genomics to phenotype are complex and not understood.[26] And, perhaps most damaging for the hopes that were vested in neurobiology to resolve the diagnostic problems of psychiatry, he questioned the view that it might ever be possible to draw clear boundaries between ill and well—it is, he thinks, better to think of "dimensions" without sharp boundaries.

This pessimistic conclusion on the current state of neuropsychiatry was echoed in the same issue of *Nature* by Krishnan and Nestler, who discussed the molecular neurobiology of depression (Krishnan and Nestler 2008). Nestler, who is Nash Family Professor of Neuroscience, chairman of the Department of Neuroscience, and director of the Brain Institute at the Mount Sinai Medical Center in New York, was one of the authors of the chapter on neurobiology in the 2002 *Research Agenda for DSM-V*. He and his coauthor were also skeptical of the prospects for GWAS in depression, not least because of the problems of comorbidities. They argued, further, that brain imaging studies were flawed because of simplistic ideas about localization (for instance, that the amygdala is the locus for fear and anxiety; the nucleus accumbens the locus for reward; and so forth). "Such artificial distinctions are of limited heuristic value," they believe, as "depressive symptoms are probably mediated by dysfunction in a diffuse series of neural networks." (ibid., 895). Earlier theories of depression, like the monoamine hypothesis, are false: "[A]lterations in central monoamine function might contribute marginally to genetic vulnerability. . . . [T]he cause of depression [is not] a simple deficiency of central monoamines." (ibid.). And the effects of drugs that act on serotonin levels, they suggested, are actually due to secondary changes in molecular and cellular plasticity. They concluded: "Collectively, these studies highlight the weaknesses of attempts to generate a 'unified theory' of depression. Mechanisms that promote depressive symptoms in response to stress differ markedly between different neural circuits and can also be distinct from changes that underlie depression in the absence of external stress ('endogenous depression'). In addition, neuroplastic events that are required for antidepressant efficacy need not function through the reversal of stress-induced plasticity . . . and might function through separate and parallel circuits" (Krishnan and Nestler 2008, 898). In short, at the neurobiological level, "enormous gaps in the knowledge of depression and its treatment persist . . . [and researchers] must follow a systems approach that acknowledges the powerful bidirectional interactions between peripheral organs and the brain" (ibid., 901).

In February 2010, the APA released the proposed draft diagnostic criteria for the fifth edition of the *DSM*.[27] Among the changes proposed were several relevant to our discussion here: the current diagnoses of autistic disorder, Asperger's disorder, childhood disintegrative disorder, and pervasive developmental disorder (not otherwise specified) were to be incorporated into a single diagnostic category of "autism spectrum disorders"; there were various changes to the categories relating to substance abuse and dependence, which were to be incorporated into a new category of "addiction and related disorders"; a new category of "behavioral addictions" was proposed, with gambling as the sole disorder; there was to be consideration of a new "risk syndromes" category, with information to help clinicians identify earlier stages of some serious mental disorders, such as neurocognitive disorder (dementia) and psychosis; and there were other changes designed to assist clinicians in better differentiating the disorders of children. Dimensional assessments would be added to diagnostic categories to permit clinicians to evaluate the severity of symptoms. And arabic numerals were to replace the previously used roman numerals—thus the new edition would be, not *DSM-V* but *DSM-5.0*, to enable future revisions to be identified as *DSM-5.2*, etc.—to make *DSM* "a living document."

Many of those who had high hopes for neuropsychiatry expressed disappointment at the way things were shaping up. One prolific commentator, Dr. Charles Parker, writing in his CorePsychBlog three days after the draft criteria were published, expostulated:

> *The Earth Is Flat! DSM 5 Points at the Tips of Icebergs— Only What You See*
>
> The tip-toe progress with these new superficial labels for office appearances misses altogether the complexity of new brain and body science. Neuroscience evidence is easily available, often paid for by insurance, and remains almost completely ignored by psychiatry. Real facts, not labels, will foretell the changes necessary for psychiatric practice, for treatment strategies to evolve with the rapidly evolving new science. . . . [T]he new DSM5 conclusions are based almost completely upon 19th century vertical thinking, insufficient feedback with patients in the office, and questions that ignore modern neurophysiology and metabolism. Psychiatrists will remain speculative with dreams and fantasies, while hard evidence from molecular and cellular physiology remains in the closet—frequently derided as quackery. Interestingly, psychiatry is held to a different standard on evidence than the rest of medicine, as SPECT brain imaging and the measurement of neurotransmitter biomarkers is still derided as non-specific, while patently non-specific

biomarkers, such as cholesterol screening, are accepted uniformly in general medicine. It's time for diagnostic change—but DSM5 is already old news.[28]

Others, however, did identify something new emerging in the proposals for the new version of *DSM*—the inclusion of some subthreshold or prodromal conditions. These were conditions where clear symptoms of disorder were not yet visible, but where there was believed to be high risk of progressing to a frank disorder—a perception intrinsically linked with the belief in the value of screening to identify those at risk at the presymptomatic stage and to intervene early to avert or mitigate the development of the disorder. Two areas were notable here: psychosis risk syndrome and minor neurocognitive disorder. Let us focus on the first.

Early intervention for those at risk of psychosis had advocates dating back at least to the 1990s, with some arguing that best results would be obtained by intervention at as early a point as possible, even before any outright symptoms had emerged (McGorry et al. 2009). During the first decade of this century, a group of international collaborators, with key investigators at the Institute of Psychiatry in London, developed a program for early diagnosis of psychosis, to be followed by intervention with atypical antipsychotics, perhaps combined with the prescription of some essential fatty acids and cognitive behavioral therapy. It appeared that general practitioners could be educated to make early diagnosis (Power et al. 2007) and that it might be possible to use imaging technologies to assist accurate diagnoses at the prodromal stage (Pantelis et al. 2003; Howes et al. 2007; Meisenzahl et al. 2008; Smieskova et al. 2010), although the efficacy of these early interventions with drugs and other measures in preventing further development of the disorder was not yet clear (Gafoor, Nitsch, and McCrone 2010). Some suggested that cost savings would be generated by such interventions with those deemed to be at high risk (Valmaggia et al. 2009). Others, however, were more skeptical, arguing that prodromal interventions were not clinically justified, generated high levels of false positives, that no significant differences arising from early intervention were found on follow-up, and that there were significant problems from the side effects of antipsychotic medication (De Koning et al. 2009; Bosanac, Patton, and Castle 2010).

More radically, Nil Kaymaz and Jim van Os,[29] writing in the journal *Psychosis*, took particular issue with the proposed diagnosis itself, arguing that the category of Psychosis Risk Syndrome "is based on the notion that labeling people with invalid diagnostic terms has more clinical relevance than simply addressing care needs, is contingent on elusive sampling strategies posing as precise diagnostic criteria, and is associated with a false-positive rate of at least 90% in the year after diagnosis. In the twenty-first century, opinion-based diagnostics continues to pose a threat to the process of diagnostic revision" (Kaymaz and van Os 2010).[30] Not, however, that these authors were opposed to early intervention in itself; indeed, although they recognized that all current examples of such intervention generate large numbers of false positives, they remained of the view that early intervention "improves subsequent course and outcome given that psychotic disorders, arising during adolescence and early adulthood, disrupt important maturational tasks and increase the risk of long term exclusion from meaningful societal participation" (ibid., 100). Despite their assertion that "the hypothesis of valid nosological entities is very weak," they look to resolve this issue not by an appeal to biomarkers,[31] but to the generic diagnosis already in *DSM-IV*: "Psychotic Disorder, Not Otherwise Specified" (Kaymaz and van Os 2010)![32]

Some leading psychiatrists, however, warned of the dangers of these proposals for early intervention. In the month before the publication of the draft criteria for *DSM-5.0*, Allen Frances, who had been the chair of the *DSM-IV* task force, had expressed his concern that the subthreshold diagnoses would "medicalize normality":

> The most appealing subthreshold conditions (minor depression, mixed anxiety depression, minor cognitive disorder, and prepsychotic disorder) are all characterized by nonspecific symptoms that are present at extremely high frequencies in the general population. These proposed "disorders" might well become among the most common diagnoses in the general population—particularly once they are helped along by drug company marketing—resulting in excessive use of medications that often have serious long-term complications associated with weight gain. Early case finding is a wonderful goal, but it requires a happy combination of a specific diagnostic test and a safe intervention. Instead, we would now have the peculiarly unhappy combination of a wildly false-positive set of criteria coupled with potentially dangerous interventions. . . . Altogether, in my view, the costs and risks of the subthreshold diagnoses far outweigh any possible current gains. (Frances 2010)

However, as we shall see in many of the areas discussed in this book, the logic of 'screen and intervene,' the identification of prodromal or subthreshold disorders coupled with interventions in the name or preventing or mitigating the condition, has proved hard to resist.

Neuropsychiatry and the Dilemmas of Diagnosis

Could *DSM 5.0* avoid those dangers of neurobiology's medicalization of normality by utilizing biomarkers, by grounding diagnostic classification on objective anomalies identified in the brain? The situation did not look very encouraging. In the draft criteria, positive references to the current utility of biomarkers are rare. One does find reference to "a positive genetic test for dominantly inherited AD, or as the field develops, evidence that certain imaging markers, atrophy of medial temporal lobe structures on MRI, temporo-parietal hypometabolism on FDG-PET, amyloid deposition on PET scanning or markers for tau and abeta in the CSF"[33] as a requirement if one is to make a prediction of Alzheimer's disease in cases of minor neurocognitive disorder, but the working group does not consider that this would be appropriate in cases where the diagnosis is Mild Cognitive Impairment (MCI).[34] But this reference to a biomarker is an exception. Each of the pathways that neuro-psychiatry has attempted to trace through the brain seems to run, not into the bright uplands of clarity, but into the murky, damp, misty, and mysterious forests of uncertainty. Where, then, do matters stand in the question of the role of neuroscience in the diagnosis and delimitation of mental disorders?

Despite the penetrating gaze of neuroscience, which has opened up the brain to vision in so many ways, as we have seen in chapter 2, psychiatric clas-sification remains superficial. This neuromolecular vision seems incapable of grounding the clinical work of psychiatry in the way that has become rou-tine in other areas of medicine. Despite the conviction of most practitioners that they deal with conditions that have a corporeal seat in the brain of the afflicted individual, psychiatry has failed to establish the bridge that, from the nineteenth century on, underpinned the epistemology of modern clini-cal medicine—the capacity to link the troubles of the troubled and troubling individuals who are its subjects with the vital anomalies that underpin them.

Why, then, has neuropsychiatry not been able to "self-vindicate"—to use Ian Hacking's term (Hacking 1992a)?[35] Of course, in part this is because it is not a laboratory science, for while its research may, in part, follow the protocols of an experimental laboratory discipline, the phenomena that it addresses—persons deemed to be ailing in particular ways—are not laboratory subjects. One might have thought that *DSM* criteria would enable such self-vindication—after all, they are used to select subjects for entry into neuropsychiatric research stud-ies, which then become the basis for hypotheses, trials, developments of drugs, labeling and treatment by relevant personnel, changes in self-perception, and so forth. And as we have seen in chapter 3, the behaviors of animals in ani-mal models of psychopathologies—what we called *models*[5]—are indeed as-sessed according to those very *DSM* criteria. But although the criteria have summoned up an image of a mental disorder in thought—a specific genetic sequence or set of sequences linked to anomalies in the neurotransmitter sys-tem or other aspects of neurobiology causally related to a particular array of cognitive, affective, volitional, or behavioral dysfunctions—this assemblage has proved impossible to summon up in reality, despite the expenditure of millions of dollars and thousands of person-hours. The phenomenon itself, that is to say, has proved resistant. Why should this be?

No doubt a certain quantum of realism is essential here. Genomics, once welcomed as a potential savior, has not lived up to its early promises. The style of thought that underpins GWAS has proved to be radically oversimple when applied to the field of mental disorder.[36] The response of neuropsychiatrists committed to this way of thinking has been to appeal to *complexity*. They suggest that many genes are likely to be involved in any psychiatric disorder. They argue that any gene sequence that might be involved is likely to be pleiotropic. They contemplate the likelihood that there may be many routes to the final common pathways that generate the phenotypical symptoms. And they try to incorporate in their explanatory systems something that, elsewhere within psychiatry, has always been well known: that the manifestation of a disorder is an outcome of complex and individual variable interactions between genomics, developmental processes, epigenetics, neuroplasticity, life experiences, and hormonal and nutritional states.

More fundamentally fatal to self-vindication has been the recognition that psychiatric diagnoses, however much they may appear to be stabilized by the circular transactions between the diagnostic manuals and the training of the clinical gaze of the psychiatrist, seem to have a kind of irreducible fuzziness, so that selection of cases and controls on the basis of phenotypes is unable to frame studies in a way that is capable of generating clear results. Of course, social scientists have argued for decades that sociocultural belief systems play a crucial role in shaping not only the recognition of unease by others, but also its shaping into phenotypical symptoms. Hence it is no surprise to them to find that there are no good routes from psychiatric phenotypes to SNPs; that in almost every case, GWAS signals have been weak in psychiatry; and that functional pathways that have been implicated are those that are basic to neurodevelopment, and not specific to psychiatric disorders.

For the same reason, the drive to discover biomarkers, though stimulated by a whole constellation of forces—commercial interests, public health concerns, employer and health service concerns, and pressure from individuals and families—has yielded no validated results. Even if some markers are found in the future, all available evidence suggests that they will, at best, indicate marginal increases in susceptibility in particular environmental circumstances, and that these will be modulated by other genetic and neurobiological factors that are related to *resilience*—a term that is itself now becoming the basis of an expanding program of research. Further, despite the fervent beliefs in the virtues of early diagnosis and intervention, there is little evidence that presymptomatic intervention is more effective than intervention at the time when problematic conduct is shown. Premature use of such biomarkers within a strategy of screen and intervene would produce large numbers of false negatives and false positives, generate stigma rather than reduce it, expand interventions and their costs and downsides, and start many unnecessarily on a psychiatric career (Rose 2008; Singh and Rose 2009).[37] Perhaps there is a lesson here about the dangerous lure of the 'translational imperative'—neurobiological work in these areas may have been valuable in exploring brain mechanisms, but it may also be wise for those working in this area to avoid premature claims about the implications of their results for humans and clinical practice.

It is clear that neuropsychiatry has a long way to go in developing a style of thought adequate to the complexity of its object, when that object is an ailment suffered by an individual human being living in a social world. Few would doubt the enormous advances in our understanding of neural processes that have been made since the 1960s. But neurobiology will not solve psychiatry's current diagnostic dilemmas. Neurobiology cannot bear the weight placed on diagnosis in psychiatry for many reasons. First, allocation to a diagnostic category does not tell one much about the severity of symptoms, their duration, the level of impairment involved, the amount of distress caused to the individual or the family, and the extent of need for care. Further, diagnosis has low predictive validity and tells us little about the course of development of the problem, or the specific social, environmental, or other factors that may shape this course. More fundamentally, neuropsychiatry won't solve the dilemmas of diagnosis, because diagnosis is not solely about the nature or causes of a disorder. As we have seen, diagnoses in psychiatry answer to many different and competing demands: statistical record keeping for institutions and clinicians, entry into research trials and studies, eligibility for treatment, choice of intervention, prognosis, predictions of need for care, eligibility for work, eligibility for compensation via insurance, and epidemiology and public health policies, including predictions of need and distribution of resources. Perhaps no single system could ever bear this weight, but it is certain that one based on the brain alone cannot.

At root, the neurobiological project in psychiatry finds its limit in the simple and often repeated fact: mental disorders are problems of persons, not of brains. Mental disorders are not problems of brains in labs, but of human beings in time, space, culture, and history. And, indeed, so is diagnosis; however well-trained and stabilized the gaze of the doctor, however regulated by criteria and augmented by neuromolecular indicators, diagnosis is a practice governed by its own rules—rules that are different in epidemiology, in cohort studies, in research, in the clinic, and in life. Has neuropsychiatry brought us any closer to resolving that centuries-old problem of how to define 'true madness'—to find what one recent author referred to as "a brain based classification of mental illness" (Miller 2010)? At present, one must answer in the negative.

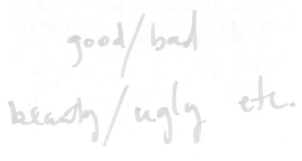

Chapter Five

· · · · · · · · · · · · · · · · ·

The Social Brain

A primate specialization which has now been amply documented
by behavioral scientists, but ignored for the most part by
neuroscientists, is the specialization for social cognition. What
I mean by the term "social cognition" is the following: *Social
cognition is the processing of any information which culminates in
the accurate perception of the dispositions and intentions of other
individuals.* While many non-primates (for example, ants) can
interact in highly specific ways with others of their kind, it appears
that primates, especially those most closely related to ourselves,
have developed a unique capacity to perceive psychological facts
(dispositions and intentions) about other individuals. This capacity
appears to distinguish primate social behavior from that of other
orders.

—**Leslie Brothers**, "The Social Brain," 1990

An . . . hypothesis offered during the late 1980s was that primates'
large brains reflect the computational demands of the complex
social systems that characterize the order. . . . There is ample
evidence that primate social systems are more complex than
those of other species. These systems can be shown to involve
processes such as tactical deception and coalition-formation . . .
which are rare or occur only in simpler forms in other taxonomic
groups. Because of this, the suggestion was rapidly dubbed the
Machiavellian intelligence hypothesis, although there is a growing
preference to call it the social brain hypothesis.

—**Robin Dunbar**, "The Social Brain Hypothesis," 1998

Philosophers and psychologists have long wondered about "this
thing called empathy." Only recently social neuroscience has begun
to provide some support in this endeavor. Initial findings are
encouraging that we will some day have a better understanding
of why, when, and how we experience empathy and whether we
can use that knowledge to increase prosocial behavior and an

intersubjectivity that is grounded in a better understanding
of ourselves and of others.

—**Tania Singer and Claus Lamm,** "The Social Neuroscience
of Empathy," 2009

Human beings, to state the obvious, are social creatures. We have platitudes
by the dozen to assure us that 'no man is an island.'[1] We live in groups—
families, communities, societies. We work collaboratively in organizations,
fight in bands and armies, take pleasure in events where we gather together
to dance, party, watch or play sports. We interact in pairs, and small and large
groups, whether in love or in hate, in teams and gangs, and in everyday activities. We care for one another and experience sympathy, empathy, or a sense
of obligation to some, though not to others. And so forth. But why? What
accounts for this apparently intrinsic and ubiquitous sociality of our species?
Is it, as one might say, in our nature, as we primates have evolved to become
human? And, if so, does that nature, that biology, that neurobiology, account
not only for the fact that we humans are social, but also for the ways that sociality has varied across history, between cultures, among different practices?

For the human sciences, the constitutive role of human sociality is undoubtedly a premise, but it is one that tends to be framed in a particular way.
What *is* in our nature is the fact that we humans are born incomplete, requiring interaction with others, and enmeshing in language, meaning, and culture, to be completed—to paraphrase Simone de Beauvoir, we are not born,
but become social.[2] Absent that, humans remain mere animals; and for some,
the repetitive stories of wild children, abandoned, isolated, living without
human contact, seem to testify to that truth (Hirst and Woolley 1982).[3] But
what if it was the case that our sociality, like that of our primate ancestors,
was a consequence of the evolved capacities of our brains, if our group formation was shaped by our neurobiology, if our empathy was not merely social
but neural? And what if our neurobiology does not just establish the *conditions* for sociality but also the *forms* that it takes? What if our friendships and
our loves, our communal ties and hates, our social and cultural lives take the
shape that they do as only a consequence of our neuronal architecture? What
then for the social and human sciences? Would such a neurobiologically
grounded conception of human sociality represent a fundamental threat to
the social and human sciences, as suggested by some of the critics discussed
in earlier chapters? Or does it represent a new opportunity for collaboration
across the 'two cultures,' one area where the 'critical friendship' we have proposed might be possible and valuable? And whether or not the claims made
about the neural basis of human sociality are true, what might be the wider
implications of such arguments being *considered truthful*—what might follow
for how we are governed or govern ourselves—how we manage our brains?

The "Social Brain Hypothesis"

In 1985, neuroscientist Michael Gazzaniga published a book called *The Social Brain* (Gazzaniga 1985b). It was based on the work that he had done, initially with Roger Sperry, on so-called split-brain animals—where the corpus callosum bridging left and right hemispheres had been lesioned—and other evidence from clinical studies of individuals with brain lesions.[4] This work seemed to show that damaging or destroying specific areas of the brain affected very specific aspects of cognition, language, and memory. On this basis he argued that the brain was not an indivisible whole but a confederation of relatively independent modules, each of which was relatively discrete and had specific functions. This modular way of thinking about the brain has been accepted by many.[5]

But it is not in this sense of a kind of confederation of modules that the phrase *social brain* has become so popular. Rather, the term has come to stand for the argument that the human brain, and indeed that of some other animals, is specialized for a collective form of life. One part of this argument is evolutionary: that the size and complexity of the brains of primates, including humans, are related to the size and complexity of their characteristic social groups. To put it simply, it is argued that the computational demands of living in large, complex groups selected for brains that had the capacities for interaction required to live in such social groups. *Social group* is used here in the sense attributed to it by ethologists—the characteristic number of animals such as primates who interact with one another in a relatively stable pattern and who differentiate themselves from other groups of the same species (see, for example, Bateson and Hinde 1976).

As far as humans are concerned, the evolutionary psychologist and anthropologist Robin Dunbar suggested that the number 150—subsequently known as the Dunbar number—specifies the optimal size of self-managing collectivities of human beings bonded together with ties of reciprocity, trust, and obligation, such as the clan or the village (Dunbar 1993).[6] He and those who followed him suggested that this group size was intrinsically related to the size and structure of the brain as it had evolved to manage the complex tasks required to collaborate with others, to solve problems, maintain cohesion, and deal with others as enemies or collaborators; even today, it is claimed, it remains significant in shaping the optimal size of the functional groupings in the military and other human organizations, and for contemporary self-assembled networks (Dunbar 1996, 1998, 2003a, 2003b; C. Allen 2004; Dunbar and Shultz 2007; Shultz and Dunbar 2007).[7]

But the social brain hypothesis is more than a general account of the role of brain size: for in this thesis, the capacities for sociality are neurally located in a specific set of brain regions shaped by evolution, notably the amygdala, orbital frontal cortex, and temporal cortex—regions that have the function of facilitating an understanding of what one might call the 'mental life' of others.

This faculty became termed *social cognition*. A key paper by Leslie Brothers in 1990 drew on evidence from work with macaques, and from humans with various neurological syndromes, to develop the hypothesis that humans, together with some primates, had a "specialization for social cognition": "*the processing of any information which culminates in the accurate perception of the dispositions and intentions of other individuals*" (Brothers 1990, quoted from Brothers 2002, 367; emphasis in original). Brothers suggested that face recognition, with a specific neural basis, was crucial for this capacity and that social cognition formed a separate module subserved by a discrete neural system. Brothers, who herself was trained in psychiatry and later in psychoanalysis, developed this insight into a popular book, *Friday's Footprint*, which argued that human brains are "designed to be social"—that is to say, to attribute mental lives to human bodies (Brothers 1997).

But she could not have anticipated the ways in which, over the next twenty years, a body of work grew up that turned increasingly toward a neurobiological account of human behavior. Indeed, she later criticized this movement on Wittgensteinian grounds, and attributed it, in part at least, to the dominance of hidden interests, such as those of the pharmaceutical industry (Brothers 2001). But despite such concerns, this way of thinking has underpinned a growing body of work that focused on the identification of these neural bases and emphasized their determining role in social behavior: a body of research that called itself variously the social brain sciences, social cognition, or social cognitive neuroscience (Cacioppo 1994; Adolphs 2003; Heatherton, Macrae, and Kelley 2004). Ralph Adolphs has described this as "an uneasy marriage of two different approaches to social behavior: sociobiology and evolutionary psychology on the one hand, and social psychology on the other." Nonetheless, he argues that "neuroscience might offer a reconciliation between biological and psychological approaches to social behavior in the realization that its neural regulation reflects both innate, automatic and COGNITIVELY IMPENETRABLE mechanisms, as well as acquired, contextual and volitional aspects that include SELF-REGULATION" (Adolphs 2003, 165; capitalization as in original), because there seems to be a relation between some biological factors that we share with other species and other psychological factors that are unique to humans.

Does neurobiology now claim to have resolved the question of how I know what is going on in your mind? This has long been the concern of philosophers, playwrights—and, of course, of psychologists—for my relations to you are shaped by my beliefs about your beliefs, what I think that you are thinking, what I take to be your motives—the phenomenon of ascription. In the late 1970s, Premack and Woodruff coined the phrase that social neuroscience would adopt to describe such ascription—the way humans and some other primates think about others and take account of what others are thinking in their own conduct—such creatures had what they termed a "theory of mind" (Premack and Woodruff 1978). It was by means of a theory of mind that we are able "to attribute mental states to other people," in particular to attribute beliefs, and false beliefs, to them (Adolphs 2003, 171).

Of course there was no reason, in principle, to seek to localize such a capacity in specific brain regions, but in the developing style of thought within the experimental neurosciences that we discuss in chapter 2, this was exactly what followed: it appeared that theory of mind entails the activity of particular brain regions, including the amygdala, the medial and orbital prefrontal cortex, and the superior temporal gyrus. Now, of course, the reciprocal of such a claim was the argument that lesions or anomalies in the sites that underpin theory of mind might produce different kinds of deficits in social abilities—such as the inability to detect social faux pas. Many are cautious about such localization, not least because while a specific area might be activated when a particular task is carried out, often a lesion in that area does not result in an impairment in carrying out that task, or else it produces only subtle and very specific changes. Indeed, because of the plastic nature of the brain, discussed in chapter 1, such deficits may take unpredictable turns. Thus Adolphs concludes: "Although the data that I have reviewed . . . converge on several key brain structures that mediate social cognition, it cannot be overemphasized that the causal role of these structures remains unclear. . . . This probably reflects the considerable redundancy and plasticity of the brain. . . . Caution should be taken in attempts to predict people's behavior from knowledge about their brains" (Adolphs 2003, 174).

Nonetheless, despite the difficulties entailed by localization, theory of mind has come to be a central concern for social brain theorists and a key element in their explanatory repertoire. One dimension has come to be termed *mentalization*—a kind of built-in intersubjectivity in human experience, perhaps shared with some other primates. Mentalization refers to the "largely automatic process by which we 'read' the mental states of others" and which helps us make predictions about their future actions (Frith 2007b, 671). Much emphasis has been placed on the apparent phenomenon termed *mirroring* that was reported in macaque monkeys in the early 1990s and in humans soon after: the existence of a so-called mirror system in the brain that was mapped in 1999 (Di Pellegrino, Fadiga, Fogassi, et al. 1992; Gallese, Fadiga, Fogassi, et al. 1996; Rizzolatti, Fadiga, Gallese, et al. 1996; Iacoboni, Woods, Brass, et al. 1999; Hari 2004; Rizzolatti and Craighero 2004).

When an individual observes a movement being carried out by another— usually a 'conspecific' but sometimes an individual of another species—a small number of neurons are activated in those areas of the brain that are also activated when the individual carries out that same movement him- or herself. If I watch you reaching for a banana, there is some activation in the same areas that would be activated more intensively if I was reaching for that banana myself—the original observations were made by chance in monkeys and initially termed "monkey see, monkey do"—a phrase that did not travel so well (Iacoboni 2003). The mirror system thus seems to embody a neural basis for imitation learning, "the capacity to learn an action from seeing it done" (the definition from Thorndyke in 1889 is quoted in Rizzolatti and Craighero 2004, 172).

Now this capacity is apparently present only in humans and some apes, leading some to believe that it has been crucial for human development, not only for the learning of complex motor skills, but for interpersonal communication and learning social skills. V. S. Ramachandran extrapolated further than most, arguing that this mirror system, and the capacity for imitation that it enables, facilitated the propagation of culture and was "the driving force behind 'the great leap forward' in human evolution" and "set the stage for the emergence, in human hominids, of a number of uniquely human abilities, such as proto-language . . . empathy, 'theory of other minds,' and the ability to 'adopt another's point of view.'"[8]

But mirror theorists developed their argument beyond their original claims about simple imitation or simulation (Gallese and Goldman 1998). It was not merely some kind of mechanical imitation of action that was at stake, but the capacity to grasp the intention behind an action—for example, to understand someone's intention to drink when reaching for a glass, but not when reaching in an identical manner for some other object. This means that I don't just understand your intentions through the mirroring mechanism: I also feel your feelings. For mirroring is not restricted to the observation of another *carrying out a movement*, but is extended to the observation of another *experiencing an emotion* such as joy or pain, at least when this other is a conspecific.[9]

Thus Chris Frith explains that the notion of a "mirror system in the brain arises from the observation that the same brain areas are activated when we observe another person experiencing an emotion as when we experience the same emotion ourselves" (Frith 2007b, 673). It appeared that this challenged the conventional, rationalist accounts of ascription—that we understand the intentions of others by creating theories about what lies behind their appearance or their acts, on the basis of our own commonsense psychology about their desires and their beliefs about how to achieve them. It was not through *theorizing* about others, but through *feeling what they feel*, that we understand the mental states of others. I really do 'feel your pain.'

By around 2005, driven largely by the work of the research groups that had formed around the original mirror neuron researchers, mirror neurons had moved on from imitation and language to account for everything from a feeling of disgust when observing others showing disgust (Wicker, Keysers, Plailly, et al. 2003; Rizzolatti 2005) to providing a unified general theory of the neural basis for interpersonal relations (Metzinger and Gallese 2003; Gallese, Keysers, and Rizzolatti 2004; Iacoboni, Molnar-Szakacs, Gallese, et al. 2005). Ramachandran, with characteristic hubris, predicted that "mirror neurons will do for psychology what DNA did for biology: they will provide a unifying framework and help explain a host of mental abilities that have hitherto remained mysterious and inaccessible to experiments."[10]

It was not long, however, before these claims for mirror neurons were being contested. Critics argued that the claims made were philosophically, conceptually, and empirically problematic and that there was little or no evidence that these mirror neurons actually could achieve the effects claimed for them—of understanding the intentions from observing actions or appearances of others, especially when it came to complex interpretations of social behavior or of the communicative acts of others—and indeed that there was actually rather little evidence for the existence of mirror neurons in humans (Jacob and Jeannerod 2005; Borg 2007; Dinstein 2008; Dinstein, Thomas, Behrmann, et al. 2008; Lingnau, Gesierich, and Caramazza 2009). Skeptical voices suggested that the enthusiasm for these claims derived, not from their scientific bases or importance, but because the neuroscientists concerned were carried away by their desires to make their research seem socially important and relevant outside their own narrow disciplinary domain.[11]

The controversy continues.[12] But perhaps because of such doubts, many of those working in social neuroscience chose not to base their arguments on mirror neurons, but on the more limited claim that the social brain consists of brain areas specifically attuned to the recognition of different aspects of conspecifics—individuals of the same species—such as faces, motion, and attribution of mental states, some of which seem present at birth, and others of which develop over time. As Sarah-Jayne Blakemore puts it: "The social brain is defined as the complex network of areas that enable us to recognize others and evaluate their mental states (intentions, desires and beliefs), feelings, enduring dispositions and actions" (Blakemore 2008, 267). Even without reference to mirroring, this is the basis of the argument that there is a neural basis for many human social phenomena, such as empathy (Decety and Meyer 2008; Singer and Lamm 2009).[13] Such arguments rapidly became central to a way of thinking about the human brain as specialized for sociality through the built-in capacity to understand the beliefs and intentions of others (Kilner and Frith 2008; Rizzolatti and Fabbri-Destro 2008).[14]

For social brain theorists, then, the human brain, as it has evolved from our primate forebears, actually embodies the capacity for intersubjectivity: when I observe your actions, or the visible signs of your inner states, areas of my own brain activate that enable me to understand the intentions or feelings that lie behind those observable features, and hence to feel a tiny trace of what it would take, what it would mean, for me to act or feel that way myself. Of course, on the one hand this proposal, whether it is neurobiologically accurate or not, is sociologically anodyne. Few would doubt that there are neurobiological conditions for our capacities to impute feelings, thoughts, and intentions to others, or for our capacities to engage in social interaction. For the human sciences, the possibility of such imputations and attributions has long been taken as a premise; what has been crucial has been the analysis of the varying forms of such attribution, and the ways in which they have been shaped by culture, history, language, and by highly variable regimes of meaning and value.

Hence, from this perspective, it is not clear what would be gained by attempts to localize such feelings in particular brain regions such as the anterior insula or anterior cingulate cortex, or in suggesting that particular brain regions provided the causal basis for such capacities. The interesting question from the human sciences would concern the ways in which, given that apparently universal basis, we might account for that historical and cultural variability in the ways in which individuals do, in fact, attribute beliefs and intentions to other entities, and the variability in the forms and limits of the empathy or fellow feeling that they exhibit to others. But the human sciences would not be alone in such concerns. When Tania Singer and Claus Lamm reviewed the neuroscience of empathy in 2009, they agreed that this research is beginning "to reveal the mechanisms that enable a person to feel what another is feeling" (Singer and Lamm 2009, 81). But they were exceptionally cautious about claims for universality, not only arguing for very careful distinctions to be made between different forms of "affect sharing"—empathy, sympathy, compassion, and so forth—but also showing that "a clearcut empirical demonstration of the link between empathy and prosocial behavior is still missing" (ibid., 84). Further, they concluded, we know little of variations across a life span, almost nothing about individual differences, have not sufficiently recognized the fact that affect sharing is highly contextually specific and far from universal, and know almost nothing about the malleability of these features and the ways in which such sharing of affects may be influenced by training and experience—the latter, they claim, having potentially "enormous implications for education and society as a whole" (ibid., 93).

As we have seen in previous chapters, such expressions of caution about what we know today coupled with optimism about future possibilities are characteristic of leading researchers in the new brain sciences. But the lure of translation is hard to resist. Whatever their consequences in other areas, a number of researchers saw, in these arguments, a way in which some important problems of mental health in humans could be reframed. Were not many human mental disorders characterized by problems in precisely those capacities explored by social brain theorists? Crucially, if these disorders were recoded as pathologies of the social brain, they appeared to offer the prospect of various forms of neurobiologically based interventions.

Pathologies of the Social Brain

Could conditions such as schizophrenia and autism be disorders of the social brain, arising from disruptions in the evolved neural circuits that implement social behavior?[15] As early as the mid-1990s, Chris Frith and his colleagues were suggesting that those diagnosed with schizophrenia had deficits in their theory of mind and problems in "mentalizing" (Corcoran, Mercer, and Frith 1995; Frith and Corcoran 1996). Over the next ten years, this idea was developed into a full-scale social brain theory of schizophrenia (Russell, Rubia, Bullmore, et al. 2000; Sharma, Kumari, and Russell 2002), largely based on neuroimaging studies that seemed to show that patients with such a diagnosis had abnormal activation patterns in regions thought to be involved in social cognition. In a characteristic reverse move, it could then be suggested that the social brain networks in the medial prefrontal cortex, the prefrontal cortex, the amygdala, and the inferior parietal lobe were the neural systems underlying disorders in theory of mind, emotion perception, and self-agency in such patients (Brunet-Gouet and Decety 2006).

For some, mirror neurons, once more, were involved. Mirror neuron theorists suggested that pathologies such as autism or schizophrenia arise because of deficits in that capacity for mind reading found in all presumably normal human beings (Iacoboni 2008). For example, mirror neuron deficits might lie behind the false attribution of auditory hallucinations to external sources found in some patients diagnosed with schizophrenia (Arbib and Mundhenk 2005). Mirror neuron deficits might provide the neural substrate for deficits in mentalization in autism—a view proposed in a popular piece in *Scientific American* by Ramachandran and Oberman under the title of "Broken Mirrors" (Ramachandran and Oberman 2006).[16]

Others, however, managed without mirror neurons. Well before the advent of the mirroring arguments, Uta Frith and Simon Baron-Cohen had argued that autism and autism spectrum disorders were linked to deficits in the capacity of an individual to mentalize, that is to say, to construct a theory of the minds of others in order to attribute beliefs and intentions to their actions (Baron-Cohen, Leslie, and Frith 1985); it was only later that they looked to neurobiology for reinforcement, although in combination with psychological arguments (Baron-Cohen, Tager-Flusberg, and Cohen 2000; Frith and Hill 2003; Hill and Frith 2003; Baron-Cohen and Belmonte 2005). Baron-Cohen and Belmonte concluded their 2005 review by hoping that it will eventually be possible to identify the way in which "large numbers of genetic biases and environmental factors converge to affect neural structure and dynamics, and the ways in which such abnormalities at the neural level diverge through activity-dependent development to produce an end state in which empathizing abilities are impaired and systemizing abilities are augmented. Such a unified understanding of the psychobiology of autism will offer targets for intervention on many levels" (Baron-Cohen and Belmonte 2005, 119): not a single mention of mirror neurons here.

There were other routes to explanations of mental disorders in terms of the social brain. Thomas Insel, who was appointed director of the NIMH in 2002, had previously adopted a different research strategy to explore the neurobiology of sociality. In his famous studies of prairie voles, he had argued that the neuropeptides oxytocin and vasopressin played crucial roles in the establishment of enduring social bonds among these creatures, and suggested that differences between the reproductive, parenting, and affiliative behaviors of the microtine voles and the asocial behavior of the closely related montane voles were related to differences in the patterns of oxytocin and vasopressin receptors in their brains (Insel and Shapiro 1992; Winslow, Hastings, Carter, et al. 1993; Insel and Young 2000).

In 2004, he linked these arguments into more general neurobiological accounts of the way in which the brain "processes social information"—as he and Russell Fernald phrased it when they were "searching for the social brain" (Insel and Fernald 2004). They singled out four dimensions of such processing of social information—epigenesis; sensory specializations for social perception, such as smell; imprinting or early social learning based on the exceptional plasticity of the infant brain; and the presence of specialized neural circuits for social recognition, also often involving the effect of olfaction on specific brain regions. Insel argued that these evolved mechanisms for processing social information or social signals might provide a way of understanding human psychopathology that offered possibilities for translation into the clinic. "Results from these studies and some recent functional imaging studies in human subjects begin to define the circuitry of a 'social brain,'" they concluded (Insel and Fernald 2004, 697). Not only does this conception of a brain evolved for sociality help us understand why isolation and social separation are linked to many types of disease, but it also suggests that conditions such as autism and schizophrenia could be viewed as neurodevelopmental disorders characterized by abnormal social cognition and corresponding deficits in social behavior. Thus "social neuroscience offers an important opportunity for translational research with an impact on public health" (ibid.).

By the time that Insel was appointed to the NIMH, others were also suggesting that the idea of the social brain could become an integrating zone for a new kind of social psychiatry. Thus Bakker and colleagues argued that the social brain provided the physiological focus that could be a unifying foundation for psychiatry "The social brain is defined by its function—namely, the brain is a body organ that mediates social interactions while also serving as the repository of those interactions. The concept focuses on the interface between brain physiology and the individual's environment. The brain is the organ most influenced on the cellular level by social factors across development; in turn, the expression of brain function determines and structures an individual's personal and social experience" (Bakker, Gardner, Koliatsos, et al. 2002, 219). While not expecting social psychiatrists to resolve the mind-brain problem, we may note, once again, the phrasing—"the expression of brain function determines and structures . . . experience"—that is called into service to bridge the explanatory gap between the neurobiological and the experiential.

The link between psychopathology and this complex notion of the social brain, in which arguments from evolution and comparative studies of animals, explanations based on mirror neurons, and findings from genomics and neurochemistry are brought together, is perhaps best exemplified in the work of Klaus-Peter Lesch. Lesch's classic paper of 1996 identified a polymorphism in the promoter region of the 5-HTT (serotonin) transporter gene—basically a region of the gene sequence that came in two version, short or long—where possession of one or two copies of the short variants led to a lower expression of serotonin in the brain and was associated with greater levels of anxiety and personality traits often linked to depression and suicidal behavior (Lesch, Bengel, Heils, et al. 1996).[17] In subsequent work, Lesch argued that the 5-HTT transporter was a kind of controller that was responsible for fine-tuning the serotonin system and hence for the regulation of emotion and disorders of emotion regulation, within which he included such conditions as depression, substance abuse, eating disorder, and autism.[18] Further studies in rhesus monkeys, where there was a different but analogous polymorphism, seemed to show that stress in early life reinforced the effect of the low-acting form, suggesting that 5-HTT was a susceptibility gene for depression.

For Lesch, these studies reveal a set of complex interactions between single nucleotide polymorphism (SNP)-level genetic variations, "embryonic brain development[,] and synaptic plasticity, particularly in brain areas related to cognitive and emotional processes," which "transcends the boundaries of behavioral genetics to embrace biosocial science" (Lesch 2007, 524). Life stress interacts with variations in the 5-HTT genotype to influence patterns of brain activation, provoking different cognitive and emotional states. And Lesch turns to Iacoboni's mirror neuron theory (Iacoboni 2005) to argue that the regions affected by allelic variations in 5-HTT genotype are none other than those "that integrate imitation-related behavior, from which social cognition and behavior have evolved" (Lesch 2007) and that epigenetic programming of emotionality by maternal behavior demonstrates the ways in which environmental influences can remodel neuronal organization in the course of development. "We are clearly a long way from fully understanding the evolutionary and neural mechanisms of social cognitive phenomena and social functioning," he concludes. "But the potential impact of 5-HTT variation on social cognition transcends the boundaries of behavioral genetics to embrace biosocial science and create a new social neuroscience of behavior" (Lesch 2007, 528).[19] Here, in perhaps its most complex formulation to date, seeking to incorporate evidence from molecular genomics, evolutionary theory, brain imaging, and cognitive neuroscience, we have the new image of the social brain.

Social Neuroscience

This emergent, though far from homogeneous, discourse on the social brain seems to escape some of the ritual criticisms leveled by social scientists at biological explanations of human conduct—reductionism, essentialism, individualism, and the like. There is certainly a genealogical link back from today's social brain to the ethological studies of human behavior that were popular in the 1960s and early 1970s, but there is little of sociobiology here, either in its scholarly or popular forms. The teleological tropes of evolution-ary psychology are, if not absent, then mitigated. There is no appeal to a fixed human nature to account for either virtuous or vicious conduct, and humans are not portrayed as naked apes. If evolution has provided humans with their behavioral attributes through selection for certain neural capacities, these are not those of unmitigated self-interest, but rather of empathy, sociality, and reciprocity, enabled by imitation learning, the ability to form a theory of other minds, and the plasticity of the human brain. All these are in turn shaped by gene expression that is modulated by biographical, social, and en-vironmental inputs in utero, infancy, and across the life-course. The human brain is now portrayed as evolved for sociality and culture—open, mutable, and in constant transactions with its milieu, whether these be at the level of the epigenetic modification of gene expression; the interactions of gene and environment in development; the interaction of evolved visual, olfactory, or other sensory capacities and relations between individuals in a milieu; or the role of mirror neurons in the attribution of beliefs, intentions, and mental states to others (Cacioppo and Berntson 2004). A discipline of *social neuro-science* has become possible.

Most of those who write the history of social neuroscience agree that the beginnings of the discipline emerge in the 1990s, if by that one means the invention of a name and the framing of a project (Cacioppo, Berntson, and Decety 2011).[20] Certainly some of the psychological ethologists working in the 1960s and 1970s discussed the social functions of human intellect and the evolutionary pressures for intellectual development (e.g., Humphrey 1976). However, the first to use the term *social neuroscience* in print seem to have been two psychologists, John Cacioppo and Gary Berntson, who published an article in *American Psychologist* in 1992 titled "Social Psychological Con-tributions to the Decade of the Brain":

> [T]he brain does not exist in isolation but rather is a fundamental but interacting component of a developing or aging individual who is a mere actor in the larger theater of life. This theater is undeniably social, beginning with prenatal care, mother-infant attachment, and early childhood experiences and ending with loneliness or social sup-port and with familial or societal decisions about care for the elderly. In addition, mental disorders such as depression, schizophrenia, and phobia are both determined by and are determinants of social pro-cesses; and disorders such as substance abuse, juvenile delinquency, child and spousal abuse, prejudice and stigmatization, family discord, worker-dissatisfaction and productivity, and the spread of acquired im-munodeficiency syndrome (AIDS) are quintessentially social as well as neurophysiological phenomena. (Cacioppo and Berntson 1992, 1020)

The term *social* was used here to index the fact that (a) neurochemical events influence social processes and (b) social processes influence neurochemical events. However, despite references in the paper to economic factors and to social systems, the social processes that the authors sought to integrate into their "multilevel analysis" were not placed under the sign of sociology but rather of social psychology. Cacioppo and Berntson used the term *social neuroscience* only in the concluding section of their paper, and it was followed by a question mark. The question mark expressed a concern about potential for reductionism and the need to recognize different levels of explanation: "[T]he brain is a single, pivotal component of an undeniably social species," but we must recognize that "the nature of the brain, behavior, and society is . . . orderly in its complexity rather than lawful in its simplicity" (Cacioppo and Berntson 1992, 1027). Two years later, Cacioppo used the term more affirmatively in the title of an article in 1994 discussing the effect of social and psychological factors on the immune system (Cacioppo 1994), and by the time that he and others published their entry under that title in the *International Encyclopedia of the Social and Behavioral Sciences* in 2001 (Cacioppo, Bernston, Neil, et al. 2001), the question mark had been firmly left behind.

A discipline needs an infrastructure (Star and Ruhleder 1996; Star 1999)—a fact of which the practitioners are usually very well aware. Cacioppo identifies the first infrastructural element in a grant to him from the Behavior and Neuroscience Division of the National Science Foundation in the 1980s to establish a five-year interdisciplinary research program, although it appears from the resulting edited volume that this research was actually focused on the neurophysiological, psychoneuroendocrinological, and psychoneuroimmunological foundations of psychophysiology (Cacioppo and Tassinary 1990). Following that, he identifies other infrastructural elements, ranging from the establishment of the Summer Institute in Cognitive Neuroscience, to the use of EEG technology and the study of "event-related potentials" (ERPs) and other measures of brain activation within experimental social psychology, and eventually to fMRI scanning, which made it possible to study neural activity by noninvasive means as humans engaged in various tasks in laboratory settings. Over the 1990s, the number of papers published that used the term *social neuroscience* gradually increased, from one or two a year at the start of the decade to around twenty per year at the end, but this number rapidly increased after the turn of the century—there were 36 in 2001, 128 in 2006, and 279 in 2009.[21] By the start of the twenty-first century, MIT Press was publishing, largely under Cacioppo's influence, a book series on social neuroscience, about which it quoted a review in *Science*: "In a very short time, social neuroscience has come from almost nothing to become one of the most flourishing research topics in neurobiology."[22]

The Center for Cognitive and Social Neuroscience at Chicago, founded in 2004 and headed by Cacioppo, provided a crucial institutional basis for social neuroscience. As it describes itself, the center focuses on the study of the biological systems that "implement"—an intriguing term—social processes and behavior.[23] In many different publications since, Cacioppo has asserted his own definition of social neuroscience as "an interdisciplinary approach devoted to understanding how biological systems implement social processes and behavior and to using biological concepts and methods to inform and refine theories of social processes and behavior . . . ," and his work explores the effects of social environment—notably social isolation—on everything from gene expression to hormonal regulation.[24] This is the direction that his own work has taken, both in his general accounts of the discipline (Norman, Cacioppo, and Berntson 2010) and in his specific studies of the physical and mental consequences of a preeminently culturally shaped human experience—loneliness (Hawkley and Cacioppo 2010). Further, Cacioppo and his coworkers made a move that opens a space for the entry of more conventional arguments from the human sciences, stressing that the effects of loneliness were not merely those of actual isolation but also those where people perceive that they live in isolation (Cacioppo and Patrick 2008). For along with the entry of human perception, rather than raw external sensory input, comes culture, language, meaning, history, and all the other dimensions known to the cultural sciences. An interdisciplinary effort, it seemed, might be on the cards. Or was it?

When the *Journal of Personality and Social Psychology* published a special section on "Social Neuroscience" in its October 2003 issue, the editors defined the field in terms that owed a lot to Cacioppo: "Social neuroscience is an integrative field that examines how nervous, endocrine, and immune systems are involved in sociocultural processes. Being nondualist in its view of humans, it recognizes the importance of understanding the neural, hormonal, and immunological processes giving rise to and resulting from social psychological processes and behaviors" (Harmon-Jones and Devine 2003, 590). However, the bulk of the papers were rather different: they used various imaging and localization technologies to identify the patterns of brain activity associated with tasks performed in the traditional experimental set-ups of social psychology (e.g., Cunningham, Johnson, Gatenby, et al. 2003) and considered the potential of brain imaging for the discipline (Willingham and Dunn 2003). Indeed, in the same collection, Cacioppo himself expressed some reservations about that direction of travel and the way it framed its research questions (Cacioppo, Berntson, Lorig, et al. 2003). But when, in 2004, a range of journals in psychology, such as *Current Directions in Psychological Science*, and neurobiology, such as *Current Opinion in Neurobiology*, published expository and review papers on social neuroscience, they mainly followed the brain localization route (Heatherton 2004; Jackson and Decety 2004; Ochsner 2004).

It is no surprise, given its focus, that when *NeuroImage* published its special section on "Social Cognitive Neuroscience" in the December 2005 issue under the editorship of Mathew Lieberman, whose own work focused on the use of functional neuroimaging to investigate social cognition, it was indeed the power of imaging that was central, notably its role in revealing the nonconscious impact of situational factors on motives, attitudes, behaviors, perceptions, beliefs, and reflective thought (Lieberman 2005).[25] And the bulk of what came to call itself social and cognitive neuroscience follows this direction, adding various versions of brain imaging to the methodological armory of experimental social psychology. Indeed, the journal *Social Neuroscience*, which commenced publication in 2006, defines its focus as "how the brain mediates social cognition, interpersonal exchanges, affective/cognitive group interactions, and related topics that deal with social/personality psychology"; the editors define social neuroscience as "the exploration of the neurological underpinnings of the processes traditionally examined by, but not limited to, social psychology" (Decety and Keenan 2006).

In the same year, the journal *Social Cognitive and Affective Neuroscience* published its first issue. It asserted that it was devoted to "publishing research on the social and emotional aspects of the human mind by applying the techniques from the neurosciences and providing a home for the best human and animal research that uses neuroscience techniques to understand the social and emotional aspects of the human mind and human behavior." However, the opening editorial, once more written by Mathew Lieberman, was quite clear about the type of work it wished to publish: "[S]ubmitted research should report new empirical investigations using neuroscience techniques (including, but not limited to, fMRI, MRI, PET, EEG, TMS, single-cell recording, genetic imaging, psychopharmacological perturbations, and neuropsychological lesion techniques) to examine social and affective processes" (Lieberman 2006, 2). Location and visualization, it seemed, were the paths to understanding the social brain.

To know is to see; and, as we have seen, many believed that to see the brain meant to localize. By 2010, localization by means of brain imaging had become the method of choice for much social neuroscience, especially within psychology departments. Although scanning was usually an intensely individual experience for any experimental subject, researchers instructed their subjects to carry out tasks in the scanner that they believed simulated social interactions, and then tried to find the regions of the brain that were activated. This style of investigation was not merely an American or European preoccupation. By 2010, the Culture and Cognitive Neuroscience Laboratory based at Peking University was receiving research funding for "Neuroimaging studies of human aggressive and affiliative behaviors" (funded by the Ministry of Science and Technology of the People's Republic of China), "Modulation of empathy by culture and social relationship—A transcultural cognitive neuroscience study" (funded by the National Natural Science Foundation of China), and a study of the "Cognitive and neural mechanisms underlying emotional influences on risk behaviors" (funded by the National Natural Science Foundation of China).[26]

Nonetheless, tensions remained within the field itself. When the Society for Social Neuroscience was established in 2010, under the leadership Cacioppo and Decety, and headquartered with Cacioppo in Chicago, it framed its mission in terms that owed much to Cacioppo: the impact that social structures—from dyads, through families, neighborhoods, and groups to cities and civilizations—have on the brain and nervous system of the individual "through a continuous interplay of neural, neuroendocrine, metabolic and immune factors on brain and body, in which the brain is the central regulatory organ and also a malleable target of these factors. . . . Social neuroscience is the interdisciplinary academic field devoted to understanding how biological systems implement social processes and behavior, and how these social structures and processes impact the brain and biology. A fundamental assumption underlying social neuroscience is that all social behavior is implemented biologically."[27]

Each of the two sessions in its first conference in November 2010 focused on one dimension. The first session examined such matters as the social regulation of brain gene expression and aggression in honey bees; measuring social behavior in the rat and mouse; neuropeptides, bonding, and social cognition in voles to man; and perspectives on stress, brain, and body health. The other featured evidence from neuroimaging in areas ranging from fronto-temporal dementia to reasoning processes in game theory.[28] Cacioppo and Decety commented that despite the agenda for integration set out in the early project for social neuroscience, "contemporary social neuroscience investigations to elucidate the genetic, hormonal, cellular, and neural mechanisms of social behavior have grown from two largely separate root systems: one based on data from humans grounded in the discipline of psychology and the other based on data from animal models grounded in biology and biomedicine. . . ."[29] With a few notable exceptions in each group, there has been little communication between these two groups of scientists. In our view, this is a gap that must be filled for social neuroscience to reach its full potential" (Cacioppo and Decety 2011). In many ways, the relation of social neuroscience to the human sciences more generally depends on which of these two paths is followed, for they point to two very different conceptions of sociality.

Social Neuroscience beyond Neuroscience

By 2005, social neuroscience was attracting the interest of those funding research on types of human social behavior considered problematic. The NIH, the U.S. National Institute on Drug Abuse (NIDA), the U.S. National Institute of Alcohol Abuse and Alcoholism (NIAAA), the U.S. National Institute on Aging (NIA), and the U.S. Institute of Neurosciences, Mental Health and Addiction (INMHA), and the Canadian Institutes of Health Research (CIHR) were calling for proposals under the heading of social neuroscience to "act as a catalyst for the emerging area of social neuroscience in order to elucidate fundamental neurobiological mechanisms of social behavior. . . . Clinical and preclinical research supported by this initiative must take a multidisciplinary, multilevel approach to a social behavioral research question (or set of questions) that is framed at the behavioral level (e.g., social cognition, social development, social interaction, and social aspects of emotion)."[30]

In the same year in the United Kingdom, the Economic and Social Research Council carried out a 'scoping' workshop on social neuroscience and issued an enthusiastic media briefing titled "Social Neuroscience Will Shed New Light on Human Behavior" that asserted this "new domain of study is set to spearhead an assault on what has been described as science's final frontier: understanding, even decoding the complex interconnected web that mediates between the brain, and the mind and human behavior. Already it is making us think again about humanity's understanding of itself and what this means for the norms and dynamics of how we behave in society. . . . The Economic and Social Research Council fully intends that the UK should not be left behind in this exciting and potentially extremely valuable field."[31] Together with the Medical Research Council, in 2007 it was calling for cross-disciplinary grants in this area, arguing that "[e]normous advances are being made in our understanding of the links between the mind, brain, innate traits, society, culture and behavior, whether normal or abnormal."[32]

Some of those within the social and human sciences were open to the opportunities presented by these new ways of thinking. At the 2008 meeting of the American Anthropological Association, a group convened a panel on "the encultured brain: neuroanthropology and interdisciplinary engagement." They argued in their proposal that neuroanthropology

> can help to revitalize psychological anthropology, promote links between biological and cultural anthropology, and strengthen work in medical and linguistic anthropology. . . . Neuroscience has increasingly produced basic research and theoretical models that are surprisingly amenable to anthropology. Rather than "neuro-reductionist" or determinist approaches, research has increasingly emphasized the role of environment, body, experience, evolution, and behavior in shaping, even driving organic brain development and function. At the same time, the complexity of the brain makes a mockery of attempts to pry apart "nature" from "nurture," or to apportion credit for specific traits. Research on gene expression, endocrine variability, mirror neurons, and neural plasticity all beg for comparative data from across the range of human variation—biological and cultural. . . . Anthropology has much to offer to and much to learn from engagement with neuroscience.[33]

At almost the same moment, in November 2008, the first Social Brain Conference convened in Barcelona, focused on "a neurocognitive approach to fairness." It included sessions titled "From Animosity to Empathy: Neuroimaging Studies on the Building Block of Fairness" and "Genetic Markers for Good and Bad Cooperators: A Biological Approach to Fairness in Economic

Exchanges" alongside those on "Sacred Values: Anthropological Perspectives on Fairness in Social Conflicts." Political figures were among the speakers, for instance, Lord John Alderdice, who played an important role in brokering the peace deal in Northern Ireland. The representatives from social neuroscience included psychologists using brain imaging, researchers on economic game theory and social groups, as well as Tania Singer, whose work on moral feelings we have already encountered, and who published the first imaging studies of pain empathy and the relation of neural signatures of empathy with cooperation and competition (Singer, Seymour, O'Doherty, et al. 2004). The organizers were modest about whether this new approach to issues from nationalism, terrorism, and suicide bombers to cooperation, fairness, and empathy would form the basis for "ruling human communities according to high standards of values and morality," as some had hoped (Atran 2009, 2), but undoubtedly the social brain hypothesis was enabling some anthropologists to reframe their research such that it seemed to address pressing social issues and perhaps even inform public policy.[34]

Anthropology has always had a biological branch, so perhaps its openness to the social brain hypothesis is not so surprising. Similarly, we should not be surprised that social neuroscience also seemed attractive to many social psychologists. One of the most articulate spokespersons was Chris Frith, for whom social neuroscience seemed to promise this fragile discipline some objectivity at last (Frith 2007a). Frith placed the social brain firmly within the forms of sociality studied within experimental social psychology—sociality as a kind of interactional nexus: "what the social brain is for" is that it "allows us to interact with other people," and specifically, "the function of the social brain is to enable us to make predictions [which need not be conscious and deliberated] during social interactions" (Frith 2007b, 671). The themes that have most engaged such researchers—empathy, trust, love, social bonding, maternal-infant relations, deception, ascriptions of beliefs and feelings to others, decision making—all fit easily within a premise that social life is, in essence, a life of interactions. The preconditions for such interactions may lie in some form of socialization or development, but the capacity for enactment of those interactions in particular ways—or the failure to enact them—is, at any one instant, located within the capacities and attributes of the interacting entities.[35] The social brain is construed as the fleshy and embodied material substrate that enables (subserves, implements, underpins) the transactions between self and other—in pairs, families, friendships, and social networks. This interactive conception of sociality is embodied in the experimental setups of those whose preferred technology is that of brain imaging.

Imaging technologies, with their focus on the localization of function and the experimental manipulations they require and utilize, impose their own constraints and logics as to what can be thought. As with a long tradition of experimental psychology, they make it seem unproblematic to draw upon evidence that occurs in the laboratory setting where such interactions

prob with experiment

can be simulated, because the other features of the laboratory—its specific properties as a location full of meaning, expectations, dense with a different kind of sociality for all the participants—may somehow be deleted from the analysis (Cohn 2008). Such premises lead, usually almost imperceptibly, to a theory of sociality that renders it in terms of the capacities required to maintain interactions between individuals in small groups, because that is all the sociality that can enter the experimenters' purview. That explicit or implicit theory is then operationalized in experiments in laboratory situations that appear to simulate such interactions, and, through the use of recording machines of one sort or another, the outcomes are turned into data that will then stand in for the experiment itself. The data produced by this technology are then used to support the theory that has enabled its production. In the process, the tool or technology has, as it were, merged with the theory itself. We can see here a reciprocal relation between concepts and technologies—a slightly more recursive example of what Gerd Gigerenzer once termed "tools to theories," in which physical tools, such as computers and scientific instruments, and intellectual tools, such as statistics or intelligence tests, start off as the means to produce data and end up providing the theories of the data that they themselves have produced. Where human mental life is concerned, scientific tools are not neutral: "the mind has been recreated in their image" (Gigerenzer 1991, 19).

Cacioppo and those who think like him seek a conception of sociality that goes far beyond interpersonal interactions: "As a social species, humans create emergent organizations beyond the individual—structures that range from dyads, families, and groups to cities, civilizations, and international alliances. These emergent structures evolved hand in hand with neural, hormonal, cellular, and genetic mechanisms to support them because the consequent social behaviors helped humans survive, reproduce, and care for offspring sufficiently long that they too survived to reproduce."[36] The work he and his colleagues have done on social isolation, for example, is placed in the context of a range of other sociological and epidemiological evidence and acknowledges the fact that humans "are an irrepressibly meaning-making species, as well, and a large literature has developed showing that perceived social isolation (i.e., loneliness) in normal samples is a more important predictor of a variety of adverse health outcomes than is objective social isolation. . . . [The] social environment, therefore, is fundamentally involved in the sculpting and activation/inhibition of basic structures and processes in the human brain and biology."[37]

But while many social psychologists have been eager to ally themselves with these new styles of thought and research, the same cannot be said for sociologists. Warren TenHouton ascribes the first use of the term *neurosociology* to a 1972 paper on the impact of different cultures on the development of the left and right hemispheres of the brain (Bogen, DeZure, TenHouten, et al. 1972).[38] Five years later, in a paper in the *Journal of Social and Evolutionary Systems*, TenHouton argued: "The case for neurosociology begins with the observation that no social relationship or social interaction can possibly be carried out in the absence of human mental activity. It follows immediately that, as scientists, sociologists must find ways to describe and measure such mental functions and cognitive structure. . . . Just as neuropsychology has far reaching implications for general psychology, so also, then, does neurosociology for general sociology" (TenHouten 1997, 8). The case was carried forward by David Franks, who described neurosociology as "the nexus between neuroscience and social psychology" (Franks 2010; cf. Franks and Smith 2001), taking up themes of emergence and neuroplasticity to argue that neuroscience can be brought into alignment with the sociology of embodiment as well as with insights from interactionist sociology and the pragmatism of George Herbert Mead and the Chicago pragmatists.[39] But he is a rather lonely voice: few sociologists seem to believe that the explanatory models of social neuroscience have implications for their often unarticulated conception of human beings as sense making creatures, shaped by webs of signification that are culturally and historically variable and embedded in social institutions that owe nothing substantial to biology.

Governing Social Brains

Gordon Allport opened his historical review of social psychology in 1954 by quoting from Giambattista Vico's phrase of 1725: "government must conform to the nature of the men governed" (Vico, quoted in Allport 1954, 1). Despite the commonly held view that governing is an essentially pragmatic affair, attempts to conduct human conduct are always based on certain premises about people, even if these are often tacit and frequently incoherent. That is to say, there are intrinsic links between 'ontological' assumptions (about what human beings are really like), 'epistemological' questions (about the kinds of knowledge practices according to which they are best understood), and 'governmental' questions (about how the conduct of human beings might best be conducted according to their nature). No doubt for much of our history, these assumptions have been derived from a mixture of folk wisdom, theological doctrines, philosophies, and the accumulated experience of those who have exercised authority in different domains. Across the twentieth century, the *psy-* disciplines came to infuse these assumptions, not just in relation to identifying and managing aberrant individuals, but also in understanding and governing groups and collectivities. Psychology, that is to say, became a 'social' science (Rose 1996). It is thus not surprising that social neuroscience too is becoming an expertise with a social vocation, claiming the capacity to provide an objective knowledge about key features of human social conduct and of human pathology that can and should underpin interventions.[40]

In his book *Mirroring People: The New Science of How We Connect with Others*, Iacoboni remarks: "Unfortunately, Western culture is dominated by an individualistic, solipsistic framework that has taken for granted the assumption of a complete separation between self and other. . . . Against this dominant view, mirror neurons put the self and the other back together again" (Iacoboni 2008, 155). And he concludes by asking how, despite the fact that we are "geared toward empathy and sharing of meaning" our world is so filled with violence and atrocities. We are at a point, he asserts, "at which findings from neuroscience can significantly influence and change our society and our understanding of ourselves. It is high time we consider this option seriously" (ibid., 271). Others share this view that significant implications follow from the recognition that the human brain is profoundly social. Sarah-Jayne Blakemore argues that understanding the social brain should help us adjust educational practices in the light of knowledge about the impact of specific types of environmental input on social cognition (Blakemore 2010). The United Kingdom's Royal Society for the Encouragement of Arts, Manufactures and Commerce (RSA) launched its Social Brain project in 2009, based on the belief stated by its director, Mathew Taylor, that we are entering an age of "neurological reflexivity."[41]

Becoming aware of the social and biological conditions that underpin our actions can, it seems, help each of us develop our competencies and understand and manage our cognitive frailties. But more important, our authorities have much to learn from the lessons of social neuroscience, notably that much of our behavior results from automatic, nonconscious, and habitual reactions to social situations, and that although our brains are inherently disposed to value cooperation and altruism, they are sadly myopic and focus on things close to us temporally, spatially, and emotionally (Grist 2009, 2010).[42] For the Social Brain project, neuroscience has shown us that "the brain is fundamentally social and largely automatic" and thus that policies based on "the notion of a rational, profit maximizing individual who makes decisions consciously, consistently and independently" was at best a very partial understanding of who we are; hence it was not surprising that social policies based on this idea were hardly successful (Rowson 2011). While recognizing that individual behavior was largely habitual and shaped by nonconscious dynamics did not lead to clear "policy levers," individuals can be aided to become more "reflexively aware" of the way their brains shape their lives, and hence move from passivity in the face of such constraints to proactively shaping their lives with an awareness of them. We can anticipate many similar initiatives over the next decade.

Elliot Aronson's social psychology textbook *The Social Animal* was first published in 1972.[43] Much praised and much translated, and now in its tenth edition, the book deals with topics from conformity to interpersonal attraction, and manages its task without any reference to the role of the human brain. But when David Brooks published his book with the same title, *The Social Animal*, in 2011, he looked first to the brain to understand human sociality, drawing wide implications from neuroscientific arguments about the myriad nonconscious processes occurring each millisecond that make humans exquisitely finely attuned to other human beings throughout their lives. Seeking to dethrone consciousness from its role in shaping human relationships, the book claimed to show "the relative importance of emotion over reasons, character over IQ, emergent, organic systems over mechanistic ones, and the idea that we have multiple selves over the idea that we have a single self" (Brooks 2011). The subtitle of the book, *How Success Happens*, points to its pedagogic ambition—if we understand these profoundly unconscious forces shaping decisions and our affects, whether as individuals, as those in authority, or as politicians, we will not only learn so much more about ourselves, we will learn why so many attempts to improve our societies have failed in the past to combat educational underachievement, achieve social mobility, and tackle welfare dependency and crime. Through social neuroscience and the recognition of the nonrational processes of our social brains, it appears that human beings—at least those in the kinds of societies envisaged by such authors—will know better how to achieve social and political goals.

Despite the radical claims of many of those who write about the implications of understanding the role of nonconscious processes of the social brain in shaping human lives, the policy implications they extrapolate are usually entirely familiar: for example, that the social environment has a role in promoting the individual; stressful and deprived environments are harmful; good parenting and family support is desirable; educational input should take account of critical periods in children's development; and social institutions need to override short-term impulses (now arising in the brain rather than the instincts or the psyche) and foster commitment and social control. Perhaps arguments from neuroscience are merely being invoked to give such proposals a sheen of objectivity—for they are often criticized as arising from hopes rather than facts. Yet the very fact that invoking the brain increases the power of a truth claim shows us something about the way we are coming to understand ourselves today.

One is reminded of an earlier movement in which a theory of the human mind with the status of truth, this time derived from dynamic psychology, grounded a program for transforming policy. This was of course the movement for 'mental hygiene,' whose many adherents and advocates argued that authorities from politicians to industrialists, and from teachers to parents, needed to reform their practices in the light of the new understanding forged by dynamic psychology of what shaped us as human beings, in order to maintain and maximize the adjustment of the individual to his or her social milieu (Rose 1985). Today, it is no longer a matter of the social shaping of the dynamic forces of instincts inscribed in our biology by evolution producing adjustment or maladjustment with personal, social, and industrial costs. Now, the inescapable nexus where these forces meet is the brain. All agree that social factors are very important. But everything that is considered social now must pass through the brain and have effects via the brain. Parenting, biography, experience, diet, alcohol, drugs, stress, and lifestyle all pass through the brain, shaping and reshaping that brain at the very same time as those capacities and attributes—cognition, emotion, conduct, disorder, resilience, and the like—are shaped by the brain. Hence we have a social brain, in the sense that it now seems that capacities that are crucial to society are a matter of brains. We have a social brain in that the brain has evolved to favor a certain type of sociality manifested in all the interactions between persons and groups that come naturally to humans in our social lives. And we have a social brain in that this organ is now construed as malleable, open to, and shaped by, social interactions—shaping sociality as it is itself reshaped by it.

We should not diagnose neuroreductionism here.[44] It is not that human beings have become conceived of as mere puppets of their brains, far from it. Human beings are not thought of as identical with their brains, or reduced to their brains or determined by their brains. Mental states have not disappeared; an interior psychological realm of thoughts, intentions, beliefs, and will is still accorded a role. Indeed, as we have seen, in the social neuroscience project, elements drawn from social psychology are frequently invoked. Not only are human beings understood as persons with mental states that are in constant transactions with their neurobiology; they also have the responsibility—presumably via an exercise of will—to nurture their mental capacities in the light of a knowledge of their brains. We, as persons, must adopt the mental states, the habits, the relationships and forms of life appropriate for this work on our brains—we must shape them as they shape us. And we must look after our brains, so that they may continue to look after us. As responsible subjects obliged to manage ourselves in the name of our own health, it seems now we have the added obligation of fulfilling our responsibilities to others by caring for our mutable, flexible, and valuable social brains.

Chapter Six

· · · · · · · · · · · · · · · ·

The Antisocial Brain

A difference between the brains of psychopaths and ordinary people has been identified. . . . Research by British scientists using advanced brain-scanning techniques has revealed that a critical connection between two regions of the brain appears to be abnormal in psychopaths. The findings are preliminary and do not show that brain abnormality causes psychopathy but they suggest a plausible biological explanation for the antisocial and amoral behavior that characterizes the condition. If the link to brain wiring could be proved it would raise the prospect of using brain scans to help in diagnosing psychopaths. . . . Professor Murphy said the findings offered the most compelling evidence yet that altered brain anatomy might be involved in psychopathy.

—**"Brains of Psychopaths Are Different**, British Researchers Find,"
  The Times (London), August 3, 2009

The exceptionally strong influence of early experience on brain architecture makes the early years of life a period of both great opportunity and great vulnerability for brain development. An early, growth-promoting environment, with adequate nutrients, free of toxins, and filled with social interactions with an attentive caregiver, prepares the architecture of the developing brain to function optimally in a healthy environment. Conversely, an adverse early environment, one that is inadequately supplied with nutrients, contains toxins, or is deprived of appropriate sensory, social, or emotional stimulation, results in faulty brain circuitry. Once established, a weak foundation can have detrimental effects on further brain development, even if a healthy environment is restored at a later age.

— **J. Cameron**, G. J. Duncan, N. A. Fox, et al., "The Timing
  and Quality of Early Experiences Combine to Shape Brain
  Architecture," 2007[1]

The headline on *Sky News* on April 26, 2010, read: "Murdered Gangster's Brain Donated to Science: Scientists Have Been Given the Chance to Get Inside the Mind of One of Australia's Most Notorious Gangsters."[2] It was reporting the fact that Roberta Williams had given "experts" permission to examine the brain of her husband, Carl Williams, who had been murdered in Melbourne's maximum security Barwon Prison less than three years into a thirty-five year sentence. "I believe it's to help with research," Mrs. Williams is quoted as saying, "and might explain why guys like Carl do the violent things they do."

A month earlier, in March 2010, *Nature* carried a news feature titled "Science in Court: Head Case" (Hughes 2010). Virginia Hughes reported on the brain imaging—using fMRI—of Brian Dugan, who was serving two life sentences in DuPage County Jail for two murders committed in the 1980s, and was facing the prospect of the death penalty for having confessed to an earlier killing. Dugan had been taken to Chicago's Northwestern Memorial Hospital "to meet one of the few people who might help him to avoid that fate: Kent Kiehl, a neuroscientist at the University of New Mexico," Hughes wrote. "Kiehl has been amassing data on men such as Dugan for 16 years. Their crimes are often impulsive, violent, committed in cold blood and recalled without the slightest twinge of remorse. They are psychopaths, and they are estimated to make up as much as 1% of the adult male population, and 25% of male prisoners." And Hughes went on to tell her readers that Kiehl had, by that time, used fMRI to scan more than one thousand inmates, often using a mobile scanner that he took to the prisons. Kiehl's studies, she reported, show that "the brains of psychopaths tend to show distinct defects in the paralimbic system, a network of brain regions important for memory and regulating emotion"—work that he believes will "eliminate the stigma against psychopaths and find them treatments so they can stop committing crimes" (Hughes 2010, 340).

Kiehl is among a number of neuroscientists who believe that advances in genomics and brain imaging are, at last, enabling researchers to detect anomalies in the specific brain regions that are responsible for the pattern of behavior shown by violent and impulsive individuals.[3] For most social scientists, these claims are just the latest episodes in a history of biological criminology dating from at least the early nineteenth century, and indeed are coextensive with the history of modern psychiatry—a history of attempts to find the roots of criminality in the criminal body itself, replete with bold claims, misplaced fears, false hopes, and sometimes tragic failures. It is, however, not just sociologists and historians who are doubtful about claims made by brain researchers for the significance of their work for the criminal justice system. ABC in Australia reported on the Carl Williams case under the headline "Carl Williams' Brain Unlikely to Reveal Secrets" and quoted Professor Brian Dean from the Mental Health Research Institute in Melbourne on the "contentious history" of examining brains and the lack of good evidence linking violence in men to a change within a particular brain structure, while commenting more optimistically about the role of brain research in depression and schizophrenia.[4]

Virginia Hughes in *Nature* (Hughes 2010) notes that some researchers, such as Ruben Gur of the Brain Behavior Laboratory at the University of Pennsylvania, who has testified in court on the basis of structural brain scans, believe that, unlike psychiatric "soft data," "the brain scan doesn't lie" where structural brain anomalies are concerned. They remain skeptical, however, about functional imaging, which they feel is not reliable enough for legal settings. Similarly, Helen Mayberg, a neurologist from Emory School of Medicine who uses brain imaging in studies of depression, has testified in several court cases *against* the use of brain scans: "It is a dangerous distortion of science that sets dangerous precedents for the field."[5] Among the reasons for criticisms are the fact that, as discussed in chapter 2, fMRI measures BOLD responses that are difficult to interpret, is based on small samples, usually compares differences between groups and is difficult to extrapolate to single cases, and tells us nothing about the mental state of an individual at the time when the crimes were committed, which may have been many years before the scan was performed.

In the Dugan case, as in the other cases to which we will refer in this chapter, the court was not convinced by the evidence from neurobiology. But Kiehl is not alone in believing that neuroscientific evidence can and should play an increasingly important role in the criminal justice system. In this chapter we will examine these arguments that claim that human antisocial behavior—notably impulsivity, aggression, and related forms of criminal conduct—have neurobiological roots. We argue that while neurobiological evidence from genomics or functional brain imaging is likely to have limited traction in the criminal courtroom itself, a new diagram is nonetheless emerging in the criminal justice system as it encounters developments in the neurosciences. This does not entail a challenge to doctrines of free will or an exculpatory argument that "my brain made me do it," as some have suggested. Rather it is developing around the themes of susceptibility, prediction, and precaution that have come to infuse many aspects of criminal justice systems as they have come to focus on questions of risk—risk assessment, risk management, and risk reduction (Feeley and Simon 1992, 1994; O'Malley 1996, 1998, 2000; Ericson and Haggerty 1997; Pratt 2000; Baker and Simon 2002).

This is an element in a wider shift in strategies and policies in many societies toward preemption of unwanted incidents, including criminal offences. In criminal justice systems, especially in the English-speaking world, this has focused particularly on certain types of offending behavior involving violence, impulsivity, repeat offences, danger, sexual predation, and the like.[6] We term the new diagram of control taking shape 'screen and intervene,' and in this chapter we trace out this new configuration and consider some of the consequences. Whatever the virtue of arguments for preventive intervention in relation to health, in the case of neurobiologically based strategies for pre-

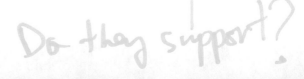

Do they support?

we can't map the mind without disturbing it, fervour

ventive crime control, we argue that policies of screen and intervene—which have a family resemblance to other strategies for 'governing the future' in the name of security—are likely to contribute to a further widening of the net of the apparatus of control to the 'precriminal' or 'predelinquent,' and to play a part in the new ways in which subjects and subjectivities are governed in the name of freedom in an age of insecurity.

Embodied Criminals

In 1890, the same year that he published his classic study *The Criminal* (Ellis 1890a), extolling the European tradition of scientific analysis of the bodies and brains of criminals to an English audience, Havelock Ellis summarized his analysis in an article published in the *Journal of Mental Science* (Ellis 1890b). He traced the pathway from the physiognomic epigrams of Homer and Aristotle to Galen and Lavater, via a brief mention of Johann Christian August Grohmann, whose 1820 work he quoted: "'I have often been impressed in criminals, and especially in those of defective development, by the prominent ears, the shape of the cranium, the projecting cheek-bones, the large lower jaws, the deeply-placed eyes, the shifty, animal-like gaze'" (Ellis 1890b, 3). He then arrived at Franz Joseph Gall, that "unquestionable scientific genius" who was the founder of criminal anthropology and for whom "the varying development of the brain was the cause of the divergent mental and moral qualities of the individual." For Ellis, it was Gall's "unceasing study of all the varieties of the brain and of the living head that he could find . . . throughout Europe, in lunatic asylums and in prisons, as well as among the ordinary population" that led Gall to foresee "the extent of the applications of the science that he was opening up to medicine and to law, to morality and to education" (ibid).

Gall had indeed studied the criminal brain, and he provided a full account of the "carnivorous instinct" or the "disposition to murder"—which was one of the twenty-seven faculties that he hypothesized were linked to different regions of the human brain (Gall [1822] 1835, 4:50–119).[7] This was based on assessments of the differences in skull shape between carnivorous and non-carnivorous animals, and on observations of the skulls of a parricide and a murderer: "[T]here was, in each, a prominence strongly swelling out immediately over the external opening of the ear" (Gall [1822] 1835, 4:50, quoted from R. Young 1970, 39). This was supported by further inspections of the heads of animals, as well as by an examination of the skulls of many famous criminals and the busts and portraits of others. These observations, together with much doleful evidence of cruel, sadistic, murderous, and tyrannical behavior of humans led Gall to the conclusion that this propensity to violence against members of his own species was innate in man and particularly prominent in certain individuals—and that the outward mark of their cruel and bloody character could be seen in this particular prominence on their skulls. "There

self-fulfilling prophesies!

can be no question," he said, "of culpability or of justice in the severe sense; the question is of the necessity of society preventing crime. The measure of culpability, and the measure of punishment cannot be determined by a study of the illegal act, but only by a study of the individual committing it" (Gall [1822] 1835, quoted from Ellis 1901, 31).

One of Gall's earliest followers in the exploration of the criminal brain was Moriz Benedikt, whose text of 1879, *Anatomische Studien an Verbrecher-gehirnen für Anthropologen, Mediciner, Juristen und Psychologen bearbeitet*, was translated into English in 1881 as *Anatomical Studies upon Brains of Criminals*. The moral sense, he believed, was located in parts of the occipital lobes that were absent in apes and criminals (Verplaetse 2004).[8] Benedikt, according to Verplaetse, was a freethinker and committed materialist—it was in the brain, not in the soul, that one should look for human moral sensibility. So it was with Paul Broca: the Society for Mutual Autopsy, whose foundation in Paris in 1876 he inspired, was organized around specifically atheist values— notably the belief that there was no soul, only brain (Hecht 2003).[9] As we saw in chapter 2, while Broca may now be remembered for his identification of the brain area that still bears his name, he was also a keen measurer of brain size, in particular seeking the differences between races and between men and women. He was also rather interested in the brains of lunatics and criminals. It might have been assumed that criminals, on the whole considered of low intellect, would have smaller than average brains. The reverse turned out to be the case—the criminal brain was large—something that Broca attributed to the manner of death itself in those whose brains were available to him—sudden death by execution, he reasoned, diminished the original deficit in brain size.[10]

Ellis, in the paper we have just discussed (1890b), selects Broca as the founder of the modern science of anthropology and credits him with initiating the special science of criminal anthropology. The next significant advance, in his view, was the conception introduced by the French psychiatrist Bénédict Morel: that the criminal was one of the "types," or forms, taken by degeneration (Morel 1857). This idea was developed by Prosper Despine in his *Psychologie naturelle* of 1868 and led to Henry Maudsley's comment in *Responsibility in Mental Disease* (1872) that "[i]t is a matter of observation that this criminal class [instinctive criminals] constitutes a degenerate or morbid variety of mankind, marked by peculiarly low physical and mental characteristics" (Maudsley 1872, 30). He considered such criminals morally insane individuals who should be treated in the same way as the intellectually insane. "Circumstances combined," Ellis writes, "to render possible, for the first time, the complete scientific treatment of the criminal man as a human variety" (1890b, 9).

In 1872, inspired by Gall, Benedikt, and Broca, and after early attempts to understand the insane by the use of various instruments of measurement, Cesare Lombroso published the results of some investigations he had made on prisoners at Padua in *Revista delle Discipline Carcerarie*. And by 1876, he published the first volume of *L'uomo delinquente*, in which he "perceived the criminal as, anatomically and physiologically, an organic anomaly" (Ellis 1890b, 9). Ellis acknowledged the many criticisms of this work, notably Lombroso's somewhat indiscriminate collection of facts, but then opined that Lombroso "was, as he has himself expressed it, the pollen-conveying insect, and the new science which he fecundated has grown with extraordinary rapidity. A continuous stream of studies from books of the most comprehensive character down to investigations into minute points of criminal anatomy or physiology is constantly pouring forth" (Ellis 1890b, 10). Ellis points to work in criminal anthropology not only in Italy, but in France, Germany, Portugal, Spain, Poland, and Russia, and to the various international congresses on criminal anthropology. "In Great Britain alone," he laments, "during the last fifteen years there is no scientific work in criminal anthropology to be recorded. . . . Fifty years ago English men sought to distinguish themselves by the invention of patent improved tread mills and similar now antiquated devices to benefit the criminal. We began zealously with the therapeutics of crime; it is now time to study the criminal's symptomatology, his diagnosis, his pathology, and it is scarcely possible to imagine that in these studies England will long continue to lag so far behind the rest of the civilized world" (Ellis 1890b, 14–15).

For in Great Britain, as Ellis noted with regret, despite Maudsley's initial commitment to criminal anthropology, the uptake of these ideas was slow. It is true that at the Exeter meeting of the British Association in 1869, Dr. G. Wilson had read a paper on "The Moral Imbecility of Habitual Criminals as Exemplified by Cranial Measurements": he measured 464 heads of criminals and found that habitual thieves presented well-marked signs of insufficient cranial development, especially anteriorly. And soon after came the study of more than five thousand prisoners that was published in the *Journal of Mental Science* in 1870 by J. Bruce Thomson, resident surgeon to the General Prison for Scotland at Perth, pointing to the large numbers who showed morbid appearance at postmortem. This was followed by a series of papers on "The Morbid Psychology of Criminals" in the *Journal of Mental Science* by David Nicholson (Nicolson 1873a, 1873b, 1874, 1875a, 1875b). But it was Italy that had become the home of criminal anthropology and of all the sciences connected with crime and the criminal. We are all too familiar with the charts of degenerate ears, jaws, and noses, along with the measurements of skulls and other physical characteristics that were created by Lombroso and his followers, initially understood in terms of atavism and later recast in the language of degeneracy. It was partly as a result of the proselytizing by Havelock Ellis that the work of the criminal anthropologists—notably of the Italians Lombroso, Ferri, and Garafalo—became much better known in the English-speaking world in the 1890s.[11]

In the United States, criminal anthropology seemed, at first, to have a hopeful future. At least a dozen books on this topic were published by American authors between 1890 and 1910, repeating Lombroso's catalogue of physical deformities and anomalies of behavior—laziness, frivolity, use of argot, tendency to adorn body and surroundings with hieroglyphics, moral insensibility, and emotional instability (Rafter 1992). Although some expressed misgivings, in America, Britain, and continental Europe, the search for a science of man and a knowledge of the criminal, the delinquent, and the degenerate based on the objectivity of numbers—height, weight, relative proportions of various body parts, and the like—generated an extraordinary flurry of activity. Heights of criminals were compared with those of other sections of society to determine whether, as some thought criminals were inferior in stature: for example, "Mr. Jacobs, in a paper communicated to the *Journal of the Anthropological Institute*," determined that the average stature of the London criminal, at 64.3 inches, was some five inches less than that of the well-to-do classes, and very nearly the same as that of the "East End Jew" (Morrison 1891, 179). Many focused on the question of impaired intelligence, which encompassed not only intellectual feebleness but moral weakness, and argued that criminals suffered from a defective moral sense—arguments that would feed into a growing concern with the problems of the feebleminded and the need for eugenic strategies to counter their threat to the race and the nation.

But while many in the English-speaking world were initially impressed with the claims of criminal anthropology, in Great Britain, most medical professionals interested in the specific links between insanity and crime expressed their doubts, as for example, did the influential Scottish psychiatrist Thomas Clouston (Clouston 1894). This is not to say that research into 'pathological' brains was nonexistent. We have already encountered James Crichton-Browne, one of the founding editors of *Brain*, and medical superintendent of the West Riding County Asylum in Yorkshire, where David Ferrier carried out his experiments on cerebral localization. As Cathy Gere has pointed out (Gere 2003), Crichton-Browne sought to turn his institution into a center for neurological research, appointing W. Lloyd Andriezen as his pathologist in 1872: Andriezen's study of more than one hundred human brains was published in *Brain* in 1894 (Andriezen 1894). Yet, as David Garland has argued, the psychiatric and medico-legal framework within which early criminological science developed in Great Britain differed from the Lombrosian tradition: it "was a therapeutically oriented discipline based upon a classification system of psychiatric disorders [within which] there were a variety of conditions which criminals were typically said to exhibit—insanity, moral insanity, degeneracy, feeble-mindedness, etc. But generally speaking, *the* criminal was not conceived as *a* psychological type" (Garland 1988, 3).[12]

Further, the medico-legal approach to the criminal was institutionally based within the prison establishment and wary of challenging the basic doctrines of legal responsibility as they were elaborated in the courts. Indeed, by the 1890s when Havelock Ellis was proselytizing, most of the leading figures—Needham, Hack Tuke, Nicolson, Maudsley—were explicitly criticizing the excesses of criminal anthropology. Maudsley, having initially been attracted by criminal anthropology, now became one of its most influential critics, arguing "first, there is no general criminal constitution, predisposing to and, as it were, excusing crime; second, . . . there are no theories of criminal anthropology so well-grounded and exact as to justify their introduction into a revised criminal law" (in his "Remarks on Crime and Criminals," quoted in Garland 1988, 4). Or, as David Nicolson put it in his presidential address to the Medico-Psychological Association in 1895:

> Writers give us a copious and precise history of the anatomical con- figuration, the physiological eccentricities, the complexion, the shapes of the ear and nose, the tattoo marks, etc., in certain criminals. We get a striking and elaborated account of their numerous fearful crimes, of their atrocious mental peculiarities and hideous moral obliquities. This analytical and biological process is applied by those who call them- selves criminalists to a comparatively small group of criminals; and by implication, and even more directly, it is made applicable to criminals generally. The whole picture is by some writers exaggerated to distor- tion as regards even the few, and it is in its main features so spurious and unfair as regards the many that it becomes impossible to regard the conclusions or assumptions to be either authentic or authoritative. (Nicolson 1895, 579)

Nicolson proposed a complex classification of types of crime and crimi- nals, linked to an assessment of the most appropriate methods of dealing with each, arguing that if the criminal brain was significant, this was only where education and training had prevented the normal development of the brain.[13] In this, his approach was typical of British criminology as it was to develop across the first three decades of the twentieth century: it was to be largely an exercise in individual classification and administrative management of the criminal around the different sites of punishment and reform, using forms of psychiatric and psychological knowledge that undoubtedly made reference to heredity, but moved away from a focus on the exterior of the body and the organic structures of the brain toward an internal mental realm—it was the criminal mind, as formed by domestic circumstances, education, training, and its moral milieu, not the degenerate criminal brain, that was at issue.

In any event, in the decades at the turn of the century, many of those with skills in neurosurgery and access to brains of criminals cast doubt on the evi- dence suggesting that there were specific anomalies in the brains of criminals. In Canada, as early as 1882, William Osler rejected Benedikt's argument: he examined the brains of two executed murderers, compared them with thirty- four controls, and found no differences (Osler 1882a, 1882b). In words that may have some resonance today, he wrote that "as society is at present consti- tuted it cannot afford to have a class of criminal automata and to have every rascal pleading faulty grey matter in extenuation of some crime" (quoted from Couldwell, Feindel, and Rovit 2004, 711). And in the United States, the anatomist Franklin Paine Mall, writing in 1909, considered that racial comparison of brains spoke volumes about the misleading nature of such claims: "It certainly would be important if it could be shown that the complexity of the gyri and sulci of the brain varied with the intelligence of the individual, that of genius being the most complex, but the facts do not bear this out, and such statements are only misleading. I may be permitted to add that brains rich in gyri and sulci, of the Gauss type, are by no means rare in the American negro" (quoted from Finger 1994, 308).

By the end of the first decade of the twentieth century, these attempts to ground differences in intellect and morality in variations in the gross features of the brain—weight, complexity of gyri and sulci and the like—had more or less run their course, at least in the English-speaking world, as evidence failed to confirm earlier reports, and pointed to some problematic implications. Nonetheless, criminal anthropology continued to flourish into the twentieth century in a number of countries in continental Europe—one of which was Germany. As Michael Hagner has shown, the work of neurologists and psychiatrists on the criminal brain was to become a significant element within what he terms "the cultivation of the cortex" in German neuroanatomy (Hagner 2001).[14] From our point of view, two essential elements are consolidated in the work of Meynert, Flechsig, Forel, and Oscar and Cécile Vogt, so well analyzed by Hagner, and to which arguments about the antisocial brain will return almost a century later. The first is the conception of the brain as a complex organ that can be anatomized into multiple functional areas, and where anomalies in one or more areas have specific and determinate consequences for normal and pathological conduct. The second is the belief that important sociopolitical issues are at stake in an understanding of the structures and functions of the human brain, together with all that contributes to their healthy or unhealthy development. As August Forel put it in 1907: "As a science of the human in man, neurobiology forms the basis of the object of the highest human knowledge which can be reached in the future. . . . It will . . . also have to provide the correct scientific basis for sociology (and for mental hygiene, on which a true sociology must be based)" (quoted from Hagner 2001, 553).

A true sociology based on the brain—these words have a renewed resonance a century later. Whatever their many differences, Meynert, Flechsig, Forel, and the Vogts mapped out a picture of the brain as an organ differentiated into distinct regions, sometimes thought of along a temporal axis running from the oldest, primitive parts wherein were located the basic drives for survival and reproduction, to the newest, higher regions of the cortex, where impulses from the sense organs were associated with one another. It was in the cortex, it seemed, where complexes developed over time as a result of experience and the progressive myelination of the nerve fibers. It was the cortex that coordinated these into images of the world and into ideas, and in the cortex where the civilizing and rational controls over those lower impulses were located. The anatomists and pathologists waged an interdisciplinary battle with the developing experimental psychologists, namely, Wilhelm Wundt and his followers. Oscar Vogt termed their position one of "neurobiology" and linked their arguments with those from what he termed "neuropsychology." At the same time, the Vogts, working with Korbinian Brodmann, undertook "the most complete structural and functional mapping of the cortex possible, and integrated neurophysiological studies and brain surgery into their research in order to find a physiological correlate for every element of consciousness" (Hagner 2001, 554).

The Vogts examined many brains of eminent men in the 1920s and 1930s, including those of Lenin, Forel, and Ferdinand Tonnies,[15] but they also collected the brains of criminals and thought they had observed "a weak development of the third cortical layer" in some feebleminded criminals. They linked this analysis to programmatic proposals for both positive eugenics ("the deliberate marriage of men and women with similar talents") and the prevention of the "pre-destined development" of some individuals into criminals (Hagner 2001, 556–57). The Vogts proposed and established the Kaiser Wilhelm Institute for Brain Research in 1914, which increasingly focused on the genetic underpinnings of brain pathology. But in 1933, Oscar Vogt was dismissed from his position as director, and according to Hagner, the Vogts themselves turned away from their ideas of cultivation through breeding and did not advocate forced sterilization or euthanasia. However, those themes were enthusiastically pursued by Oscar Vogt's successor as director of the institute, Hugo Spatz, and his colleague Julius Hallervorden. Spatz and Hallervorden received many brains from the Aktion T4 extermination campaign. However, despite the many measures enacted by the Nazis to target, sterilize, incarcerate, and murder habitual criminals and others thought to be psychopaths or asocial, the focus of these Nazi brain explorers was not directly on the criminal brain, but on the brains of those deemed feebleminded or mentally ill.

Inside the Living Brain

In June 1989, the Max Planck Society announced plans to give a respectful burial to some ten thousand glass slides with sections of brain tissue that had been collected by Hallervorden; the slides were cremated and buried in Munich a year later (Dickman 1989). There was little interest in research on the criminal brain in the period following the end of the Second World War, no doubt, in part as a reaction to such research on those murdered in the Nazi period. Yet four decades later, at the same time as the fate of the Hallervorden collection was being debated in Germany, the relation between the brain and criminality was to be reopened within another style of thought, in a very different sociopolitical climate, with other technical means, and in living subjects. In 1987, the *British Journal of Psychiatry* published a study by Nora

Volkow[16] and Lawrence Tancredi from the University of Texas of four individuals "with a history of repetitive purposeless violent behavior," using three neuroimaging techniques: EEG, CT scan, and PET. They reported the following results: "Three patients showed spiking activity in left temporal regions, and two showed CT scan abnormalities characterized by generalized cortical atrophy. The PET scans for the four cases showed evidence of blood flow and metabolic abnormalities in the left temporal lobe. Two patients also had derangement in the frontal cortex. The patients showing the largest defects with the PET scans were those whose CT scans were reported as normal" (Volkow and Tancredi 1987, 668).

A year earlier, when Thomas Grisso, a leading forensic psychiatrist at the University of Massachusetts Medical School, had reviewed the tools available to those conducting assessments for forensic purposes, the repertoire seemed limited to traditional symptom description and diagnosis, together with history-taking about head injuries, brain insults, and epilepsy, and psychological tests of intelligence and personality, including some specifically developed for forensic use (Grisso 1986). In this context, and given the many criticisms of the validity and forensic utility of the psychological tests, it was perhaps not surprising that Volkow and Tancredi's study, probably the first to use such neuroimaging techniques in relation to violent individuals, was hailed by Daniel Martell, an American forensic neuropsychologist,[17] as foreshadowing a larger role for the neurobehavioral sciences in criminal law: "[T]he particular appeal of neuropsychological evidence in the criminal context [is] the expert's ability to bring quantified, normative data on brain-behavior relationships to bear in support of what have traditionally been professional opinions based on mental status examinations and clinical interview techniques." (Martell 1992b, 315).

Martell noted that such imaging studies nonetheless raised the thorny issue of how to relate specific brain abnormalities to individual behavior in particular cases: "while neuroimaging technologies can provide powerful evidence of the presence, type, location, size, and angle of brain lesions, they provide the trier of fact with little information about the behavioral sequelae of a given lesion" (Martell 1992b, 324). Hence, he suggested, it is precisely in this area that the neuropsychologist, using a battery of other tests, is best prepared to provide evidence to inform the decision-making process in the court: the "very best neuropsychological assessments represent the highest level of contemporary objective psychological evaluation [and, with] the highest standards of thoroughness, accuracy, and care in the administration, scoring, interpretation, and reporting of findings . . . speak adequately to the presence or absence of mental disease or defect, and its relationship to specific criminal-legal referral issues" (ibid., 318).

The early 1990s saw a series of papers from Martell and others, arguing for a new neurobiology of violent crime and pointing in particular to the results of research using EEGs and brain scanning of violent offenders (Kahn 1992; Martell 1992a; Volavka, Martell, and Convit 1992; Pincus 1993; Rojas-Burke

1993; Dolan 1994). The direction of the brain scanning research itself, here as elsewhere, was toward localization. Within the style of thought that had taken shape in brain imaging, if there was something anomalous in the brains of violent offenders, it had to be locatable in a specific brain region. But where?

In 1990, Antonio Damasio and colleagues at the University of Iowa Hospitals and Clinics published a paper drawing on the clinical observation that adults with previously normal personalities who sustain damage to the ventromedial frontal cortex develop defects in decision making and planning revealed in abnormal social conduct (Damasio, Tranel, and Damasio 1990). They suggested that this might be a result of inability to activate the somatic states related to previous experiences of punishment and reward, hence depriving the individual of the automatic devices regulating conduct; they claimed to find confirmation in the abnormal autonomic responses of patients with frontal lobe damage. The belief in a link between regional (frontal) brain abnormalities and problems of impulse control seemed to be firming up.

In 1991, Joseph Tonkonogy published the results of head CT and MRI scanning studies that had been undertaken since 1987 at the Worcester State Hospital. Reporting on five patients with violent behavior and local lesions in the anterior-inferior temporal lobe, he argued that "violent behavior in psychiatric patients may manifest as a primary symptom of tissue loss in the areas adjacent to the amygdaloid nuclei or in the amygdala itself, especially when combined with a history of seizure disorders. . . . To the best of our knowledge, this is the first report of local tissue loss in the anterior temporal lobe in the group of psychiatric patients with violent behavior"—although the author noted that such patients formed only a relatively small percentage of all violent psychiatric patients in the Worcester State Hospital (Tonkonogy 1991, 189–90, 193).

As this work gathered pace, one of the first to reflect on its more general implications was Adrian Raine, a former airline accountant turned experimental psychologist, who had worked for a spell in high security prisons prior to his academic career, first at Nottingham University and then at the University of Southern California. In 1993, Raine published *The Psychopathology of Crime: Criminal Behavior as a Clinical Disorder*. He reviewed the evolutionary, genetic, and neurobiological evidence and sought to rehabilitate biological investigations of criminal behavior. His aim was to free them from their associations with racism, suggesting that many of the critics were attacking a straw man: he argued that there were good grounds for considering at least some types of offending behavior as a clinical disorder. What was still lacking, he contended, was a concerted effort to integrate biological and social theories with empirical evidence, in particular to resolve currently tangled issues concerning the relations of causality between the biological, psychological, and social factors involved (Raine 1993, 288).

Across the 1990s, the apparent insights to be gained from using the technology of brain imaging, which had now become relatively easy for researchers to access, encouraged many other investigators to use PET and fMRI to study the brains of individuals with a history of violence or aggression, and indeed to extend the groups studied to those diagnosed with personality disorder (Goyer, Cohen, Andreason, et al. 1994). Raine himself continued his studies of the brains of murderers. He argued that "seriously violent offenders pleading not guilty by reason of insanity or incompetent to stand trial are characterized by prefrontal dysfunction" and suggested that "deficits localized to the prefrontal cortex may be related to violence in a selected group of offenders," although he accepted that the group that he studied—those accused of murder or manslaughter undergoing examination in relation to a potential plea of not guilty by reason of insanity (NGRI) or incompetence to stand trial—might not be typical of violent offenders in the community (Raine, Buchsbaum, Stanley, et al. 1994). By 1997, he and his colleagues were arguing that "[t]he advent of brain imaging research has recently made it possible for the first time to directly assess brain functioning in violent individuals," and their research followed up the suggestion that frontal brain regions were implicated in addition to the temporal cortex (Raine, Buchsbaum, and LaCasse 1997, 496). Using a larger sample group of forty-one people (thirty-nine men and two women) accused of murder or manslaughter who were undergoing examination in relation to a potential plea of NGRI or incompetence to stand trial—a group the researchers designated somewhat misleadingly as "murderers"—Raine and his colleagues concluded that significant brain abnormalities were present. In their discussion, the researchers conceded that this was both a relatively specific subgroup of offenders, and also one that was itself clinically heterogeneous. They emphasized that the findings did not demonstrate that violence was caused by biology alone, nor that murderers pleading NGRI were not responsible for their actions, nor that PET could be used diagnostically, nor that their findings "could speak to the issue of the cause (genetic or environmental) of the brain dysfunction," nor could they be taken as specific to violence in particular, since a nonviolent criminal control group was not used (Raine, Buchsbaum, and LaCasse 1997, 505). But their abstract stated their findings in less cautious terms: "Murderers were characterized by reduced glucose metabolism in the prefrontal cortex, superior parietal gyrus, left angular gyrus, and the corpus callosum, while abnormal asymmetries of activity (left hemisphere lower than right) were also found in the amygdala, thalamus, and medialtemporal lobe. These preliminary findings provide initial indications of a network of abnormal cortical and subcortical brain processes that may predispose to violence in murderers pleading NGRI" (ibid., 495).

In 2000, Raine and his team published their much-cited study claiming to find reduced prefrontal gray matter volume and reduced autonomic activity in people diagnosed with antisocial personality disorder (Raine, Lencz, Bihrle, et al. 2000). By this time, they were able to refer to more than a dozen studies that had used different brain imaging techniques to explore anomalies in the brains of alcoholics and other substance abusers, those diagnosed with socio-

pathic behavior or antisocial personality disorder with convictions for violence, as well as those diagnosed with mood disorders or schizophrenia, and of
course their own studies of "murderers." Their conclusion was that prefrontal
structural deficits may underlie the low arousal, poor fear conditioning, lack of
conscience, and decision-making deficits in antisocial, psychopathic behavior.

Some forensic psychiatrists and neurobiologists expressed caution about
the claims that frontal lobe dysfunction was strongly associated with, or predictive of, violent behavior (Brower and Price 2001). But despite such doubts,
in the first decade of the twenty-first century, an expanding research program in what Raine termed "neurocriminology" took shape in many labs and
in many countries, using brain imaging techniques to seek the neurological
basis of psychopathy, antisocial behavior, impulsive violence, and aggression
(Blair 2004, 2005; Blair, Newman, Mitchell, et al. 2006; Rogers, Viding, Blair,
et al. 2006; Crowe and Blair 2008). Biological criminology was no longer, it
seemed, linked inseparably with the pseudoscience of criminal anthropology and its focus on the exterior of the bodies of those forming a criminal
class. The gaze of the neurocriminologist could now plunge into the interior
of the living criminal brain and discover therein the roots of the violence
and failures of impulse control in each dangerous individual. But with what
consequences? For if the roots of such socially undesirable behavior were
neural, was that exculpatory or inculpatory? Was it grounds for mitigating
the responsibility and the culpability of offenders because their capacities to
constrain their conduct were diminished? Or did it increase the need for their
incarceration and exclusion from society on the grounds of their incorrigibil-
ity? On these crucial matters, the leading researchers remained ambivalent.[18]

Neurolaw?

What, then, would be the impact of this research in neurocriminology within
the legal system? When the *neuro-* prefix came to law in the 1990s, neurolaw
was initially focused on the question of traumatic brain injury and the role of
neuropsychologists as expert witnesses in the increasing number of cases involving claims for damages (Taylor, Harp, and Elliott 1991; Taylor 1995, 1996).
But by 2004, when the Dana Foundation convened a meeting of twenty-six
neuroscientists for a conference on law and neuroscience, the focus of the debate was firmly on criminal law and the relations of the brain to the criminal
justice system (Garland 2004). The commissioned papers marked out what
was to become the territory for neuroethical discussions in this area for the
next decade.

What, if any, were the implications for the legal doctrines of free will and
responsibility? asked Michael Gazzaniga and Megan Steven. Stephen Morse,
writing under the title "New Neuroscience, Old Problems," rehearsed his
argument that legal systems operated with their own "folk psychological"

conception of human beings as intentional persons possessing mental states that were the causes of their actions, and hence could be held responsible for them, and that neuroscientific evidence was not likely to pose a radical challenge to this view. And Henry (Hank) Greeley discussed issues of "Prediction, Litigation, Privacy and Property," focusing, presciently, on the likelihood that the real impact of neuroscience would probably not be in the courtroom, where legal reasoning and evidential rules were likely to retain hegemony, but would lie outside the courtroom, in the pretrial and posttrial practices of the criminal justice system. This overall verdict was reiterated a couple of years after that event by the editor of the DANA volume: "Much discussion has focused on 'ultimate' issues such as free will, determinism (genetic or mechanistic), and their effect on whether the concept of criminal culpability will be undone by new scientific discoveries. This seems unlikely, at least in the near future" (Garland and Frankel 2006, 103).

This has not deterred those who continue to speculate about these themes. However, as predicted by most of the contributors to the DANA debate, the courts or juries, at least in adversarial criminal justice systems, have proved resistant to changing their own conceptions of responsibility, or their ways of attributing guilt, in the light of neurobiological evidence (Rose 2000, 2007c, 2008).[19] They do not seem to find evidence from the neurosciences about factors predisposing to a criminal act any more relevant to legal notions of responsibility than arguments or evidence from the other 'positive sciences' of human conduct—psychiatry, psychology, sociology—despite the frequent suggestions by commentators that the rhetoric of objectivity surrounding brain scans and other neurobiological evidence will radically enhance their plausibility. Anglo-American legal systems conceptualize their subjects— with specific exceptions—as individuals with minds, or mental states, who intend the acts they commit, and who foresee their outcome to the extent that any reasonable person could so do (Morse 2004, 2006, 2008; Morse and Hoffman 2007).[20] If they are to mitigate this presupposition, they require compelling evidence that some state or event—an illness, sleepwalking, coercion from a third party—was directly and causally related to the commission of the act in question. Probabilistic arguments, to the effect that persons of type A, or with condition B, are in general more likely to commit act X, or fail to commit act Y, hold little or no sway in the process of determining guilt.[21]

Further, as we have seen in chapter 2, evidence from brain scans or other neurobiological tests does not 'speak for itself'—it must be spoken for. Images do not convince by means of their own intrinsic plausibility: interlocutors are required, and expert witnesses such as the clinicians who have requested and interpreted the brain scans of the accused must make the claim that the scan shows relevant anomalies.[22] As in other areas of criminal trials in adversarial systems, these claims will usually be contested by other expert witnesses, who argue that it shows no such thing. Judges advise juries to take into account many factors, including their own judgments as to what might be a reasonable and plausible account of the circumstances that led to the commission of the act in question. It is no surprise, therefore, to find that despite the repetitive discussion of this issue by philosophers and ethicists, there is no evidence, to date, of neurobiological testimony concerning *functional* abnormalities in the brain of the accused affecting the verdict in any recent murder trial in the United States or the United Kingdom (Rose 2007c, chap. 8).[23]

But this is not to say that neurobiological arguments about the neural basis of violent or antisocial conduct will have no traction in the criminal justice system or in crime control policy.[24] Initially, as might be expected, the effects are likely to appear in relation to those kinds of persons who have long been treated differently by the legal system—notably children and women,[25] but also those who are thought to be incapable of understanding the nature or meaning of their acts, and hence, in the criminal law at least, not capable of bearing responsibility. As far as children are concerned, the age of legal subjectivity for the purpose of attributing criminal responsibility has always been somewhat arbitrary, contested, and variable across jurisdictions. There is some evidence that neurobiological arguments may have an impact here, for example, in the much-discussed judgment of the U.S. Supreme Court in *Roper v. Simmons*, which barred capital punishment for juvenile offenders under the age of eighteen.[26] The Court was presented with an amicus brief drawn up by a group of New York lawyers at the request of a number of key American professional organizations, including the American Medical Association, the American Psychiatric Association, the American Society for Adolescent Psychiatry, and the National Mental Health Association (Haider 2005). Taking advice from a panel of scientists, the brief argued that juvenile brains are anatomically different and that the frontal lobes, "the part of the brain responsible for reasoning, impulse control, cost-benefit calculation and good judgment is not fully developed. This means that adolescents are inherently more prone to risk-taking behavior, and less capable of governing impulses, than adults. . . . These studies were presented to the Court as evidence that adolescents are biologically different" (Haider 2005, 371–72).

We should, however, be wary of reading too much into this case. The judgment itself did not explicitly refer to neurobiological evidence, or the adolescent brain, although it did refer to some sociological research on adolescence, and of course to commonsense knowledge about the impulsiveness of youth. The brief was submitted in the context of a fairly major mobilization of liberal U.S. academics, lawyers, and medical professionals against the death penalty, seeking any opportunity to mitigate or limit its use.[27] Nonetheless, despite the well-known cultural and historical specificity of the notion of adolescence, and the relative recency of the association of this period in life with risk taking and poor impulse control—which would rule out any underlying evolutionary changes in the rate of brain development—a number of U.S. researchers argue that such characteristics are a consequence of 'the developing brain' and the neural mechanisms responsible for processing emotions and incentives (Casey, Jones, and Hare 2008). It is doubtful whether such arguments will have much influence on questions of *culpability* in the courtroom, but they are being referred to in debates on the age of criminal responsibility. And in the United States, there are suggestions that they will increasingly impact on sentencing decisions for adolescents, especially those concerning the legitimacy of the death penalty and the sentence of life imprisonment (Maroney 2009, 2011).

Will neurobiological arguments reshape those other categories in which persons are released from the category of legal subject—for example, in relation to the question of unfitness to plead? It is too early to tell, but in any event, such cases are rare.[28] It is more likely that neuroscience, like the other positive sciences since the nineteenth century, will gain its traction outside the agonistic space of the courtroom. We will see its impact in criminal justice policy, notably policies aimed at prevention, in the identification of risky persons, and in the agonized debates over the existence and management of psychopathy. Further, we will see the effects of neuroscience in the investigatory process and in the procedures following arrest and leading up to trial—these are crucial, since only a small percentage of cases actually reach trial, as in many instances the suspect is either released or pleads guilty prior to trial, sometimes to a lesser charge.

And we will also see the impact after the verdict has been delivered, in the determination of sentence and the appropriate institutional disposition, in decisions as to release from prison, and of course, in the probation service. Indeed, in the DANA event mentioned earlier, one of the neuroscientists expressed concerns over just these issues: "[A significant fear], both in the behavioral genetics area and with [neuroscience], is that there are no [rules of evidence] that control the use of these kinds of technologies in the preformal stages of criminal processes. When it gets to the formality of sentencing, the cry will come up, but the ability of judges and prosecutors to make decisions about whether they're going to initiate charges, [whether] they're going to accept diversion [from criminal prosecution] for people, et cetera—using [neuroscience tests] that haven't been validated—is a serious risk that the technology poses" (quoted in Garland and Frankel 2006, 106).

But before turning to examine some areas where notions of the antisocial brain are indeed entering these arenas,[29] let us turn from evidence about brain development to examine the challenges posed by research in behavioral genomics.

The Genetics of Control

Adrian Raine's 1993 book, *The Psychopathology of Crime*, contained one chapter on the genetics of crime. It was largely concerned with defending those who were undertaking this work at the time against the charges of genetic reductionism, political bias, racism, or eugenics; from allegations that they rule out environmental factors; and from accusations that they believe that genetics is destiny. Nonetheless, he argued: "Overall, data from genetic studies [he is referring to twin and adoption studies] supports the prediction made by sociobiological theory that heritable influences in part underlie criminal behavior" (Raine 1993, 79). Brain scanning studies of violence and personality disorder in the 1990s made little reference to genetic research, and, when they did, generally noted that they were unable to determine whether the anomalies that they reported had a genetic base or resulted from environmental insults. But a new genomic approach to antisocial conduct was taking shape that would prove more compatible with the emerging image of the impulsive brain.

This is not the place to rehearse the depressing history of research on the genetics of criminality.[30] From about 1990, something began to change in the way in which the relations between genetics, neurobiology, and crime were construed. It was no longer a question of the search for a gene for crime, or even of a gene for aggression (even though this was the way that it was sometimes reported in the popular press). Indeed, these studies no longer sought to explain criminality in general, but focused on the impulsive behaviors that were thought to be involved in many criminal or antisocial acts, in particular, acts of aggression. These behaviors seemed to involve something that would once have been understood in terms of the will but that now fell under the rubric of *impaired impulse control*. And those searching for a genetic basis for such impairment now conceptualized the issue in neuromolecular terms— that is to say, they tried to identify variations in SNPs in the coding sequences that affected neuronal circuits in the brain, notably in the neurotransmitter system.

At the risk of repeating a familiar story, let us begin with the paper by Han Brunner and his team—based in Nijmegen in the Netherlands—published in *Science* in 1993 under the title "Abnormal Behavior Associated with a Point Mutation in the Structural Gene for Monoamine Oxidase A" (Brunner, Nelen, Breakefield, et al. 1993). For if there was one paper that can be said to mark a mutation in the thought style of neurocriminology, this was it. Like much genetic research at the time, Brunner's team was conducting a *kindred study*—that is to say, the researchers studied a problem that seemed to pass down a lineage. In this case, it was "a large kindred in which several males are affected by a syndrome of borderline mental retardation and exhibit abnormal behavior, including disturbed regulation of impulsive aggression" (ibid., 578). It seems that the kindred in question came to light because an unaffected maternal great-uncle, aware that there was a pattern of violent behavior among his male relatives, perhaps going back to the eighteenth-century, had constructed a detailed family history of what he thought was an inherited mental disability; this led a sister of one of the affected males to come to Brunner's genetic clinic seeking a genetic test to see if she was a carrier of this hereditary trait.[31]

Brunner and his team collected samples and histories from fourteen affected males and ten nonaffected individuals in the extended family; their first report, initially submitted to the *American Journal of Human Genetics* in 1992, focused largely on the question of X-linked mental retardation. It did, however, relate this to previous research on the neurotransmitter anomalies that might underlie disturbed aggression regulation, notably the work by a veteran of earlier struggles over the legitimacy of biological psychiatry, Herman van Praag (van Praag, Asnis, Kahn, et al. 1990; van Praag 1991; van Praag, Brown, Asnis, et al. 1991).[32] The later paper in *Science*, however, specifically made the connection between human impulsive aggression and low or absent monoamine oxidase A (MAOA); it placed this in the context of a dozen studies of aggressive behavior in animals and humans published over the previous decade that had implicated altered metabolism of neurotransmitters—serotonin, and to a lesser extent dopamine and noradrenaline—and suggested an as yet unknown genetic basis.

Brunner and his colleagues were aware that their kindred was rather special; all the affected males had a single point mutation, in which a C was replaced by a T at nucleotide 296 in the coding sequence, changing a glutamine (CAG) codon to a termination (TAG) codon; this resulted in complete deficiency (that is to say, total absence) of MAOA in affected males. MAOA is specifically implicated in the breakdown of serotonin, norepinephrine, and epinephrine, and therefore it seemed reasonable to postulate that a lack or deficiency in this enzyme could be related to an increase in available levels of these neurotransmitters in the central nervous system (and indeed in the other organs where it is present, such as the liver and gastrointestinal tract). Actually, although this was not a concern of Brunner and his group, one might wonder how, in the absence of such a key enzyme for neurotransmitter functioning, these unfortunate individuals survived at all—a testament to the redundancy within human neurotransmitter systems and its homeodynamic mechanisms.

Brunner and his colleagues reported that impulsive, rather than premeditated, aggression in humans had been linked to low concentrations of the breakdown product of serotonin, 5-HIAA,[33] in cerebrospinal fluid, and that "this may be caused by absent MAOA activity in these subjects"; they also noted that "inhibition of MAOA activity has not been reported to cause aggressive behavior in humans," although they speculated that "deficiencies throughout adult life might have different consequences" (Brunner, Nelen, Breakefield, et al. 1993, 579). They concluded their paper with the suggestion that would spark a whole research program. For while their kindred was marked by complete absence of MAOA, they pointed to the wide range of MAOA activity in the normal population: "one could ask whether aggressive behavior is confined to complete MAOA deficiency" (ibid., 590). Could this, then, be some kind of dimension, where levels of aggression were linked to variations in MAOA activity, rather than the complete absence of the enzyme?

This paper attracted great interest, not only among researchers but also in the international press, who dubbed the mutation that had been identified "the aggression gene," and in the legal profession—very soon after publication the defense lawyers in the murder trial of Stephen Mobley, in the U.S. state of Georgia, sought unsuccessfully to introduce it as evidence to mitigate their client's responsibility for the crime for which he had been convicted.[34] Brunner himself, clearly troubled by the implications, attempted to cast doubt on extrapolations from this rare kindred to more general genetic explanations of aggression and violence. By 1996, in a CIBA Foundation event on the genetics of crime, he was arguing that his research gave no support for the notion of an aggression gene, despite having been interpreted in this way by the popular press: "[T]he notion of an 'aggression gene' does not make sense, because it belies the fact that behavior should and does arise at the highest level of cortical organization, where individual genes are only distantly reflected in the anatomical structure, as well as in the various neurophysiological and biochemical functions of the brain. . . . [A]lthough a multitude of genes must be involved in shaping brain function, none of these genes by itself encodes behavior" (Brunner 1996, 16).

Brunner soon turned away from this area of work,[35] but this paper sparked a veritable industry of research, both in humans and in animal models, on the relations between aggression and monoamine oxidase, and more generally on neurotransmitter anomalies that might be correlated with aggression. As far as the animal studies are concerned, one of the leading labs was that of Klaus Miczek at Tufts University.[36] Miczek, who we mentioned briefly in chapter 3, had been working on animal models of aggression since the early 1970s, initially examining the effects of various drugs of abuse—cannabis, amphetamines, alcohol—and later focusing on the role of the monoamines. He was also a member of the National Academy of Sciences's somewhat controversial panel on "Understanding and Preventing Violence" from 1989 to 1992.[37] Throughout the 1990s, he and his research group at Tufts published a series of papers arguing for the importance of experimental ethological studies of aggression in animals, mostly in rodents, to pinpoint the most appropriate selection of pharmaceutical interventions in clinical situations to modify aggressive behavior in humans. They suggested that such studies would enable one to select precise receptor sites that could be targeted by drugs to reduce situationally specific forms of aggressive, defensive, and submissive behavior (e.g., Miczek, Weerts, Haney, et al. 1994).

Reviewing their own and other relevant research almost a decade later, Miczek and his team asserted: "Aggressive outbursts that result in harm and injury present a major problem for the public health and criminal justice systems, but there are no adequate treatment options" (Miczek, Fish, DeBold, et al. 2002, 434). It seemed that obstacles "at the level of social policy, institutional regulation, and scientific strategy in developing animal models continue to impede the development of specific anti-aggressive agents for emergency and long-term treatments" (ibid.). There were, they conceded, no such things as specifically pathological *mean genes*, as "aggressive behaviors serve mostly adaptive functions in all phyla, rather than being behavioral aberrations"; however, the authors nonetheless argued that "[b]ehavioral and neurobiological analyzes of aggression in various animal species contribute to understanding human violence and to the development of therapeutic interventions" (ibid., 435). Their aim was to identify the neural circuits underlying different types of aggression, so as to ascertain the particular serotonin, dopamine, and GABA receptor sites that could be targeted by psychopharmaceuticals in order to modulate human aggression.

The authors noted that a usual definition of aggression in humans entails an *intent* to do harm. However, like others working in this field, they were not detained for long by the problems that this invocation of a mental state might entail. They proceeded on the basis that certain behavioral patterns in animals can be used as models for aggression in humans, supporting their argument with reference to studies of higher primates; laboratory studies with humans, for example, those involving simulation of violent or aggressive encounters with fictional opponents; and the effects of various forms of reinforcement or drug treatment. They also pointed to neurobiological evidence concerning the effects of varying levels of neurotransmitter activity in humans, primates, and rodents, which seemed to demonstrate a homology between the neurobiological mechanisms in human and animal aggression. This enabled them to draw upon their rodent models to explore the specific neurotransmitters and receptors that seem to be entailed in either activation or reduction of different types of aggression, with a special interest in individual differences in escalated—that is, non–species typical—aggressive behavior. The premise of the argument was that this neuromolecular dissection of the bases of aggression might lead to the development and use of pharmacological interventions to control escalated aggression in humans.

This argument for a "pharmacotherapeutic approach for managing violent and aggressive [human] individuals" was made more explicit in a paper published two years later, in which the authors lamented the fact that

> [t]he development of selective anti-aggressive agents, once referred to as "serenics," [here they refer to the work of Berend Olivier who proposed this term (Olivier, Mos, Raghoebar et al., 1994)] has been sporadic and, at present, appears to be inactive. There are many reasons for this calamitous situation, including a lack of enthusiasm by regulatory agencies to approve medicines for a non-disease and the reluctance of targeting neurobiological sites for an ostensibly social dysfunction. The latter issue derives from a conceptual problem that divorces aggressive experiences from neurobiological mechanisms. (Miczek, Faccidomo, de Almeida, et al. 2004, 344)

Despite this "calamitous" situation and the reluctance of authorities to license pharmaceuticals whose main indication for use was behavioral control, the authors conclude: "It can be anticipated that currently developed tools for targeting the genes that code for specific subtypes of serotonin receptors will offer new therapeutic options for reducing aggressive behavior, and the 5-HT1B receptor appears to be a promising target" (ibid., 336).

By 2007, this time writing with Adrian Raine, Miczek noted with regret that the 2002 WHO *World Report on Violence and Health* completely ignored neuroscience; nonetheless, they argued, neurobiological research on aggressive behavior was emerging from its shameful past of eugenics and lobotomies, and that "the schism between these approaches [social and biological] promises to be overcome by advancing our knowledge of the molecular events through which social experiences sculpt future aggressive acts. Insights into the gene-environment interactions are critical for the way in which the criminal justice and the public health systems deal with aggression and violence. . . . [E]merging successes in behavioral and molecular biology of aggression research . . . may have important implications not only for diagnosis, prevention, and treatment but also for guidance of public and judicial policies" (Miczek, de Almeida, Kravitz, et al. 2007, 11,803).

Yet despite this apparent openness to social and environmental factors in the genesis of aggressive and violent conduct, they summed up their argument in their abstract, cited in chapter 3, suggesting that genetic and imaging research on aggression in fruit flies and feral rats leads to the hypothesis that "antisocial, violent, and psychopathic behavior may in part be attributable to impairments in some of the brain structures (dorsal and ventral PFC, amygdala, and angular gyrus) subserving moral cognition and emotion" (ibid.). Despite the fact that this argument would have most human scientists, and many neurobiologists working with animal models, reeling at the bizarre jumps of logic entailed in the connections between flies, rats, and humans, they concluded by arguing for a "*neuroethical* discussion" of the implications among representatives of the criminal justice system, psychiatry, neuroscience, and the social sciences (ibid., 11,805). One might be forgiven for thinking that critical scientific analysis of the conclusions drawn by these researchers might preclude the necessity for discussion of their supposed implications.

Despite the publicity around Han Brunner's study discussed earlier, the specific role of MAOA in aggression was not central to animal studies in the 1990s, with the exception of the work of Isabelle Seif and colleagues based at the CNRS in Paris. Referring directly to Brunner's work, Seif and her group created a line of transgenic mice with a deletion in the gene encoding MAOA, thus providing an animal model of MAOA deficiency. This increased serotonin concentrations in the brains of the mouse pups by up to ninefold, and also increased levels of norepinephrine concentrations in pup and adult brains. It appeared that (perhaps unsurprisingly given such an uplift in neurotransmitter levels at an early stage in brain development), behavioral abnormalities followed: "Adults manifested a distinct behavioral syndrome, including enhanced aggression in males," and the authors concluded that their study "supports the idea that the particularly aggressive behavior of the few known human males lacking MAOA is not fostered by an unusual genetic background or complex psychosocial stressors but is a more direct consequence of MAOA deficiency" (Cases, Seif, Grimsby, et al. 1995, 1763, 1766). However, by 1999, Schuback, and a group that included one of the researchers on Brunner's original study, Xandra Breakefield, screened a large sample, including males with mental retardation, those with a history of sexually deviant behavior, and other patient groups, and found no evidence at all for mutations disrupting the MAOA gene; they concluded that such mutations were not common in humans (Schuback, Mulligan, Sims, et al. 1999). One might have thought that this would spell the end, or at least the beginning of the end, for MAOA-linked hypotheses concerning impulsive and violent criminal behavior. But this was not to be the case.

Lacking much else to go on, researchers would continue to search for MAOA links whenever they had the chance. And MAOA was to come back into prominence via another route, with the first significant publication from Avshalom Caspi and Terrie Moffitt and their team from their work on the Dunedin Multidisciplinary Health and Development Study. This was a cohort study of 1,037 people born in a coastal region of the South Island of New Zealand between April 1, 1972 and March 31, 1973, and still living in the area three years later, who were assessed every few years up to and beyond the age of twenty-six.[38] Caspi and Moffitt and their team drew on the earlier studies of the links between deficiencies in MAOA activity and aggression in mice and humans—including of course, that of Brunner's group—and argued that although "[e]vidence for an association between MAOA and aggressive behavior in human populations remains inconclusive, . . . [c]ircumstantial evidence suggests that childhood maltreatment predisposes most strongly to adult violence among children whose MAOA is insufficient to constrain maltreatment-induced changes to neurotransmitter systems" (Caspi, McClay, Moffitt, et al. 2002, 851).

They supported this claim with reference to animal studies showing that early stress alters levels of norepinephrine, serotonin, and dopamine in ways that can be long-lasting (Francis and Meaney 1999; Bremner and Vermetten 2001) and reports linking variations in norepinephrine and serotonin levels to aggressive behavior in humans, although they did acknowledge that no study had shown that MAOA plays a role in these variations. Nonetheless, they proceeded to test the hypothesis "that MAOA genotype can moderate the influence of childhood maltreatment on neural systems implicated in antisocial behavior" by seeking to correlate the number of tandem repeats in the promoter region of the MAOA gene and response to childhood maltreatment. Dividing the sample on the basis of repeat numbers into "low MAOA activity" and "high MAOA activity"; classifying childhood maltreatment as severe, probable, and none; and assessing outcomes in terms of *DSM-IV* conduct disorder, convictions for violent crimes, personality assessments of disposition toward violence, and anecdotal reports of antisocial personality disorder, they generated the finding that was to launch a host of studies of 'gene-environment interaction.' Maltreated males with high levels of MAOA activity did not have elevated antisocial scores, while maltreated males with low levels of MAOA activity were more likely to have adolescent conduct disorder, violent convictions, and to score high on measures of antisocial conduct. Hence the conclusion: "a functional polymorphism in the MAOA gene moderates the impact of early childhood maltreatment on the development of antisocial behavior in males" (Caspi, McClay, Moffitt, et al. 2002, 853). It appeared that some genetic variants could both *increase* risk of antisocial behavior in the face of stressful childhood experiences, while other variants might be protective. The team speculated that the findings had practical implications for drug treatment. "Both attributable risk and predictive sensitivity indicate that these findings could inform the development of future pharmacological treatments" (ibid.).

In the subsequent eight years, this study was cited some 1,700 times. Some tried to replicate it and failed; others criticized various aspects of the methodology or the analysis. However, from our perspective, these problems are not the key issue. More to the point is that, as with the Brunner study, this argument seemed to encapsulate a whole style of thought concerning the role of genetic variation in generating behavioral phenotypes, not just violent and antisocial conduct but a whole range of psychiatric conditions and problematic conduct. Caspi and Moffitt made this explicit in a paper published four years later. Rather than proceeding by trying to link genetic variations to disorders, or even seeking to find endophenotypes, they suggested that the "gene-environment interaction approach assumes that genes moderate the effect of environmental pathogens on disorders" (Caspi and Moffitt 2006, 584). Given that we know a great deal about the environmental pathogens and risk factors for many mental disorders from depression to antisocial behavior, and heterogeneity of response between individuals characterizes even the most overwhelming of environmental traumas, we can now look to genetics to help us understand why some succumb and others do not: "The hypothesis of genetic moderation implies that differences between individuals, originating in the DNA sequence, bring about differences between individuals in their resilience or vulnerability to the environmental causes of many pathological conditions of the mind and body." (Caspi and Moffitt 2006, 584). And they suggested that we should go beyond genetic epidemiology: "Neuroscience can complement psychiatric genetic epidemiology by specifying the more proximal role of nervous system reactivity in the gene-environment interaction" (ibid.).

In particular, Caspi and Moffitt's group focused on three genes where high and low activity variants could be linked to susceptibility or resilience. The first concerned variants of the MAOA genotypes and their link with susceptibility to childhood maltreatment, which we have already discussed. The second concerned the long and short versions of the promoter region of 5-HTT, the serotonin transporter gene, where individuals carrying one or two copies of the short version were believed to be more susceptible to depression following stressful life events (Caspi, Sugden, Moffitt, et al. 2003).[39] The third was the influence of a polymorphism in the catechol-O-methyltransferase (COMT) gene, in which individuals carrying the valine allele were thought likely to develop severe psychiatric symptoms following cannabis use, while those with two copies of the COMT methionine allele seemed to exhibit no such adverse effects (Caspi, Moffitt, Cannon, et al. 2005).

We were, it seems, ready for a new relationship between psychiatric genetics and neuroscience, in which they would work together to identify the neurobiological pathways that accounted for the variations in susceptibility or resilience between individuals to environmental insults, in order to clarify "[h]ow an environmental pathogen, especially one that is psycho-social in nature, get[s] under the skin to alter the nervous system and generate mental disorders" (Caspi and Moffitt 2006, 589). Despite its neuromolecular form, the basic form of this argument is reminiscent of debates in the mid-nineteenth century on hereditary dispositions, predisposing and exciting causes, and the like in the inheritance of insanity and all manner of socially problematic conduct. So what, today, does this imply for understanding of, and intervention into, problems of antisocial and aggressive conduct?

In 2008, Joshua Buckholtz and Andreas Meyer-Lindenberg summed up current thinking on the neurobiology of antisocial conduct along the following lines (Buckholtz and Meyer-Lindenberg 2008). Antisocial and violent conduct is a major social problem, which also generates significant economic costs, and is hence an important target for government intervention. Such conduct runs in families, and a small proportion of families in any community account for a large proportion of its violent crime. Hence risk for antisocial aggression is transmitted intergenerationally. While twin studies have shown significant heritability, they have been unable to identify the specific genetic bases of risk. However, they say, Brunner's study began to clarify this, by demonstrating the link between MAOA deficiencies and human aggression. This link has been further demonstrated in rodents, where MAOA knockouts demonstrate increased aggression, which is correlated with elevated levels of serotonin and noradrenaline. MAOA has a key role in regulating release and clearance of serotonin and noradrenaline in monaminergic synapses in brain, suggesting that aggression is also regulated by such levels. Levels of MAOA at or before birth have long-term effects into adulthood. While the complete deficiency of MAOA in the population is rare, there are polymorphisms in the MAOA gene—basically long and short versions of a VNTR (variable-number tandem repeat). This determines whether an individual has a low level of monoamine oxidase activity (MAOA-L) or a high level (MAOA-H). In vitro studies have failed to demonstrate the relevance of these differences for in vivo levels of MAOA-activity in adults; nonetheless, much research has sought to link these differences to aggression and impulsivity in animals and humans.

Buckholtz and Meyer-Lindenberg note that direct association studies have been inconclusive but argue that gene-environment studies have shown a consistent relationship between MAOA-L and the risk of impulsive violence in those males who have experienced early-life adversity. Other studies in animals have also shown a link with variations in other genes that modulate serotonin levels in brain, such as the promoter region of the serotonin transporter gene. Neuroimaging studies are beginning to demonstrate that there are differences between MAOA-L and MAOA-H alleles in cortical circuitry relevant to emotional regulation, notably in the medial prefrontal cortex, which has a role in modulating amygdala function, which is well known to be crucial in fear extinction, emotion regulation, and temperamental variation; the medial prefrontal cortex is also well known to be modulated by serotonin levels. These variations in neural circuitry can be shown empirically to be associated with individual differences in temperamental traits, notably increased reactivity to threat cues; increased tendency to experience anger, frustration, and bitterness; and reduced sensitivity to cues for prosocial behavior. These alterations in corticiolimbic structure, function, and connectivity may underlie MAOA's role in genetic susceptibility for aggression as a result of the crucial role played by the implicated circuitry in social evaluation and decision making. Thus, they conclude, "although MAOA is not a 'violence gene' per se, susceptibility alleles might bias brain development towards alterations in function and structure which, in combination with other factors, predispose the development of antisocial behavior." Yet they also admit that in the absence of such factors, inheritance of the low-activity version of MAOA "is completely compatible with psychiatric health" (Buckholtz and Meyer-Lindenberg 2008, 127).

Once again, we can see the logic of 'bioprediction' that is taking shape here: screen and intervene. This is not a search for the biological roots of crime but for the biological susceptibilities that increase the likelihood of manifesting undesirable behavior involving violent, impulsive, or aggressive conduct, and the apparent weakening of the conventional workings of guilt and remorse. The focus is on risk identification, intervention, precaution, and prevention. The neurobiological basis here is not conceived of as genes for crime or even genes for aggression, but rather as specific variations at the level of SNPs, in the sequences that are believed to code for the various neurobiological components of behavior control and the neurochemical anomalies that correlate with low impulse control or aggression. This genomic analysis is combined with imaging research on localization of functions, seeking to identify the brain regions or neural circuits activated in aggression or impulsivity, carried out using brain scanning techniques to identify characteristic patterns of brain activity. This neurobiological approach allies itself well with its sociopolitical milieu, for it has taken shape in a context in which crimes of this sort—impulsive, aggressive, antisocial—are understood not merely as infractions of the law but as problems of public insecurity and generators of economic burden. Within these control strategies of precaution, prevention, and preemption, the question shifts from that of response to the offense after the act, to programs of prediction and prevention that identify those at risk on the basis of a kind of algorithm that combines genetic and neurobiological factors with those relating to family life, parental behavior, poverty, housing, and other environmental factors. For these violent or impulsive behavior at least, crime, is reframed as a problem of public health.

The stage is set for the reinvention of strategies for governing childhood and family life in the name of crime prevention. On the one hand, we can find approaches that focus on the neurobiological underpinnings of those at risk of antisocial conduct or psychopathy and advocate screening to identify children at risk, initially using well-known scales, but hopeful of the possibility of identifying biomarkers in brains or genes. On the other, we can see approaches that seek to act upon the developing brain of the child who is at risk by shaping his or her interpersonal milieu; they aim to maximize the potential of each child by targeting those 'incredible years.' But, as we will see, strategies developed in the second approach merge with those proposed from the first—and for both, the key actors are not the child, but the parents.

Nipping Budding Psychopaths in the Bud

Should one screen children to identify those at risk of developing antisocial behavior or even psychopathy?[40] In the 1990s, the case for this approach was made by Donald Lynam, first at the University of Kentucky and later at Purdue University (Lynam 1996, 1998; Lynam and Gudonis 2005).[41] But if screeing is desirable, what methods might be appropriate? Population-level screening is controversial in many regions, for example, in the United Kingdom, though some argue for its cost-effectiveness and utility even at the preschool age, given certain conditions (Wilson, Minnis, Puckering, et al. 2009). In some areas of the United States, screening programs for children's behavioral problems using non-neurobiological forms of assessment have already been tried. The most controversial was the Texas Childhood Medication Algorithm Project (TCMAP)—a development of the adult version (TMAP), which was designed to screen adults for major psychiatric conditions.[42] The version for children proved particularly contentious, with claims that it was acting as a powerful marketing tool for pharmaceutical companies: it was suspended in September 2008.[43] Other U.S. screening programs for children have included Teenscreen, employed for the early identification of mental illness in teenagers, notably screening for depression to reduce suicide risk.[44]

While there are a multitude of screening programs that seek to predict future violence on the basis of current behavior or symptoms, at the time of writing, there seem to be no screening programs in general use for antisocial conduct that involve biomarkers (genetic profiling, brain imaging, etc.) or endophenotypes, such as hormone levels.[45] Indeed, as we discuss in chapter 4, despite the hopes raised by developments in genomics and in brain imaging, it has proved difficult, if not impossible, to develop validated and reliable biomarkers for *any* psychiatric condition that can be used to make a diagnosis, even in a clinical context. Thus, those researching the possibilities of screening for future antisocial behavior have tended to rely largely on a behavioral assessment of features, notably those termed "callous and unemotional traits" (J. Hill 2003; Moran, Ford, Butler, et al. 2008), derived ultimately from Cleckley's *Mask of Sanity*—published in 1941 (Cleckley [1941] 1976), operationalized by Robert Hare in *The Hare Psychopathy Check List—Revised* (Hare 1991; cf. Hare 1999), and then reformatted for children (Frick and Hare 2001).[46] But endeavors to go beyond the behavioral and the symptomatic, to diagnose through the brain itself, are certainly at the research stage.

Let us take one extremely influential example to illustrate this line of thinking: the program of research funded by the United Kingdom's Medical Research Council, the Department of Health, and the Home Office, and carried out at the Institute of Psychiatry, King's College London (Viding, Blair, Moffitt, et al. 2005; Odgers, Moffitt, Poulton, et al. 2008; Viding, Jones, Frick, et al. 2008; Viding, Larsson, and Jones 2008).[47] The research has argued as follows: First, that one particular subtype of antisocial behavior, that with "callous and unemotional traits" (AS-CU) is a precursor of adult psychopathy. Second, that the characteristics of childhood AS-CU and adult psychopathy map onto one another: in both children and adults we see lack of empathy, lack of guilt and remorse, and shallow affect and manipulative conduct, and in such children we can also see precursors of adult violence, such as cruelty to animals and enjoyment of aggression. Third, if we look at studies of psychopaths, psychological tests show them to be poor at empathizing with others and bad at recognizing their fear and sadness, and brain scans show them to have a disruption of one particular part of the brain, the amygdala, which is, so it seems, implicated in these affective defects. Fourth, if we take twin pairs aged around seven years, some monozygotic and some dizygotic, who have been assessed for AS-CU and carry out fMRI scans on them, we find the same deficits in the brain's "affect circuitry" in the AS-CU subjects as we see in adult psychopaths. Finally, if we use classical twin study methods, taking environmental factors into account, we find a much higher correlation of these brain patterns in the genetically identical twins than in nonidentical twins—that is, the trait is highly heritable and suggests a strong genetic vulnerability. Hence, it is argued, we should search for the precise SNP-level variations that underlie this vulnerability. We should uses SNPs and scans to search for budding psychopaths.

As is now obvious, what these researchers are searching for is a vulnerability, a *susceptibility*. It is not surprising, then, that they are closely linked to the

Caspi and Moffitt research group and hope to identify the gene-environment interactions that provoke vulnerability into frank psychopathy (Caspi, Mc-Clay, Moffitt, et al. 2002; Kim-Cohen, Moffitt, Taylor, et al. 2005; Odgers, Moffitt, Poulton, et al. 2008). Their message, they insist, is *not* fatalism. This is how Viding and McCrory put it in a presentation to a 2007 conference commissioned by the United Kingdom's Department of Health; the Department for Children, Schools and Families; the Youth Justice Board; and the Cabinet Office: "the goal of early identification is successful intervention."[48] And the means to be used were multiple—perhaps behavior therapy, cognitive therapy, and even psychopharmaceuticals—for the aim is to reshape brain mechanisms in order to nip those budding psychopaths in the bud.

In fact, the path most emphasized is one that addresses the problem indirectly. The route to the problematic child, as so often in the past, is to be through the parents. For research now claims that these undesirable traits can be ameliorated by such means (Dadds, Fraser, Frost, et al. 2005; Hawes and Dadds 2005, 2007). That is to say, one can modulate the traits, and perhaps even their neural underpinnings, by training and 'empowering' parents—or rather 'caregivers'—in parenting skills. Perhaps, some say, what is required is "multisystemic therapy" that will improve parental disciplinary practices, increase family affection, divert children from antisocial peers to prosocial ones, improve schoolwork, engage children in positive recreation, improve family-community relations, and empower the family to solve future difficulties.[49] But in any event, the familiar answer is that one should govern the child at risk by governing the risky family.

Sculpting the Brain in Those Incredible Years

Who could question that the child is a most incredible human being?[50] And that the child needs everything that he or she can receive in order to withstand the emotional and social challenges that life will inevitably bring, especially in the teenage years? And who could doubt the virtue of interventions that seek both to ensure and sustain resilience in the child, to prevent a future of delinquency, drug abuse, and violence, with all the accompanying misery to child, family, and victims, not to mention the economic costs? Thus, in the United States more than a decade ago, Mark Cohen calculated the various lifetime costs associated with a typical criminal career, and based on the fact that such careers start young, and in an attempt to provide a basis for the cost-effectiveness of early preventive intervention programs, estimated that the "monetary value of saving a high-risk youth" was between $1.7 million and $2.3 million (Cohen 1998, 5). In the United Kingdom, in 2006, Stephen Scott, Martin Knapp, and their colleagues estimated that individuals who show conduct disorder at the age of ten will, by the time they are twenty-eight, have "cost society" up to ten times more than controls without this diagnosis (Scott, Knapp, Henderson, et al. 2001; Romeo, Knapp, and Scott 2006). All seem agreed about the implications: surely we need strategies for early identification and intervention to safeguard our children from such miserable futures, and our societies from such a social and economic burden (Harrington and Bailey 2003; Kim-Cohen, Moffitt, Taylor, et al. 2005; Margo and Stevens 2008).

No wonder, then, that we see a whole array of individuals, pressure groups, and organizations seeking to promote early intervention in the name of prevention—many of which can be found on the pages of the website Prevention Action.[51] These programs are often pioneered by individuals with an almost evangelical zeal and commitment. There is Carolyn Webster-Stratton with her "Incredible Years" training series, which has expanded from its start in the United States to England, Wales, Ireland, Norway, Australia, and New Zealand. There is Clay Yeager, a onetime probation officer, turned Pennsylvania oilman, with his consultancy working to introduce evidence-based programs for prevention such as "Nurse Family Partnerships" and "Communities That Care" into child welfare and youth justice systems in Florida and several other U.S. states. There is Steve Aos in Washington, advising the legislature on issues such as "the economic costs and benefits of investment choices in juvenile justice." Of course, much of this work does not explicitly refer to brains, let alone to neurobiology—Webster-Stratton, for example, relies on theories of social learning—but others, such as Tom Dishion, combine their practical work promoting the importance of early intervention with brain imaging research—the article on his work found at Prevention Action is titled "Putting Brain Science Back on the Streets of Los Angeles."[52]

In the United Kingdom, this agenda was proposed most passionately by Iain Duncan Smith's Institute of Social Justice. Duncan Smith, a former leader of the Conservative Party, teamed up with Graham Allen, a member of Parliament from the Labour Party, to produce a report in 2008 titled *Early Intervention: Good Parents, Great Kids, Better Citizens* (Allen and Duncan Smith 2008). After outlining the problems and costs of social dysfunction, violence, drugs, alcohol, and family breakdown, they argued that from birth through age three was the vital period when the right social and emotional inputs must be made to build the human foundations of a healthy, functioning society and that the key agents to provide those inputs were not the state but the parents. Why this age range? How many times have we heard this period emphasized in social policy, most recently in the arguments in the 1970s concerning the cycle of deprivation and the need for policies of intervention in the preschool years, such as the Head Start program in the United States? The report refers to many longitudinal studies to make its case that early neglect and lack of appropriate parental input leads to these later problems, and that youngsters who do not have the right kind of emotional and familial input at this age grow up to be antisocial adults who go on to raise antisocial children—thus perpetuating the cycle.

While the conceptual bases of Head Start and analogous schemes for pre-school intervention in the United Kingdom were psychological, the ratio-nale is now articulated in terms of the brain—because the brain's capacity for change is believed to be at its greatest in this period: "the human brain has developed to 85 per cent of its potential at age three (and 90 per cent at age four)" (Allen and Duncan Smith 2008, 48). Because the human being is born with the brain underdeveloped, and its completion depends on input after birth, "sculpting the young brain" and programming the development of the baby from the first moment after birth is of the highest importance. "The structure of the developing infant brain is a crucial factor in the cre-ation (or not) of violent tendencies, because early patterns are established not only psychologically but at the physiological level of brain formation" (ibid., 57). The mechanism is the growth and pruning of synapses in response to experience—synapses that become "hard-wired" by repeated use.

But not only that—the brains of deprived children are actually smaller, or so it is claimed: "The brain of an abused or neglected child is significantly smaller than the norm: the limbic system (which governs emotions) is 20–30 per cent smaller and tends to have fewer synapses; the hippocampus (respon-sible for memory) is also smaller. Both of these stunted developments are due to decreased cell growth, synaptic and dendrite density—all of which are the direct result of much less stimulation (e.g. sight, sound, touch) than is required for normal development of the brain"; this claim is supported by the reproduction of images from Bruce Perry's work at the Child Trauma Academy (Perry 2002).[53] The role of cortisol, neurotransmitter modulation, sensitive windows, and the whole repertoire of neurobiological accounts of child development is deployed to produce the objectivity effect that in previ-ous periods was generated by reference to dynamic and developmental psy-chology. Even attachment, that stalwart of developmental psychopathology, is now inscribed in the brain: "Babies who are healthily attached to their carer can regulate their emotions as they mature because the cortex, which exer-cises rational thought and control, has developed properly. However, when early conditions result in underdevelopment of the cortex, the child lacks an 'emotional guardian'" (ibid., 62). It seems obvious that intensive intervention on early parenting, on the parents of today, and on those at risk of being the dysfunctional parents of tomorrow is the path required to break the cycle of antisocial and violent conduct that destroys lives and costs our societies so dearly. Who could disagree?

Well, as it happens, in April 2010, this question of whether there was a link between child deprivation, brain development, and future social prob-lems became rather controversial. Dr. Bruce Perry reproached Iain Duncan Smith—who was about to become a minister in the Conservative-led coali-tion government that came to power in the United Kingdom in May 2010—for using his brain scan images of children who had suffered extreme neglect to illustrate a more general claim that neglect or family breakdown changes brain development.[54] Perry himself had written extensively on the neurode-velopmental consequences of ill treatment and neglect of children, notably opening one of his more cited pieces with the arresting phrase "Children are not resilient, children are malleable" (Perry 1997). In this article, he drew on neurobiological research and studies of neglected children, including the Romanian orphans who were discovered living in appalling conditions in Romania in the years following the fall of the Ceausescu regime in 1989, to argue for the malign consequences of early violence and neglect on different brain areas and functions, and in particular, claiming that the consequences were that those who had suffered in these ways were themselves more prone to violence and aggression in adolescence and adulthood as a result.[55] Perry called for policies that provided parents with the resources and information to optimize their children's developing brains, an argument he developed in a number of other scientific publications (Perry 2002, 2008), and in a popular book arguing for the crucial role of empathy, its roots in the developing brain, and its social and personal consequences—apparently under threat from cur-rent cultural and technological changes (Szalavitz and Perry 2010).

These disputes about whether one could extrapolate from the neurobio-logical consequences of cases of extreme neglect did not deflect those who argued for early intervention in the name of the brain. In the U.K. debate, Duncan Smith was backed by Camila Batmanghelidjh, charismatic founder of one of the country's leading early intervention charities, Kids Company, arguing that lack of care does affect brain development, and can be com-bated by intensive interactions in preventive interventions. A month later, Kids Company launched "Peace of Mind," a major campaign to raise £5 mil-lion to fund research into vulnerable children's brains. Supporters were asked to go online to access a virtual brain with one million neurons; for each £5 they donated, they received a "virtual neuron," and they were asked to invite their social networks to do the same—the virtual neurons they purchased would connect together to form clusters in the virtual brain. Supported by a section of the website summarizing genetic and neurobiological research on child maltreatment and the brain, the conclusion was that "[s]ustained child abuse and neglect has a devastating impact on a child's brain often creating a lifetime of vulnerability, tragically increasing the risk of suicide, violence, anti-social behavior, depression and drug addiction."[56] This was, it seemed, a nonpartisan issue: in January 2011, Graham Allen, still a Labour member of Parliament, was asked by the Conservative British Prime Minister David Cameron to produce a report on early intervention. Allen used the very same brain images on its cover and deployed essentially the same brain-based ar-guments in his proposals for intensive early intervention for preschool chil-dren (Allen 2011).

These arguments for the benefits of early intervention seem hard to deny. As in earlier programs, such as Head Start, they have generated considerable research seeking to evaluate preventive programs for intensive intervention for those thought to be at risk. As we have noted, currently, children at risk are not identified using biomarkers but because either they, or their parents, have come to the notice of social workers, teachers, doctors, or other authorities. And the intervention programs do not use interventions targeted directly at neural processes. As we have seen, the key is thought to be the training of parents—and teachers—in managing the behavioral difficulties of their children. Consider, for example, the randomized controlled trial by Stephen Scott at the Institute of Psychiatry in London, evaluating an intervention using the Incredible Years program on six-year-old children identified by screening all children of this age at eight schools for antisocial behavior. The study concluded that, at the cost of £2,380 ($3,800) per child, a twenty-eight-week program showed that by targeting multiple risk factors (ineffective parenting, conduct problems, ADHD symptoms, and low reading ability), and by training parents, "[e]ffective population based intervention to improve the functioning of [children] with antisocial behavior is practically feasible. . . . Including halving the rate of oppositional defiant disorder . . . there is the prospect that children's long term mental health will be improved, including better school attainments and less violence" (Scott, Sylva, Doolan, et al. 2010, 48, 56).[57]

By way of the brain, then, we reach a conclusion that does not differ greatly from arguments reaching back to the late nineteenth century about the effects of the early years on later propensities to problematic conduct (Platt 1977; Donzelot 1979; Rose 1985). From Mary Carpenter's campaigns for colonies for dangerous and perishing children (Carpenter 1851), through the social problem group and "the submerged ten percent" in the early twentieth century, via the mental hygiene movement in the 1930s, and the arguments for setting up child welfare services in the years after the Second World War (Rose 1985), to the contentious concept of the "cycle of deprivation" in the 1970s (Joseph 1972; Jordan 1974), and the interventions from the Head Start program to the Sure Start program—we find the repeated arguments that one should minimize a host of social ills, including criminal and antisocial conduct, by governing the child through its family. In each generation, unsurprisingly, these arguments are made on the basis of whatever happens to be the current mode of objectivity about the development of children—habits, the will, instinct theory, psychoanalysis, and today the brain. Social justice, it seems, lies not in tackling the causes of structural inequality, poverty, poor housing, unemployment, and the like, but in managing parents in the name of the formation of good citizens.

Governing Antisocial Brains

It may seem odd to have concluded our discussion of antisocial brains not by debating questions of free will and criminal responsibility, but by exploring proposals to train parents in ways of rearing their young children. But if there is one key lesson that should be learned from the genealogy of crime-control strategies, it is that the criminal justice system itself has rarely been considered the most important strategic site for interevention (D. Garland 1985; Rose 1985). The most significant changes occurring today at the intersection of crime-control strategies and neuroscience do not take the form that concerns many neuroethicists. They do not entail a challenge to doctrines of free will in the courtroom—despite the flurry of books claiming and discussing such a shift (Libet, Freeman, and Sutherland 1999; Dennett 2003a; B. Garland 2004; Spence 2009). Nor do they undermine the notion of the autonomous legal subject. Indeed, the reverse is perhaps closer to the truth—strategies seek to create the kinds of persons who *can* take responsibility for their actions, and they attempt to enhance such self-control by acting on the brain.

Despite the concerns of those who raise neurobiological doubts about autonomy and free will, these kinds of interventions go entirely with the grain of folk psychology, with its long-standing emphases on the importance of the early years—creating the adult through shaping the child. Indeed, they are entirely consonant with programs, strategies, and techniques since the second half of the nineteenth century that have sought (and usually failed) to address the problem of crime by early intervention to shape the family into a mechanism for creating civilized citizens (Donzelot 1979; Rose 1989; Parton 1991). Today, it is *brains at risk*—at risk of becoming risky brains—that are to be addressed indirectly and preventively. As in earlier strategies, two senses of risk are brought into alignment. The first is the desire to identify *risky individuals*—that is to say, those who will present a future risk to others—before the actual harm is committed. The second is the hope that one might be able to identify *individuals at risk*—those whose particular combination of biology and life history makes them themselves susceptible to some future condition—here personality disorder, impulsivity, aggressivity, or whatever, but more generally susceptibility for any psychiatric disorder. And of the many options available once one has identified such children, the strategies that are emerging most forcefully, at least in the United States and the United Kingdom, are strategies of *optimization* and strategies of hope—strategies that aim, in the words of the Kids Company Peace of Mind campaign, to address both "[t]he individual cost to the young person in terms of emotional wellbeing, mental and physical health, and education" and "[t]he cost to society of antisocial behavior in terms of police and criminal justice resources, public safety, NHS resources, and the continuation of the cycle of abuse and violence."[58]

It is too early to say whether neurogenetics and brain imaging will play an important role in the screening programs to identify those at risk. And it is too early to say whether interventions directly addressing the brain, perhaps by drugs, will come to supplement or supplant those based on parent, teacher, and nurse training. But it is undoubtedly the case that the reference to neurobiological consequences of childhood neglect will continue to provide such programs with crucial support, that research into such consequences will expand, and that the brain will increasingly be seen to lie at the root of anti-social and dangerous conduct, even if its effects are manifested psychologically in poor impulse control and failures of the will. Whatever the fate of the high-profile endeavors to identify anomalies in the brains of adult murderers, or to discover the neural anomalies that might finally resolve disquiets over the category of psychopathy, we can certainly anticipate future strategies to govern antisocial citizens by acting upon their developing brains.

Chapter Seven

· · · · · · · · · · · · · · · · · ·

Personhood in a Neurobiological Age

In the course of centuries the *naïve* self-love of men has had
to submit to two major blows at the hands of science. The first
was when they learnt that our earth was not the centre of the
universe but only a tiny fragment of a cosmic system of scarcely
imaginable vastness. This is associated in our minds with the
name of Copernicus, though something similar had already
been asserted by Alexandrian science. The second blow fell when
biological research destroyed man's supposedly privileged place
in creation and proved his descent from the animal kingdom
and his ineradicable animal nature. This re-evaluation has been
accomplished in our own days by Darwin, Wallace and their
predecessors, though not without the most violent contemporary
opposition. But human megalomania will have suffered its third
and most wounding blow from the research of the present time
which seeks to prove to the ego that it is not even master in its own
house, but must content itself with scanty information of what is
going on unconsciously in its mind. We psycho-analysts were not
the first to utter this call to introspection; but it seems to be our
fate to give it its most forcible expression and to support it with
empirical material which affects every individual.

—**Sigmund Freud**, *Introductory Lectures on Psychoanalysis*, 1916

Humans are probably the only animals that have made conjectures
about what goes on inside the body. . . . Today we are engaged
in the quest of understanding our brain. When we have finally
worked out the details, this remarkable organ, for all of its lofty
pretensions—seat of intellect, home of the soul—will very likely
join other remarkable pieces of biological machinery. Remarkable,
certainly, but not mysterious or possessed of any supernatural
qualities.

—**Mark Bear and Leon Cooper**, "From Molecules to Mental States,"
1998

Sometime in the twenty-first century, science will confront one
of its last great mysteries: the nature of the self. That lump of flesh
in your cranial vault not only generates an "objective" account
of the outside world but also directly experiences an internal
world—a rich mental life of sensations, meanings, and feeling.
Most mysteriously, your brain also turns its view back on itself to
generate your sense of self-awareness.

—**V. S. Ramachandran**, *The Tell-Tale Brain: Unlocking the Mystery of
Human Nature*, 2011

The self an illusion created by our brain.[1] This is the message of a host of recent semipopular books published by neuroscientists over the past decade.[2]
Not long ago, few would have had the temerity to approach this issue in such
a direct manner. But now many neuroscientists are willing to take up the
challenge posed by Freud at the start of the previous century: to find the neural bases of the processes that he lodged, in his view only provisionally, in the
psyche (Pribram and Gill 1976). Like Freud, today's neuroscientists challenge
the view that consciousness is the master in its own house and argue that
much that is conventionally ascribed to the realm of the mental, of conscious
deliberation and willed intention, actually occurs somewhere else—now not
in some immaterial unconscious but in the brain.

In one of the earlier and most provocative formulations, Francis Crick asserted: "You, your joys and your sorrows, your memories and your ambi-
tions, your sense of personal identity and free will, are in fact no more than
the behavior of a vast assembly of nerve cells and their associated molecules.
As Lewis Carroll's Alice might have phrased it: 'You're nothing but a pack of
neurons'" (Crick 1994). In similar vein, Patricia Churchland claimed: "Bit
by experimental bit, neuroscience is morphing our conception of what we
are. The weight of evidence now implies that it is the *brain*, rather than some
non-physical stuff, that feels, thinks, and decides. That means there is no soul
to fall in love. We do still fall in love, certainly, and passion is as real as it ever
was. The difference is that now we understand those important feelings to be
events happening in the physical brain" (Churchland 2002, 1).

Rather less boldly, this is how Joseph LeDoux opens his book *The Synaptic
Self*: "The bottom-line point of this book is 'You are your synapses.' Synapses
are the spaces between brain cells, but are much more. They are the channels of
communication between brain cells, and the means by which most of what the
brain does is accomplished" (LeDoux 2002, ix). For the psychologist-turned-
neuroscientist, Chris Frith, in his book *Making Up the Mind: How the Brain
Creates Our Mental World*, it is the brain that "creates the world," producing
not only "the illusion that we have direct contact with objects in the physical
world" but also that "our own mental world is isolated and private": "through

these two illusions we experience ourselves as agents, acting independently upon the world" (Frith 2007a, 17). For the philosopher Thomas Metzinger, writing under the title *The Ego Tunnel: The Science of the Mind and the Myth of the Self*, despite the way we think and talk, "nobody has ever *been* or *had* a self": conscious experience is a creation of our brains, an internal construct in what he terms "the ego tunnel," which also creates "an internal image of the person-as-a-whole," which is "the phenomenal ego, the 'I' or 'self' as it appears in conscious experience." But "you are constitutionally unable to realize that all this is just the content of a simulation in your brain . . . not reality itself but an image of reality . . . a complex property of the global neural correlate of consciousness" (Metzinger 2009, 1–12).

How should we assess such claims? Many from the social and human sciences regard this neurobiologization of the self as the most challenging feature of contemporary neuroscience. It seems to threaten the very conception of the human being that lies at the heart of their work: the idea that personhood is a matter of internal mental states—consciousness, intention, beliefs, and the like—existing in a uniquely human psychological realm of mind, embodied in a self-conscious subjectivity, and created in a world of meaning, culture, and history. Thus the French sociologist Alain Ehrenberg, writing in 2004, suggested that some neuroscientists were going beyond explorations of the neuronal machinery, and hopes for a new neuropsychiatry, to argue for a philosophical identity between knowledge of the brain and knowledge of the self: the "sujet parlant" was giving way to the "sujet cérébral"—a "cerebral subject" (Ehrenberg 2004).[3]

The "Brainhood Project," established by Fernando Vidal and colleagues in 2006, took up this idea: "Starting in the second half of the 20th century, the quality or condition of being a brain ranked as the defining property of human beings. . . . From science fiction in writing and film to neurophilosophy and the practices of intensive care and organ transplantation, humans came to be thought of not merely as having a brain, but as being a brain. Of course, many other properties of human persons were recognized. Nevertheless, whole or in part, the brain emerged as the only organ truly indispensable for the existence of a human self and for defining individuality."[4]

In this chapter we take a different approach. The conception of the person that is emerging in neurobiology does not quite match the version attacked by critics from the *Geisteswissenschaften*.[5] Indeed, neurobiological arguments in this area are beset with slippages and confusions, both linguistic and conceptual, between different notions of person and self, of consciousness and its relation to selfhood, of identity and personhood, of the relevance of questions of will and intention to self and consciousness, and much more. Further, the social and human sciences have not been so committed to the psychological ideal of the person or the self as is sometimes suggested. Before turning to neuroscience, then, it is worth beginning by reminding ourselves of some earlier challenges from within the human sciences to the idea of the bounded, conscious, constitutive self.

The Challenged Self

In our 'individualist' times, almost all aspects of political, ethical, and cultural thought hinge upon the things we believe about persons: their nature, their capacities, their freedoms and self-actualization, their rights and duties. These ideas about ourselves are central to debates about law, morality, justice, and the rights of minorities; about sexual relations, lifestyles, marriage, parenthood, and bioethical issues related to the beginning and end of life; about consumption, tastes, marketing and economic life; about politics and conflicts over national identities; and much more. They are entangled with other notions such as autonomy, agency, individuality, control, choice, self-realization, self-esteem, free will, and liberty. Human beings who inhabit our contemporary Western cultures understand themselves in terms of a kind of 'inwardness,' as endowed with a personal interiority, to which only they have direct access, in which lies their deepest truths and their truest thoughts, feelings, and wishes. It is this inwardness, inscribed in memories, capacities, and habits, that assures human beings that they *are*—that they have an enduring identity rather than being an ever-changing matrix of fleeting thoughts, feelings, and desires. This individualized inwardness, with its interiority and its continuity, is both unique to each and common to all (it was not always so, consider slaves, women, 'primitives,' children, idiots, the mad): it grounds the claim of each individual to membership in the human race and to the consideration that such membership implies.

In the 1970s and 1980s, many social historians, anthropologists, and sociologists turned their attention to the question of 'the person' and 'the self' and their historical and cultural variability. In the last essay he published—delivered in French as the Huxley Memorial Lecture in 1938—Marcel Mauss asserted his belief that "there has never existed a human being who has not been aware, not only of his body, but at the same time also of his individuality, both spiritual and physical." But, he argued, the *notion* or *concept* of the person or the self was "a category of the human mind" that had evolved over time, and has taken a different form in different societies, according to their "systems of law, religion, customs, social structure and mentality": the "cult of the self" and the respect for the self was very recent (Mauss [1938] 1985, 3).[6] Almost four decades later, Clifford Geertz argued something similar, writing that "[t]he Western conception of the person as a bounded, unique, more or less integrated motivational and cognitive universe, a dynamic center of awareness, emotion, judgment and action, organized into a distinctive whole and set contrastively against other such wholes and against a social and natural background is, however incorrigible it may seem to us, a rather peculiar idea within the context of the world's cultures" (Geertz 1974, 31).

In the years that followed, in the growing literature on the historical and cultural variability of conceptions of personhood and ideas of the self, most historians, philosophers, anthropologists, and sociologists agreed with Mauss and Geertz, as did many social psychologists engaged in comparative research on 'self-concepts' in different cultures (Dodds 1968; Shweder and LeVine 1984; Marsella, De Vos, and Hsu 1985; Roland 1988; Sampson 1988; Markus and Kitayama 1991). In different ways, they argued that ideas of the self or the person in Japanese society, among the native people of North America, in ancient Greece, in premodern Europe were different: they did not share our sense of personal uniqueness, of individual autonomy, of bodily integrity and biographical continuity. The contemporary Western conception of the self— individualized, bounded, with interior depth and temporal continuity, self-possessed, autonomous, free to choose—was not natural, given, or universal, it was a historical and cultural achievement.

More radically, some social theorists returned to Sigmund Freud and to the blow that psychoanalysis struck to human narcissism, arguing that the sense of oneself as having full insight into one's own interior psychological world, let along mastery over it, is an illusion. As Freud put it: "the ego . . . is not even master in its own house, but must content itself with scanty information of what is going on unconsciously in its mind" (Freud [1916] 1953–74, 285). Jacques Lacan, interpreting Freud in the spirit of French postwar structural linguistics, asserted that the self is radically excentric to itself: "the philosophical *cogito* is at the centre of the mirage that renders modern man so sure of being himself, . . . but yet I am not what I think I am or where I think I am" (Lacan [1957] 1977, 165, 171). A critical evaluation of a certain narcissistic mirage, image or 'ideology' of the universal, individuated, self-aware, self-conscious person, as both eternal fact and unimpeachable value, thus lay at the heart of much critical social thought in the closing decades of the twentieth century (Rose 1996).

In this light, the character of the current debate between the human sciences and the neurosciences seems paradoxical. While the human sciences have been happy to accept that our current conception of the 'naturally' autonomous, self-conscious person is an artifact of history, culture, meaning, and language, and that consciousness is far from sovereign in human affairs, most react with hostility when practitioners of the new sciences of the brain announce, usually with an irritating sense of daring, the death of our illusions about this same conscious, self-identical, autonomous person. It would seem prudent, therefore, to abstain from both celebration and critique, in order to explore the extent to which a new image of the self or person is taking shape, where, and with what consequences.

In his magisterial survey, *Sources of the Self*, the philosopher Charles Taylor traced the formation of the notion of the self, our modern sense of self, through the canonical works of Western philosophy (Taylor 1989). He argued that while there were deep and distributed supports in contemporary culture for the forms of thought that underpin psychological notions of selfhood— autonomy, depth, the capacity to choose and bear responsibility for those choices—this was nonetheless a function of a historically limited mode of self-interpretation, which had spread from the West to other parts of the globe, but which had a beginning in time and space and may have an end.

From a different perspective, that of the genealogy of personhood, we come to a similar conclusion. But, we suggest, we need to look beyond philosophy if we are to discern the signs of any such 'end' of this mode of self-interpretation today. For philosophers do not live in a separate world, at least when it comes to arguments concerning human nature or ideas of self and identity. It is not only that their arguments need to be located 'in context' as the saying goes (Tully 1988, 1993), but that philosophers draw, sometimes explicitly but usually implicitly, upon the empirical knowledge and cultural beliefs of the times and places they inhabit. Their apparently abstract considerations of the nature of humans have been bound up in all kinds of interesting ways with endeavors to shape, train, and govern those human beings, both individually and collectively. Descartes drew upon his experiences witnessing dissections and on contemporary anatomical knowledge of the body and brain to articulate his ideas for the location of the soul and his doctrines of healthy living; Locke, Condorcet, and others drew upon images and stories of wild children in their thought experiments about the existence of innate ideas and the role of the sensations (Rose 1985). Despite their often expressed disdain for 'folk psychology,' philosophers since at least the mid-nineteenth century have, explicitly or implicitly, taken up ideas from the knowledges about human beings articulated in the positive sciences—biology, psychology, neurology, and the brain sciences—even when apparently in dispute with them.[7]

If we are considering the transformations immanent in our present, then, we would do better to turn our gaze away from the great thinkers of our age and to explore the changes in language, technique, and judgment that arise in more mundane settings, in the face of more pragmatic concerns about the government of conduct. That is to say, we should examine mutations in knowledges with a lower epistemological profile—not so much in philosophy as in the human sciences and indeed in neuroscience itself. As for neuroscience, we will find once more, to paraphrase Canguilhem, that when it comes to matters of our ontology, while the normal seems prior to the abnormal, it is historically secondary. Again and again, in what follows, we will see the extent to which it is through the analysis of pathology—accidental, temporarily induced, or experimentally produced—that neurobiology has made its assault on the question of the mind and the self.

From the Pathological to the Normal

Early on in his genial book *Making Up the Mind: How the Brain Creates Our Mental World*, Chris Frith wrestles with the question of dualism and materialism. "There can," he writes, "be changes in the activity in my brain without any changes in my mind. On the other hand I firmly believe that there cannot be changes in my mind without there also being changes in brain activity"—a sentence that he annotates with a footnote declaring, "I am not a dualist" (Frith

2007a, 23). However, he immediately follows this with the following somewhat dualist pronouncement that "everything that happens in my mind (mental activity) is caused by, or at least depends upon, brain activity"—a sentence that he annotates with a footnote declaring, "I am a materialist. But I admit that I sometimes sound like a dualist" (ibid.). He says that he talks of "the brain 'not telling me everything it knows' or 'deceiving me' . . . because this is what the experience is like" (ibid.). So we can ask questions like "How does my brain know about the physical world?"—that is to say, we can ask what he terms "a mind question" as if it was a brain question. It is not clear how far this gets us in resolving those complex and slippery ways in which we are coming to shuffle the ways we speak of minds, brains, bodies, selves, persons, conscious intentions, unconscious decisions, and so forth. Frith's book is conducted as an imaginary debate with a mythical professor of English who points out the obvious error that brains can no more 'know' things than an encyclopedia that contains information about the world 'knows' about the world: even if the information is encoded in neurons rather than letters, we still seem to require some entity to 'read' it.

But nonetheless, his book is devoted to demonstrating that it is indeed the brain, not the mind, nor the conscious self, that creates the mental world in which humans live, including their capacity to act, interpret the actions of others, and inhabit a realm of meaning and culture. The route to this conclusion is an experimental one, making use of evidence from neuropathologies, or of the creation in the laboratory of some kind of artificial pathology, where damage to an area of the brain creates anomalies in what seems to be the normal experience of selfhood. In each case, that is to say, the technology for generating evidence depends upon breaching of the normal ways in which we inhabit our worlds. Breaching here, not in the sociological sense, where breaching is a method to render visible the taken for granted rules and presuppositions that make the micropractices of everyday life possible. But breaching the usual relations that hold between the brain and the world. And in so doing, the accidents and experiments that are drawn upon seem to demonstrate that certain central capacities of normal personhood are, in fact, creations of nonconscious neural mechanisms and that it is these nonconscious mechanisms that create the illusions within each individual's self-awareness—the illusions of coherent, conscious, self-controlling capacities that are attributed to the normal person.

Lessons from Lesions

"By studying patients . . . who have deficits and disturbances in the unity of self, we can gain deeper insights into what it means to be human" (Ramachandran 2011, 288). Perhaps it is obvious that the pathway from mind to brain should begin with lesions. What could give a clearer indication of the role of brain in mental life than the lacks, distortions, and other strange consequences of physical damage to particular regions of the brain? As we have

seen in previous chapters, contemporary histories of brain localization ritu-
alistically invoke Phineas Gage and his unfortunate accident with that tamping iron (Macmillan 2000), and our current knowledge of functional brain
regions still bears the names of neurologists (Broca, Wernicke, Fleschig) who
studied lesions to cognitive functions such as speech and language. The 1987
edition of the translation of the Soviet neurologist, A. R. Luria's *The Man with
a Shattered World* (Luria 1973), the account of a man who loses his speech
ability following a brain trauma during the Second World War, was introduced by Oliver Sacks, whose own romantic stories of the strange, sometimes
wonderful, sometimes poignant worlds of those with damaged brains have
become so popular (Sacks 1976, 1984, 1985, 1991). And yet, at least up to the
1990s, these authors did not frame their research, their findings, or their narratives in the way that has become routine today: in terms of the lessons that
we could learn from such episodes about the ancient problem of the relations
between mind and brain.

What, then, are the contemporary neurobiological 'lessons from lesions'
about the self—the study of what the neurologist and psychiatrist Todd Feinberg terms "neuropathologies of the self," where "a brain lesion causes a profound and specific alteration in the patient's personal identity or personal
relationships between the self and the world" (Feinberg 2011, 75)?[8] We can
start with the famous split-brain experiments dating back to the early 1960s
conducted by Michael Gazzaniga, with Roger Sperry, Joseph Bogen, and
P. J. Vogel (the history of these experiments is reviewed in Gazzaniga 2005).[9]
These *callostomies* involved severing the corpus callosum in human patients
suffering from intractable epilepsy in an attempt to control the seizures.
While it was initially claimed that they had no cognitive effects in humans,
Roger Sperry and his colleagues showed that similar operations in animals
led to very severe and profound deficits. Much research followed on the consequences in humans, for instance, in relation to the hemispheric localization
of different aspects of language, vision, and causality, but only one aspect is
relevant here—that bearing on issues of the self.

When a patient has his or her corpus callosum severed, Gazzaniga asks,
"does each disconnected half brain have its own sense of self . . . could it
be that each hemisphere has its own point of view, its own self-referential
system that is truly separate and different from the other hemisphere?" He
argues that there is indeed a lateralization, although not a complete one, in
which "the left hemisphere is quick to detect a partial self-image, even one
that is only slightly reminiscent of the self, whereas the right brain needs an
essentially full and complete picture of the self before it recognizes the image
as such. . . . [T]he data indicate that a sense of self arises out of distributed
networks in both hemispheres" (Gazzaniga 2005, 657, 658).[10] But however it
may be affected by the operation, Gazzaniga's premise is clear: that it is indeed
the brain that, through its patterned networks of neurons, produces the sense
of self.

Others draw their evidence for the neural creation of the illusion of self from disturbances concerning the boundaries and components of the body. One that is regularly cited is the 'rubber hand' phenomenon (see, e.g., Metzinger 2009). Here individuals observing a rubber hand will sometimes attribute sensation to that hand rather than to their own. For example, experimental subjects will make this false attribution if they see the 'alien hand' being stroked by a brush while their own equivalent hand is stroked in the same way but is hidden from sight. After a few minutes, they will 'feel' the stroking on the rubber hand, even though it is separate from their body, rather than in their own, hidden, hand. It seems that the brain has attributed the sensation to a physically distinct object within its field of vision, and in the process, it has somehow incorporated that alien object into the body (Botvinick and Cohen 1998; Tsakiris and Haggard 2005).

Another lesson along these lines comes from the phenomenon of 'phantom limbs.' It has long been recognized in the literature that patients who have had a limb amputated often continue to experience sensations from that missing limb, sometimes believing that they can still move the absent hand or fingers, or that the amputated limb is present but locked in space, or, worse, feeling excruciating pain arising from that absent part of the body. This evidence shows, suggests Frith, that these are "phantoms created in the brain"; a conclusion that is further strengthened by the reverse situation, where individuals who have suffered damage to a particular area of the brain feel that they have had a limb amputated—that they have 'lost' their hand, arm, leg, or foot.[11] This shows, for those who think like Frith, that "you do not have privileged access to knowledge about your own body"—that our feeling of effortless conscious control over our physical being in the world is an illusion created by the brain by "hiding from us all the complex processes that are involved in discovering about the world" (Frith 2007a, 72, 81).

Hence, as Gazzaniga and Cooney put it, neurological disorders illuminate the structure of human consciousness: "subjective awareness emerges from the interactions of specialized, modular components in a distributed neural network. In addition, the activity of such components is united cognitively by an interpretive process that occurs in the left hemisphere of the human brain. . . . [T]he specialized nature of the component neural processors suggests a potential explanation for how specific lesions can selectively alter or eliminate the contents of particular aspects of subjective experience, with no resulting experience of deficit for the brain-damaged individual" (Cooney and Gazzaniga 2003, 164). These, then, are the lessons from lesions.

Irreal Visions of a Real World

Frith, provocative as ever, titles chapter 5 of his book "Our Perception of the World Is a Fantasy That Coincides with Reality." "The remarkable thing about our perception of the physical world in all its beauty and detail is that it seems so easy. . . . But this very experience that our perception of the physical world is easy and immediate is an illusion created by our brains" (Frith 2007a, 111). He points to various contributions to this way of understanding the work of the brain—information theory, for example, which shows that messages can be recreated from physical events by eliminating noise and errors, and by the use of redundancy. He also notes the evidence that expectations and prior knowledge shape perception and that some of this prior knowledge is "hard wired into the brain through millions of years of evolution"—for example, that light comes from above (the sun), and so our interpretation of patterns of shadows on objects interprets them as convex or concave on that assumption, as many optical illusions show (ibid.).

Then there is Bayesian theory and what it tells us about the ways the brain creates beliefs or models of the world on the basis of successive predictions that are then compared with the patterns of activity detected from the senses, predictions that are linked to our actions in the world and the feedback they provide. What I perceive, then, is not the world, but "my brain's model of the world," which combines the ambiguous cues from all my senses with a picture built up by past experience to generate predictions of what ought to be out there in the world, which I constantly test and refine through my actions (Frith 2007a, 132). This, it seems, is what is demonstrated by all those familiar optical illusions, from the Ames room through the Necker cube to the Rubin vase figure: "our perceptions are fantasies that coincide with reality"—or perhaps we should say, that "usually" coincide with reality. It is not so much a question, says Frith, of whether the brain's model of the world is true, but whether it works—enables us to take appropriate actions in order to survive. And all this happens effortlessly, outside of any consciousness. Our perception of the physical world is, in short, a fantasy constructed by our brains. And as for the physical world, so for the mental world—we create models of the mental world of others in the same way, recognizing the motion of dancing figures from patterns of spots of light alone, inducing the goals and intentions of others into their movements and the direction of their eyes, imitating others, actually or mentally, and automatically imputing to them the same intentions as we ourselves would have if we were acting in this way (the line of thought extrapolated in the rhetoric of mirror neurons discussed in chapter 5) and much more.

Frith, as we have noted, conducts his book as an imaginary debate with a skeptical professor of English, into whose mouth he places the supposed objections from the human sciences. As one would anticipate, she is distressed by the absence of meaning, language, and culture from these accounts, and the attribution to brains of capacities that should properly only be attributed to persons—thinking, feeling, wanting, knowing, and so forth. But, beyond such objections, how should the human sciences respond to the fundamental proposition that the world we humans live in is not an immediate perception of some external reality? As we have seen, this has, in fact, been the found-

ing presumption of much contemporary social thought, enabling a historical and cultural interrogation of both the form and the function of changing
beliefs about persons and selves. If one was of a radical political bent, one
could look to the French Marxist philosopher Louis Althusser, to whom the
autonomous self is the central element in contemporary ideology, "a representation of the imaginary relationship of individuals to their real conditions
of existence" (Althusser 1971, 153). Or if one was of a more philosophical bent,
one could look to the American philosopher Nelson Goodman's notion of
"irrealism"—which he describes as a kind of radical relativism with rigorous
constraints, in which many world versions—some conflicting and irreconcil-
able with others—are equally right (Goodman 1978, 1983). But it seems that,
in the matter of the creation of worlds, dethroning the self-possessed self and
its illusions of reality by culture and language is one thing, and dethroning
the perceptual capacities of the self-possessed self by the nonconscious, automatic, evolutionarily shaped functioning of neurons and their hard- and
soft-wired interconnections is quite another.

Libetism

In 1983, Benjamin Libet and his colleagues published an article in *Brain* titled
"Time of Conscious Intention to Act in Relation to Onset of Cerebral Activity (Readiness-Potential): The Unconscious Initiation of a Freely Voluntary
Act"—an article that has become a standard reference for almost every account of the implications of neurobiology for mind-brain relations and the
idea of free will. Perhaps it is not too much to describe the way of thinking
that was engendered by this article as *Libetism*. What then is Libetism?

Libet's experiment relied on a finding that a certain pattern of brain activity
in prefrontal motor areas that is termed the "readiness-potential" (RP), which
can be measured by electrodes placed (externally) on the scalp, had been observed to occur up to a second or more before an individual is aware of having initiated an act. Libet and his colleagues worked with six college students,
divided into two groups of three, each sitting in a "partially reclining position
in a lounge chair" with recording equipment attached to their heads and a
cathode ray oscilloscope in front of them into which they were to stare without
blinking. They instructed each subject "when he felt like doing so, to perform
the quick, abrupt flexion of the fingers and/or the wrist of his right hand." In a
second experiment, they instructed each subject " 'to let the urge to act appear
on its own at any time without any preplanning or concentration on when to
act,' that is, to try to be 'spontaneous' in deciding when to perform each act;
this instruction was designed to elicit voluntary acts that were freely capri-
cious in origin" (Libet, Gleason, Wright, et al. 1983, 625). The time of initiation
of the RP was recorded using the monitoring equipment attached to the scalp.

Why were the subjects asked to stare without blinking at the oscilloscope?
Well, this had a revolving beam like the second hand of a clock, with spots

marked on the circumference, and the subjects had to remember, and subsequently report, the "clock position" on that oscilloscope. In some trials each had to report the position at "the time of appearance of his conscious *awareness of 'wanting' to perform* a given self-initiated movement. The experience was also described as an 'urge' or 'intention' or 'decision' to move, though subjects usually settled for the words 'wanting' or 'urge.'" In other trials they had to report "the time of subject's *awareness* that he/she *'actually moved'*" or, in a third series, where the movement was triggered by an external stimulus, the "time of *awareness of the sensation* elicited by the near-threshold stimulus pulse to the back of the hand, delivered at randomly irregular times unknown to the subject" (Libet, Gleason, Wright, et al. 1983, 627). Of course, there were many other technical details and much statistical analysis of the results. And the conclusion?

"It is clear that neuronal processes that precede a self-initiated voluntary action, as reflected in the readiness-potential, generally begin substantially *before* the reported appearance of conscious intention to perform that specific act. This temporal difference of several hundreds of milliseconds appeared fairly consistently" (Libet, Gleason, Wright, et al. 1983, 635). Thus, it seems, "the brain decides" to act before awareness of, or conscious initiation of, that act:

> Since onset of RP regularly begins at least several hundreds of milliseconds before the appearance of a reportable time for awareness of any subjective intention or wish to act, it would appear that some neuronal activity associated with the eventual performance of the act has started well before any (recallable) conscious initiation or intervention could be possible. Put another way, the brain evidently "decides" to initiate or, at the least, prepare to initiate the act at a time before there is any reportable subjective awareness that such a decision has taken place. It is concluded that cerebral initiation even of a spontaneous voluntary act, of the kind studied here, can and usually does begin *unconsciously.* The term "unconscious" refers here simply to all processes that are not expressed as a conscious experience; this may include and does not distinguish among preconscious, subconscious or other possible nonreportable unconscious processes. . . . These considerations would appear to introduce certain constraints on the potential of the individual for exerting conscious initiation and control over his voluntary acts. (Ibid., 640)

So, if the brain decides to act prior to conscious initiation of an act, where then lies the possibility for free will—a retrospective rationalization of the act, giving the subject the illusion that he or she has actually consciously willed it, whereas in fact it was the brain that decided? Libet and his colleagues allow two ways out—the subject might decide to veto or abort the performance—for both the RP and the awareness of the decision appeared *before* the actual movement in question. Or "[i]n those voluntary actions that are not 'spontaneous' and quickly performed, that is, in those in which conscious deliberation (of whether to act or of what alternative choice of action to take) precedes the act, the possibilities for conscious initiation and control would not be excluded by the present evidence" (Libet, Gleason, Wright, et al. 1983, 641). This is what we have termed *Libetism*.

"The result," observes Frith, "had such a vast impact outside psychology because it seemed to show that even our simplest voluntary actions are predetermined. We think we are making a choice when, in fact, our brain has already made the choice" (Frith, 2007a, 67). Of course, to most people who pause to consider this experiment, the idea that this tells us anything about the exercise of human will in any of the naturally occurring situations where individuals believe they have made a conscious choice—to take a holiday, choose a restaurant, apply for a job, vote for a political party, approach a potential girlfriend or boyfriend, write a letter of complaint to some authority—is absurd. Not only is the task involved in Libet-like experiments trivial, and the perspective of the subjects of the experiment on their role ignored, but the results simply meant that those subjects reported their awareness of that decision—the "urge"—a few milliseconds after a brain change occurred—a fact that seems to be about the process of reporting rather than the process of deciding. Actually, one has to be a strange kind of dualist, first to distinguish, and then to reconnect these two in a relation of causality. And, of course, even experimental psychologists find flaws in the welter of conclusions about free will that Libet and others built on such findings. For example, Frith points out that, to put it mildly, the freedom of these subjects was somewhat constrained—they had a task to perform and an experimenter to satisfy: "The participants are not actually making free choices about action. They are playing a complex game with the experimenter" (ibid., 187).

Although current neurobiological understandings of volition are significantly more complex and subtle (see Haggard 2008 for an excellent review), Libetism is alive and well, especially in debates with neuroscientists and philosophers about the idea of free will (Frith 2002; Wegner 2002).[12] When the minds (or brains) of those who think like this turn to the implications for the law and legal notions of responsibility, Libetism is often coupled with some of the other arguments we have reviewed in this section and presented without the qualifications that Libet himself makes. Here, for instance, was the British neuropsychiatrist Sean Spence, whose brain presumably initiated the writing of this in the mid-1990s:

> The purpose of this paper is to demonstrate that a conscious free will (in the sense of consciousness initiating action) is incompatible with the evidence of neuroscience, and the phenomenology described in the literature of normal creativity, psychotic passivity, and the neurological syndrome of the alien limb or hand. In particular the work of Libet and others who have directly stimulated the brain and measured activity

from its surface leads to the conclusion that subjective states (be they sensory or intentional) are preceded by predictive neural activity . . . "decisions" to act arise prior to our conscious awareness of them. Thus our "decision" or "freedom" is illusory (if by these terms we mean conscious phenomena). It occurs after activity has been initiated. . . . The decision does not make itself. It is a product or correlate of preceding neuronal events. (Spence 1996, 75, 83–84)

Michael Gazzaniga is a little more cautious, writing in *Neuron* in 2008 under the title "The Law and Neuroscience":

One thing we are certain of is that the "work" in the brain happens before we are consciously aware of our mental struggles. Researchers have, since as early as 1965, advanced our understanding of the fact that much of the work is done at the subconscious level (Kornhuber and Deecke, 1965; Libet et al., 1979; Soon et al. 2008). A decision, for example, can be predicted several seconds before the subject consciously decides. If it is simply the brain, working up from its unconscious neural elements, that causes a person to act (even before he or she is aware of making a decision), how can we hold any person liable for his or her mental decisions? To hold someone responsible for his or her actions, one must find a "there" there. Is a little guy pulling the levers in your head producing a free-floating you? Modern neuroscience, of course, tells us the answer is "no." (Gazzaniga 2008b, 413)

Where, then, does this leave the question of the self and the person? In the brief epilogue at the end of his book, Frith turns his mind, or his brain, to this issue. With all this talk of the brain, and how much it achieves without any conscious or willed engagement, without thought or awareness, how is one to account for the feeling of "I"—that I think, that I intend, that I act? Perhaps there is an area in the brain that corresponds to this "I"ness—but surely not a kind of homunculus, for that would merely be to replicate the question that it was intended to solve. No, if the brain does so much that we usually attribute to consciousness, then the perception that one is a conscious "I"—and that others are also such "I"s—must itself be produced by the brain as a result of evolution: "This is my brain's final illusion: . . . [to] create an autonomous self." But he goes on to tell us that his is not a book about consciousness, although, as he points out "[a]fter the age of about 50, many neuroscientists feel they have sufficient wisdom and experience to set about solving the problem of consciousness" (Frith 2007a, 189). He adds the acerbic footnote: "Whether or not they have ever done any experimental work on the topic."

Perhaps it would be unfair to think of Thomas Metzinger in the context of this remark, for although he is a theoretical philosopher, he works closely with cognitive neuroscientists. Nonetheless, his argument is put provocatively: human beings "do not have selves." He argues that "we live our conscious lives in our own Ego Tunnel, lacking direct contact with outside reality but possessing an inward, first person perspective." This ego tunnel, which has evolved for controlling and predicting our behavior and understanding the behavior of others, creates a robust feeling of being in contact with an outside world, and a sense of immediate contact with our selves—"the self is not a thing but a process . . . we are 'selfing' organisms: At the very moment we wake up in the morning, the physical system, that is—ourselves—starts the process of 'selfing'" (Metzinger 2009, 207–8). But do those who have done the experimental work on this topic share this view of the constitutive illusion of the human self?

The Self: From Soul to Brain

"An Ape with a Soul" is the title that V. S. Ramachandran gives to the penultimate chapter of his popular book *The Tell-Tale Brain: A Neuroscientist's Quest for What Makes Us Human* (2011).[13] It opens it with an account of a patient called Jason, who suffered a serious head injury in a car accident, followed by a long period in a semiconscious state of akinetic mutism. Jason seemed alert, could move his eyes, but could not recognize anyone, talk, or comprehend speech—except in one context. If his father, whom he neither recognized nor spoke to directly, telephoned him from another room, he became alert and talkative on the phone. For Ramachandran, Jason was a person when he was on the phone but not otherwise—a person, he suggests, requires "the ability to form rich, meaningful metarepresentations, which are essential not only to our uniqueness as a species, but also our uniqueness as individuals and our sense of self" (Ramachandran 2011, 246).

It is, it seems, *metarepresentations* that are the key. Like other neuroscientists, despite some difference of terminology, Ramachandran distinguishes between what he terms first-order sensory representations of external objects—which evolved earlier, and humans share with many other animals—and metarepresentations—representations of representations, created by a part of the brain that evolved later and works "by processing the information from the first brain into manageable chunks that can be used for a wider repertoire of more sophisticated responses, including language and symbolic thought" (Ramachandran 2011, 246). He distinguishes seven dimensions, which together create a sense of self "in the normal human brain," each of which may be disrupted or eliminated where a region of the brain is damaged or nonfunctional: *unity*—a myriad of sensory experiences somehow cohere to give the sense that one is a single individual; *continuity*—the ability to project oneself back and forward in time with a continuing identity; *embodiment*—the sense that one's body, and only parts of one's own body such as one's limbs, belong to oneself; *privacy*—that one's experiences of sensations, qualia, and one's mental life are one's own, alone; *social embedding*—

that we are part of, and can interact with, a social environment; *free will*—that one is able consciously to "choose between alternative courses of action with the full knowledge that you could have chosen otherwise" (ibid., 252); and *self-awareness*—an aspect of selfhood that is, for Ramachandran, axiomatic.

It is hardly necessary to point out that Ramachandran's sense of self consists precisely in those features of selfhood that Charles Taylor suggests arise from a historically limited mode of self-interpretation. For the human sciences, the time of such a history is that of human culture. For Ramachandran, the time is that of evolution: these features are indeed understood as the outcome of evolved features of human neurobiology, creating the sense of unified selfhood arising from the human brain itself—seemingly independent of its historical or cultural moorings. At the root of his analysis, as for many other neurobiologists, is the view that neural organization is not a unitary whole but a multiplicity, an assemblage that has come together over the course of evolution.

As Francisco Varela puts it: "brains are not logical machines, but highly cooperative, nonhomogeneous. . . . [T]he entire system resembles a patchwork of subnetworks assembled by a complicated history of tinkering, rather than an optimized system resulting from some clean unified design . . . networks whose abilities are restricted to specific, concrete, cognitive activities as they interact with each other" (Varela 1992, 321). Whether it is Michael Gazzaniga's modular conception of the brain as "organized into relatively independent functioning units that work in parallel" apart from our conscious verbal selves, contributing to consciousness but often producing behavior without conscious intervention (Gazzaniga 1985b, 4), or Marvin Minsky's idea of mind as a society of multiple agents with limited and circumscribed abilities (Minsky 1986), the message is clear: mental activity is made up from the interconnections and interactions between many relatively distinct components, each with its own mode of action. And yet, out of this assemblage, for humans at least, unless interrupted by pathology or tricked by experiment, a sense of a punctual, unified, continuous, autonomous, and self-conscious self always emerges.

The neurobiological self, then, is not to be equated with consciousness. Presenting a paper at an event he co-organized under the auspices of the New York Academy of Sciences titled "The Self: From Soul to Brain," Joseph LeDoux sought to characterize an emerging body of thought within neuroscience that articulated this "broader view of the self" as a multiplicity with several distinct aspects, many of which are nonconscious (LeDoux 2003, 296–97). He argued that the notion of the self can be thought of along an evolutionary continuum: while only humans have the unique aspects of the self that are made possible by their brains, other animals have the kinds of selves made possible by *their* brains.[14] As summarized by Henry Moss, in his review of the conference, for LeDoux: "our 'selves' are our brains, or, more accurately, the synaptic connections and systems that capture our experiences and memories, and express our habits and dispositions" (Moss 2003, 2). "Because you are a unique individual, the particular multifaceted aspects of the self that define 'you' are present in your brain alone. And in order for you to remain who you are from minute to minute, day to day, and year to year, your brain must somehow retain the essence of who you are over time. . . . [T]he self is essentially a memory, or more accurately, a set of memories" (LeDoux 2003, 298).

But LeDoux is at pains to stress that this way of understanding our selfhood is not at the expense of all others—the view of the self as psychological, social, or spiritual in nature and not neural: "My assertion that synapses are the basis [of] your personality *does not assume that your personality is determined by synapses; rather, it's the other way around.* Synapses are simply the brain's way of receiving, storing, and retrieving our personalities, as determined by all the psychological, cultural, and other factors, including genetic ones. . . . A neural understanding of human nature in fact broadens rather than constricts our sense of who we are" (LeDoux 2003, 302–3; emphasis added). The self *is* the brain, the self is *based on* the synapses, the synapses are ways of storing a multitude of inputs that constitute our personality, and the self is a set of memories. One suspects this constant slippage of terminology masks a deeper confusion about what it is that is being written about here. And the matter is made even less clear when LeDoux invokes the preeminently psychological concept of *personality*, which he does not define, as the locus where nature and nurture interact in the "wiring of synapses" that makes possible the self, while still arguing that the source of identity and continuity of the person remains nonconscious—the brain itself.

Despite these diverse, terminologically slippery, and somewhat confused accounts, neuroscientists such as Ramachandran and LeDoux consider that *coherence* is the very condition of normal selfhood. LeDoux is typical in suggesting that when such coherence fails, mental health deteriorates—indeed mental illness, it seems, centrally involves the failure of that sense of coherent selfhood. But if the self is, or arises from, or is based upon, a multiplicity, how does the *impression* of unity and coherence necessary for normality emerge? LeDoux is somewhat vague, suggesting that all the different systems in the brain experience the same world, and that all the systems are interconnected and use the same monoamines, and that this, somehow, enables most of us, most of the time, to hold our self together.

In her contribution to the event, Patricia Churchland, however, addresses the problem of coherence directly. She focuses on what she terms the "self-concept." "[I]t is useful to start," she says, "with the idea that one's self-concept is a set of organizational tools for 'coherencing' the brain's plans, decisions, and perceptions," and it turns out that there are remarkably diverse ways in which individuals refer to themselves; thinking of these in terms of "the self-representational capacities of the brain thus deflates the temptation to think of the self as a singular entity and encourages the idea that self-representing involves a plurality of functions, each having a range of shades, levels, and degrees" (Churchland 2003, 32). This suggests that various levels of such co-herencing can be found in other species, as a result of the nature of their particular brains. She draws on diverse evidence for her suggestion that there is a physical basis in the brain for the capacities for self-representation—not just evidence of the neural bases for the representation of the functions of the internal milieu of the body, but also split-brain studies, developmental stud-ies of the capacity for self-representation in children and the fading of such capacities in dementia, studies of amnesic subjects with various lesions who seem to have lost all their autobiographical information, confusion about self/nonself boundaries in schizophrenia, patients with lesions in various re-gions who show limb-denial, alien hand syndrome, personality change, and the like.

And she follows Damasio in arguing that the self/nonself distinction pro-vides the basis for consciousness, as it builds capacities to distinguish be-tween internal and external representations, creating a metarepresentational model of the relation between inner and outer worlds: "as wiring modifica-tions enable increasingly sophisticated simulation and deliberation, the self-representational apparatus becomes correspondingly more elaborate, and therewith the self/not-self apparatus. On this hypothesis, the degrees or levels of conscious awareness are upgraded in tandem with the self-representational upgrades" (Churchland 2003, 37). "The self thus turns out to be identifiable not with a nonphysical soul, but rather with a set of representational capaci-ties of the physical brain" (ibid., 31).

Now this is clearer, and indeed accords with the kinds of arguments made in the human sciences—the self is an imaginary representation of the real relations that any human being has to its internal world and external milieu. Leaving aside the *re-* in representation—which always misleadingly directs us to asking what is it that is present prior to its re-presentation—we can now ask the interesting questions about how experience, culture, language, and so forth shape these representations. But that is not the direction pursued by most neuroscientists. Instead, some, entranced by the apparent capacities of neuroimaging to localize anything referred to by a 'mind' kind of word in some brain region activated during a task carried out in a scanner, ask where in the brain this representational self can be found (Feinberg and Keenan 2005; Northoff, Qin, and Feinberg 2011). One way to approach this is familiar: study those who appear to have a deficit in some aspect of self-representation (Sinigaglia and Rizzolatti 2011, 71).

Thus Baron-Cohen and others committed to the importance of theory of mind have suggested that the "mindblindness" or theory of mind deficits they find in those diagnosed with autism spectrum disorders may also affect the ways in which the individual becomes aware of his or her own experience and can use that self-awareness in understanding the experience, or the minds, of others (Carruthers 2009).[15] Some suggest that perhaps this might be located in an anomaly in the right temporo-parietal junction, which may be part of a distributed network of brain regions that is 'recruited' for mentalizing about both self and other (Lombardo, Chakrabarti, Bullmore, et al. 2010; Lombardo and Baron-Cohen 2011).[16] Others have taken different routes to discover the neural correlates of the various components of such metarepresentational capacities, seeking to identify and localize the neural systems of subsystems responsible for each aspect considered crucial for the emergence of the self: first-person perspective in spatial navigation, first-person narrative comprehension, self-face recognition, self-memory, awareness of the boundaries of one's own body, awareness of relation of self to social environment, and much more. These studies are well reviewed by Gillihan and Farah, who conclude that "many of the claims for the special status of self-related processing are premature given the evidence and that the various self-related research programs do not seem to be illuminating a unitary, common system, despite individuals' subjective experience of a unified self" (Gillihan and Farah 2005).

This is not surprising given the fact that most of those carrying out these studies are either confused or imprecise about what they mean by 'the self,' and this imprecision is manifested when they redeploy brain imaging tasks and procedures devised for other purposes to try to localize 'selfhood' in a particular brain region (Zahavi and Roepstorff 2011). Paradoxically, however, the weaknesses of attempts to 'locate' the self in the brain serves to support the general thesis—that the conscious feeling of selfhood, as self-awareness, coherent identity, willful agency, and the like emerges out of the transactions among a whole array of evolutionarily distinct and functionally separate subsystems, whose modes of operation not only exceed that self-awareness and constitute it, but are, in principle, unavailable to it.

There is, as we have already mentioned, a long tradition of research on cultural variability in self-concepts. There was a revival of interest in this work in the sociology and anthropology of the 1980s: most argued that cultures shaped the self by acting on collective beliefs, practices of socialization, and individual psychologies. Can the brain take on the role that the psyche once occupied—can these brain-based conceptions of the self open themselves to culture? Today, it seems, the answer is yes: for as we have seen in previous chapters, the twenty-first century brain is not an organ whose fate is fixed at the moment of conception or birth, but is plastic and open to external influences throughout development and indeed throughout life.[17]

For LeDoux, as we have seen, selves are memories, and memories are patterns of synaptic interconnections. How are these patterns of synaptic interconnections established? In a tradition that goes back to Donald Hebb's maxim that "what fires together, wires together" (Hebb 1949), it is now 'experience' that shapes the patterns of neural connections that are strengthened—and what is experience, if not cultural? This, for example, is the argument developed by Naomi Quinn in her contribution to *The Self: From Soul to Brain*, titled "Cultural Selves": "Thus, to LeDoux's picture of how neural processes engage multiple neural systems to endow each person with a self, must be added a description of how a very different set of processes, those that enter the cultural patterning of experience, shape and enhance that self. . . . The cultural patterning of the world we live in is so far-reaching and so all-encompassing that it cannot help but pattern our synaptic connections" (Quinn 2003, 146).In this neural version of the social shaping of the self, all culturally variable aspects of selfhood familiar from psychological accounts—habit, comportment, distinctions between good and bad behavior, practices for self-management, "cultural schema"—are inscribed into the developing brain itself through the patterns of synaptic connections formed during child rearing and socialization.

Yet despite this belated entry of culture, we are left with something of a paradox. For while evolved neural processes and pathways provide the *capacities* for self-representation, for this metarepresentation of representations that becomes conscious and amenable to awareness, what accounts for the *form* of such representations? By this we mean, not the 'content' of the self—for this is what is addressed by the claims about the role of cultural schema and the like—but the *sense of self* itself. Is that form—unitary, punctual, continuing, attributing autonomy and free will to itself—given in the neural capacities and hence universal? Or is it socially and historically variable, as argued by the cultural and historical sciences?

Perhaps this is merely a new version of that double-faced conception of the self that was articulated by Marcel Mauss many decades ago. Mauss, as we have seen, argued that the *category* of the self or the person is historically and culturally variable—by which he means, he says, the notion or concept that people in different ages have formed of it. Yet, to repeat, he asserts that "there has never existed a human being who has not been aware, not only of his body, but at the same time also of his individuality, both spiritual and physical."[18] In other words, while the *form* of the self is variable across cultures and epochs, he believes that the *sense of self* is universal. It is worth quoting the remarks with which he continues: "The psychology of this awareness has made immense strides over the last century, for almost a hundred years. All neurologists, French, English and German, among them my teacher Ribot, our esteemed colleague Head, and others, have amassed a great deal of knowledge about this subject and the way this particular awareness is formed, functions, deteriorates, deviates and dissolves. . . . My subject is entirely different and independent of this" (Mauss [1938] 1985, 2).[19]

Should we be content with this distinction between what we owe to nature and what we owe to culture? Should we delegate the task of understanding the biological conditions of awareness, consciousness, and the sense of self to neurobiology and allocate to an "entirely different" enterprise the task of exploring the ways that humans at different times and places come to speak about selves, allocate statuses and capacities to them, mark them culturally, differentiate among them, and so forth? These questions undoubtedly inhabit distinct registers. But the distinction between an esoteric scientific discourse and an everyday lay understanding is itself a fact of history—scientific truths and cultural truths are not always so distinct. What happens, then, when these two dimensions transect and intertwine? What happens when the quotidian conception that human beings have of themselves is itself shaped in part by the truth claims made in an esoteric language of science? This certainly was the case for psychological conceptions of personhood across the twentieth century. Is it also our contemporary condition in relation to neurobiology?

A Mutation in Ethics and Self-Technologies?

Despite their many differences, it is possible to identify a general picture of 'the self' and 'the person' emerging from contemporary developments in the brain sciences. In this picture, consciousness, with its image of the person as unified, purposive, intentional, and self-aware, is by no means the master in its own house. Consciousness is an effect, a metarepresentation, which creates an illusory, but probably evolutionarily advantageous, sense of coherence and self-direction. In fact, however, our conscious selves are not aware of, let alone in control of, the multitude of neural processes that enable us to live our lives: to receive, interpret, and respond to sensory inputs, to create an internal sense of our embodiment, to manage our motor skills and comportment, to remember and sometimes to recall previous experiences, to respond to many features in our environment and to humans and other animate entities, and to select among, and carry out, various options for action. It is not just that these, and many other processes essential to survival happen outside of our consciousness. It is that they are *in principle* unavailable to consciousness in normal circumstances, although some can become amenable to conscious control under particular conditions or with appropriate training.

There is not much in this account that is new or startling to those from the *Geisteswissenschaften*. The novelty of contemporary neurobiological arguments lies elsewhere, in the claims about the processes that enable all this to occur. But does this renewed recognition of the limits of our awareness and conscious control over the neural infrastructure that enables us to consider ourselves as selves have any relevance to the way in which we are, to use Foucault's phrase, "constituted as moral subjects of our own actions" (Foucault 1984, 49)? Is it now these nonconscious neural processes that are considered as the constitutors of moral action, the origin of morally relevant conduct, the subject of moral evaluation, the target of self-improvement? Has the new prominence of the brain in these truth discourses effaced older forms of selfhood, rendering personhood exclusively, or largely, a matter of brains? The answer to the first set of questions is a qualified yes—at least for some of our authorities, though perhaps not yet for many individuals themselves, those neural processes, now attributed names, locations, and mechanisms, have achieved salience and acquired a seeming practicability. But the answer to the latter is certainly no: personhood has not become 'brainhood.'

The brain has certainly become a rich register for narratives for self-fashioning.[20] There are now many neurobiological technologies of self-hood, that is to say, practices that seek to mold, shape, reform, or improve aspects of one's person—moods, emotions, cognition, desire—by acting on or through the brain. But with what consequences for personhood? Let us briefly consider perhaps the longest established of these contemporary neurotechnologies—psychopharmaceuticals.[21]

Despite the claims of some that such drugs promise a painless modulation of personality (Kramer 1992), and the reciprocal worries of some who believe the drugs offer a shallow and facile chemical manipulation of one's selfhood (President's Council on Bioethics (U.S.) and Kass 2003), there is little evidence here of a reduction of personhood to the brain. It is not only that the drugs do not work as advertised—they do not have the remarkable therapeutic properties once claimed for them, and their mode of operation is far less specific than initially thought (Moncrieff and Cohen 2005; Moncrieff 2008). It is also that what such drugs promise is not designer moods or designer selves; they do not offer a limitless and painless recrafting of one's personhood through a simple chemical redesign of the brain (Rose 2003, 2007).[22] The image that they offer is different, and it appeals to those who believe that, for whatever reason, they have lost touch with themselves, that they fail to recognize themselves in the person that they have become. The promise they offer is not that you will become someone new, but that you may be able to get your life back, to become yourself again, to feel, once more, like the "real me" or "better than well" (Kramer 1992, x). And to get your life back, to return to the 'real you' that you were once, that you that your loved ones remember, you must do more than consume a pill: this requires work on the self—introspection, self-monitoring, perhaps cognitive behavior therapy—all implying the existence of persons with brains, not persons who are their brains.

The two most powerful empirical studies of the way in which individuals engage psychopharmaceuticals in their everyday lives and practices of self-fashioning are David Karp's *Is It Me or My Meds?* (Karp 2006) and Emily Martin's *Bipolar Expeditions* (Martin 2007; see also Martin 2006). Each contains compelling firsthand accounts of individuals seeking to manage psychiatric conditions that they consider, at least in part, to be biological, based upon, arising from, or otherwise linked to, anomalies in their neurobiology. But in no case do these individuals consider themselves to be mere fleshly puppets of their brains or their neurotransmitters: they are persons with brains, and persons who believe that they experience their brain states by means of what we might loosely call a psychological sense of their own consciousness, and who need, as a matter of their psychological and subjective responsibilities, to manage their moods through the modulated use of psychoactive medication. It is true that they employ a vocabulary in which their personhood is shaped by their biology, and altered by their medication. And they wonder if their changing moods are caused by the condition itself, by the interaction of the medication upon the condition, or by their own actions and their interactions with family, friends, or professionals.

While it has certainly become possible for human beings today to understand their actions, emotions, desires, and moods in a neurobiological language, their sense that they are inhabited by a psychology remains, and remains crucial. For it is through that psychological selfhood that they are required to enact the responsibilities that they have now acquired over the consequences of their troubled and troublesome brains. While one might now have the opportunity of modulating specific aspects of personhood by neurochemical means, it remains the responsibility of the person to manage himself or herself by acting on the brain. Indeed this reminds, us, if we need reminding, that our own time is not unique in having multiple and often inconsistent and contradictory repertoires for describing and managing personhood and selfhood, which are activated and deployed in different contexts and practices, and in different departments of life. If over the past century a psychological sense of the person has come to dominate in our ways of thinking, speaking, and acting in so many practices, and interpreting the actions of others, it is unlikely easily to be effaced by neurobiology.

But drugs are not the only neurotechnologies. There are now multiple attempts to persuade individuals to adopt or use technologies that claim to act directly to modulate the brain and its development. For example, we have seen the emergence of 'brain education' or 'brain awareness programs.' The Dana Foundation, in partnership with the Society for Neuroscience, runs Brain Awareness Weeks, described as "the global campaign to increase public awareness about the progress and benefits of brain research. Every March BAW unites the efforts of organizations worldwide in a week-long celebration of the brain. During BAW campaign partners organize activities to educate and excite people of all ages about the brain and brain research. Events are limited only by the organizers' imaginations. Examples include open days at neuroscience laboratories; museum exhibitions about the brain; lectures on brain-related topics; displays at malls, libraries, and community centers; and classroom workshops."[23] They involve downloadable coloring sheets for children, based on the Dana Alliance's Mindboggling booklet series; downloadable puzzles for all ages, based on the content of Dana Alliance publications; tests of neuroscience know-how; tools for educating policymakers; and activities organized by many universities and organizations in the United Kingdom and the United States, as well as in many other countries, such as Malaysia.

Plasticity plays an explicit rhetorical role in arguments for such interventions on the brain. Thus the International Brain Education Association supports brain education across the world, basing its philosophy on neuroplasticity:

A few decades back, scientists thought that people could have very limited influence over their brains. It was assumed that by the time people reached adulthood, their brain connections were permanently and indelibly in place. . . . Recently we have discovered the neuroplasticity of the brain. Right up to the end of your life, you can restructure and adapt your brain according to your needs. Your capacity to change your brain depends on your choice, nothing is predetermined. . . . The brain is the only organ that exerts influence on the world outside your body as well as inside your body. . . . IBREA's message is that if we all chose to use our brains to pursue the common good and bring humanity together under a common system of life values, our lives and our experience in this Earth will become much more meaningful and hopeful.[24]

Or consider Power Brain Education, which disseminates programs that "teach how to use the brain well, through physical conditioning and balance, sensory awareness, emotional regulation, concentration, and imagination."[25] There is also the Brain Gym movement founded by Paul Dennison, although its promoters are honest enough to explain: "The investigation of the neuroscience that underpins Brain Gym is an eagerly awaited project for the future. We can only at the present time hypothesise about why without actually knowing."[26] And this is not to mention all the organizations offering techniques for brain fitness, brain health, or brain improvement, such as SharpBrains, whose 2011 summit was titled "Retooling Brain Health for the 21st Century,"[27] or Neurobics, "a unique new system of brain exercises based on the latest scientific research from leading neurobiology labs around the world . . . the first and only program scientifically based on the brain's ability to produce natural growth factors called neurotrophins that help fight off the effects of mental aging."[28] And, closely related, one sees the invention or rebranding of an array of nutritional products or other devices purporting to improve brain fitness—brain foods, brain drinks, brain-boosting games for Nintendo, and so forth.

But one can also act on one's brain by acting on one's mind. This is most clearly exemplified in the intersections of neuroscience and Buddhism, both in its pure versions in the Mind and Life Institutes and in the somewhat debased version of 'mindfulness' being taught in sites ranging from psychotherapy to business management.[29] But there is much more. A few minutes on the Internet will lead to dozens of philanthropic organizations, quasi-religious evangelicals for peace, and profit-seeking enterprises organized around this premise—your brain is amazing; it is flexible; it can be trained, developed, improved, optimized: learn to use it well for your own benefit and for that of your society, perhaps even for the world.[30] It is not that you have *become* your brain, or that you are identical with your brain, but you can act on your brain, even if that brain is not directly available to consciousness, and in so acting, you can improve yourself—not as a brain, but as a person.

Caring for the Neurobiological Self

Our argument in this chapter has been that the emerging neuroscientific understandings of selfhood are unlikely to efface modern human beings' understanding of themselves as persons equipped with a deep interior world of mental states that have a causal relation to our action. Rather, they are likely to add a neurobiological dimension to our self-understanding and our practices of self-management. In this sense, the "somatic individuality" (Novas and Rose 2000; Rose 2007c), which was once the province of the *psy-* sciences, is spreading to the *neuro-* sciences. Yet *psy* is not being displaced by *neuro*: neurobiological conceptions of the self are being construed alongside psychological ones. We extend our hopes and fears over our biomedical bodies to that special organ of the brain; act upon the brain as on the body, to reform, cure, or improve ourselves; and have a new register to understand, speak, and act upon ourselves—and on others—as the kinds of beings whose characteristics are shaped by neurobiology.

This is not a biology of fate or destiny—although we may sometimes describe ourselves as fated, for example, when diagnosed with a neurodegenerative disorder—but a biology that is open for intervention and improvement, malleable and plastic, and for which we have responsibility to nurture and optimize. If our contemporary ethical practices are increasingly 'somatic,' that is to say, organized around the care of our corporeal selves, aided by rules for living articulated by somatic experts from doctors to dieticians, we should not be surprised if care of the brain becomes one aspect of this 'somatic ethics' (Rose 2007c). But in all these neurotechnologies, the self that is urged to take care of its brain retains the form of selfhood that took shape in the late twentieth century—the commitment to autonomy and self-fulfillment, the sense of personal responsibility for one's future, the desire for self-optimization, the will, often the obligation, to act on our selves to improve ourselves over the whole of our lives. That is to say, we are not seeing personhood replaced by brainhood, but the emergence of a new register or dimension of selfhood, *alongside* older ones—a dimension in which we can understand and take care of ourselves, in part, by acting on our brains.

Added those other obligations, we are now acquiring a new one: the obligation to take care of our brain—and of the brains of our families and children—for the good of each and of all. For as we have seen throughout the book, brains are now construed as persistently plastic, molded across the life-course and hence requiring constant care and prudence. There is clearly an "elective affinity," to use Max Weber's term, between this emphasis on plastic, flexible brains and more general sociopolitical changes that prioritize individual flexibility across the life span to accommodate to rapidly changing economic demands, cultural shifts, and technological advances—and that demand a constant labor of self-improvement on the part of today's citizens. Indeed, that link is made explicitly in the marketing of many of the neurotechnologies we have discussed. But it would be too easy to merely map the neural onto the social in this way. Such socioreductionism would be a short-cut, sidestepping the work of analysis required to link these developments in the neurosciences and neurotechnologies with their sociopolitical context. The wish to fashion the self is not a recent phenomenon, nor is the belief that the continuous work of improving the self is a virtuous exercise of freedom. In the liberal societies of the West, from around the 1960s, at least for some of the middle classes and for many young people, such self-fashioning became no longer the privilege of the elite, the philosopher, the dandy, or the aesthete. The radical democratization of self-fashioning over the closing decades of the twentieth century has been taken into new territory with the spectacular di-versification of authorities of the self in the age of the Internet. What is novel, then, is not the aspiration to shape, improve, fashion oneself, but the source of authority that underpins it, the technologies that it deploys, and the target or substance upon which it operates—the brain itself.

Conclusion
. . . .
Managing Brains, Minds, and Selves

Are developments in the neurosciences transforming our conceptions of what it is to be a human being? If so, how have they come to achieve this role, what forms are these transformations taking, and with what consequences?[1] These questions, with which we began this book, lead to a subsidiary one, for if such transformations are occurring, how should the social and human sciences respond, given that individual and collective human beings are their own privileged objects of investigation? Of course, it is far too early to reach any definitive diagnosis. As we have argued, while investigations into the brain and nervous system can be traced back many centuries, *neuroscience* is barely half a century old. And it would be misleading to search for any single figure of the human that inhabits all practices at any time and place, let alone across an epoch: many different conceptions of personhood can and do coexist, often incompatible and sometimes contradictory. Further, we need to be wary of the tedious tendency to proclaim the novelty of our own times, or to suggest that we are at the end of one age or on the verge of something new. Yet it is hard to ignore the pervasiveness of references to the brain and neuroscience, the growth of research and scientific publishing that we have documented, and the frequency of popular accounts of new discoveries about the brain in the mass media and in popular science books. Even more significant, we are beginning to see neuroscientific arguments deployed in relation to questions of policy and practice in areas ranging from child rearing to behavioral change. How, then, should we assess these developments?

A Neurobiological Complex

In this book, we have described some of the key mutations—conceptual, technological, economic, and biopolitical—that have enabled the neurosciences to leave the enclosed space of the laboratory and gain such traction in the world. In this process, the human brain has come to be anatomized at a molecular level, understood as plastic, and mutable across the life-course, believed to be exquisitely adapted to human interaction and sociality, and open to investigation at both the molecular and molar scales in a range of experimental setups, notably those involving animal models and those utilizing neuroimaging technologies. However, the hopes that advances in basic neurobiological knowledge would translate into radical improvements in our capacities to understand and treat troubled and troublesome individuals have largely been disappointed. Nor has it yet proved possible to articulate clearly the way in which brain states give rise to mental states, and still less how they enable conscious experience: despite the proliferation of terms to describe this relationship, the explanatory gap remains.

The avalanche of neuroscientific research and its popular presentations are generating a growing belief, among policymakers and in public culture, that human neurobiology sets the conditions for the lives of humans in societies and shapes human actions in all manner of ways not amenable to consciousness. Yet, contrary to the predictions of many neuroethicists and social critics, neurobiological conceptions of personhood do not portray humans as mere puppets of their brains and have not effaced other conceptions, notably those from psychology, of who we are as human beings. Nor have they presented a monotonous picture of 'neuronal man' as isolated, individualized, and asocial.[2] Instead, they have argued that the human brain, like that of some of our primate forebears, is evolved for a collective form of life, to make the attributions of internal mental states, beliefs, and intentions to our conspecifics that enable interaction among conspecifics, and the formation of groups both small and large to pursue common aims. And, as a corollary, neuroscientists have argued not only that human interactions in dyads, groups, and larger collectivities depend on these features of our neurobiology, but also that some of the difficulties suffered by those diagnosed with psychiatric disorders arise from problems within the neurobiological circuits that underpin and sustain these capacities for interaction with others. In these and many other ways, then, the normal functioning of the human brain has been constituted as crucial for sociality, and all manner of problems of sociality have been linked back to anomalies in neurobiology. This has enabled neurobiological ways of thinking to infuse the analyses of problems of individual and collective human conduct in the many sites and practices that were colonized by the psy- disciplines across the twentieth century.

The proliferation of the neuro- prefix is, no doubt, often merely a matter of fashion. Yet when 'neuro-drinks' promise to improve our brain power, or when 'neuro-trainers' claim that their courses will enable us to take full control of our lives, we can recognize the growing power of the belief, however banal, that by acting on our brains, we can improve our mental life and enhance our everyday capacities to meet the ceaseless demands for self-development and self-management that characterize our contemporary form of life. When various kinds of neurotherapy offer to make invisible mental processes visible, for example, via neurofeedback, and to give us the power to transform them by specialized techniques, we see something of the new configuration of personhood, in which the human mind can know and act on the brain that gives birth to it, in order to transform itself. When legal debates have to confront arguments from neuroscience, and neuroscientists argue that their research has relevance for issues from the sentencing of offenders to the age of criminal responsibility, we can see the growing salience of this new topography of the human being, even where brain-based claims do not displace psychological accounts and fail to convince judges and juries in the courtroom.

When some authorities look to neuroscience to help resolve the intractable difficulties of risk predictions—whether it be risk of developing a mental disorder or risk of committing a violent offence—we can see signs of the rise of new forms of expertise whose apparent access to objective knowledge of the determinants of human conduct can legitimate the exercise of social and political authority. When contemporary proponents of long-standing arguments about the importance of a child's earliest years in determining its character and shaping its future life chances try to influence policy by reframing these in terms of our growing knowledge of the vulnerable developing brain, and supplement their arguments with images from brain scanning and data from neuroscientific experiments, we can see signs of the potential emergence of a new regime of truth about our nature as human beings.

It is too early to diagnose the emergence of a full-blown 'neurobiological complex,' or a radical shift from *psy-* to *neuro-*. But neurobiology is undoubtedly reconfiguring some of the ways in which individual and collective problems are made intelligible and amenable to intervention. The passage of neuroscience from the seclusion of the laboratory to the unruly everyday world, and the new styles of thought concerning the intelligible, visible, mutable, and tractable brain that characterize the new brain sciences are beginning to reshape some of the ways in which human beings, at least in advanced liberal societies, are governed by others. They are also providing those of us who live within their purview with new techniques by which we can hope to improve ourselves, manage our minds, and optimize our life chances by acting on our brains.

Brains In Situ?

How should the social and human sciences respond to these developments? The most common response has been to argue that neuroscience is reductionist. Of course, there are many sophisticated philosophical discussions of reductionism, but this book is not a philosophical treatise, and our discussion here is not intended to be a contribution to philosophy. Rather, we want to consider some of the different ways in which this allegation has been made from the social and human sciences and consider where we stand today, in the light of our discussion in previous chapters.

A first question of reductionism concerns the relation between the laboratory and the world, between the necessary simplification of the experimental setup in neuroscientific research and the inescapable complexity and messiness of the world in which humans and other animals live their lives. Of course, all experimental setups, even 'naturalistic,' observational ones, require a certain simplification, and there are many virtues in the quest of the laboratory sciences for the elimination of extraneous variables and the rigorous paring down that is required to demonstrate causal relations. But we have suggested that the persistent failure to translate laboratory findings to real-world situations—for example, in animal models of psychiatric conditions or in brain imaging experiments of human volition—may represent problems that are not technical but fundamental.

For, as so many ethnographic studies of laboratory life have shown, the laboratory is not a 'non-space': it is itself a real configuration of persons, devices, and techniques, which imposes its own characteristics on the data generated, as well as the interpretations made of them, and all the more so when the subjects are individual living creatures, be they humans or other animals. This is not a critique of artifice or artificiality, merely a recognition that the things that are made to appear in these laboratory setups are not those that appear in non-laboratory spaces, and their relation to what occurs in such other places and spaces is far from simple. All too often, as we have pointed out, these issues are ignored or set to one side by laboratory researchers. There are certainly some within the research community of neuroscience who recognize these problems and are seeking ways—both conceptual and methodological—to address them. Sometimes they can be addressed in part by seeking more ecological validity in the lab itself. Sometimes they can be addressed by trying to explore the ways the laboratory situation itself shapes the data that are generated. The first step on this journey, however, is to recognize the issue and to acknowledge it, however slightly, in the reports and interpretation of experimental results.

Among science policymakers, and among many researchers, there is now a ritual repetition of the theme that 'translation' from laboratory findings to clinical applications is the major challenge for the twenty-first century. We would agree. But we have also argued that the current pressures for neuroscientists to make premature claims of applicability may have perverse effects. Neuroscientists might well be advised to be frank about the conceptual and empirical questions that translation entails, rather than suggesting that the outcome of a series of experiments with fruit flies or feral rats has something to tell us about human violence, or that brain scans of individuals when they are exposed to images of differently colored faces in an fMRI machine has something to tell us about the neurobiological basis of racism.

The *translational imperative* under which all are now forced to work induces such reckless extrapolations. Perhaps one can have some sympathy for the researchers' dilemma, caught between the necessary simplifications of the experimental design and the promises of rapid applicability demanded by those who fund the research. There is a role for the social sciences here in defending scientists from these demands. For there is a wealth of historical and sociological research on the relations between scientific findings and social innovations that undermines these naive views of linear translational pathways and their associated 'roadmaps.' Social research also suggests that the translational imperative, leading as it does to cycles of promises and disappointments, is likely to fuel just that distrust in science that it is intended to alleviate.

It is not entirely misleading to suggest that the move from the lab to the social world is a move from the simple to the complex—although the lab itself is complex in its own way. We will address this issue later, for it goes to the heart of the criticism that neuroscience 'reduces' the social and cultural. However, there is another criticism of reductionism that concerns complexity. It addresses the approach that we described in our account of the early work carried out under the neuromolecular style of thought—the idea that one should first observe the basic neuromolecular processes in the simplest possible systems and then, scale up to the level of intracellular transactions, circuits, networks, regions, and finally to the scale of the whole brain. This kind of reductionism has taught us a great deal. However, such a bottom-up approach also presents enormous difficulties in moving up the scales, and in addressing the kinds of phenomena that are created when tens, dozens, thousands, millions of neurons are in communication and transaction with one another and with their milieu.

Some neuroscientific projects are directly addressing this issue, for example, in the work on phenomics that we cited in an earlier chapter, where it is framed in terms of the challenge of relating phenotypes to genotypes, and seeks to go beyond the impoverished explanatory architecture of interaction ritually invoked in the expression GxE.[3] But the issues are not merely those concerning the technical complexity of mapping multiplicities of transactions, cascades, circuits, and feedback loops. They are also conceptual. At what scale *should* different processes or functions of the brain be conceptualized? This raises a rather fundamental issue about the kind of biology that is appropriate here. We have much sympathy with the view of the microbiologist and physicist Carl Woese, when he argued that we need a new biology for a new century: criticizing mechanistic models that have transformed biology into an engineering discipline focused on applications at the expense of understanding, he argued that "biology is at the point where it must choose between two paths: either continue on its current track, in which case it will become mired in the present, in application, or break free of reductionist hegemony, reintegrate itself, and press forward once more as a fundamental science. The latter course means an emphasis on holistic, 'nonlinear,' emergent biology—with understanding evolution and the nature of biological form as the primary, defining goals of a new biology" (Woese 2004, 185).[4]

Yet despite the great interest in studying the emergent properties of networks within complex systems that one can see in many scientific domains—atomic physics, information science, and of course in neuroscience—these remain exceptionally difficult to specify with precision. It has to be said, however, that the social and human sciences are hardly exemplary here: too often, appeals to complexity are a sign of explanatory failure, not explanatory success. Here, then, is a further challenge for genuine conceptual collaboration across the human sciences and the life sciences.

There is a further dimension to the charge of reductionism from the social and human sciences. This is the argument that neuroscience, in isolating brains and their components as the focus of investigation and explanation, ignores or denies the fact that, in humans and other animals, brains only have the capacities that they do because they are embodied, organs in a body of organs with all its bloody, fluid, and fleshly characteristics. There is, of course, a long philosophical debate about brains in vats (Putnam 1981), but the problems of solipsism and the like are not our concern here.[5] Rather, like our colleagues from the social and human sciences, we are troubled by the misleading assumption that one can understand neural processes outside those of the bodies that they normally inhabit.[6] To put it simply, as far as vertebrates are concerned at least, brains in situ function within complex pathways of afferent and efferent nerves reaching both the interior and the periphery of the body; brains are infused with blood containing all manner of active chemicals that carry nutrition, hormonal signals, and much else. And all of this and more is shaped by time—at levels from the microsecond to the hour, by the times of circadian rhythms, and by the times of cell growth and cell death. Stomachs, intestines, lungs, kidneys, and livers—let alone limbs—do these play no part, or merely a peripheral one, in the faculties and functions studied by neuroscience—in memory, in thought, in feelings, in desires? Psychiatric conditions shape and are shaped by mobilizations of feelings and functions across the body. Emotions course through the veins, engage the heart and the lungs, the bowels and the genitals, the muscles, the skin and face. As for cognition, do we not think, literally here, with hands and eyes? As Elizabeth Wilson puts it, an exclusive focus on the brain "narrows the geography of mind from a diverse, interconnected system (a mutuality of moods-objects-neurotransmitters-hormones-cognitions-affects-attachments-tears-glands-images-words-gut) to a landscape within which the brain, as sovereign, presides over psychological events" (Wilson 2011, 280). This not a plea for some kind of mystical recognition of wholeness: to criticize the isolation of the properties of the brain from those of the organism of which it is a part is to demand that neurobiology addresses more adequately the biological reality of the processes it seeks to explain.

A further problem of reductionism, perhaps the most crucial challenge from the social and human sciences, concerns sociality, culture, and history. This is the argument that the neurosciences, even when they make reference to the environment, and even when they accept a role for social and cultural factors, perhaps even when they ask how experience gets under the skin, reduce this complex extraneuronal milieu—of practices, meanings, symbols, swarming collectivities, multiplicities of historically sedimented human practices—to a small number of discrete 'inputs' into neurobiological systems, ideally able to be represented in the form of quantifiable variables. Hence the repeated, not always playful, challenge to sociologists to identify 'the socio-ome' to be set alongside the genome, the proteome, the metabo-lome and all the other '-omes' so beloved of the biosciences at the end of the twentieth century.

The belief that one might be able to specify a stable, objectively given, fac-torially representable, quantitatively measurable socio-ome, generalizable across time and space, is anathema to most from the social and human sci-ences. It is, however, possible to identify some ways in which a more lim-ited engagement might be possible. We have already contrasted two ways of thinking about sociality in social neuroscience: On the one hand, we find the impoverished conception of sociality that can be found in most research that utilizes brain imaging in laboratory settings—where 'social' seems to refer to sequential transactions between isolated individuals or simulations of such transactions in scanners. On the other, we encounter the more subtle arguments of those who seek to understand sociality in terms of the multiple networks of collective association that characterize human existence, in fami-lies, groups, villages, towns, nations, and beyond, and the neurobiological consequences of isolation for humans evolved to live in such ecological set-tings. We have argued that the latter, for all their current weaknesses, open a pathway for collaboration beyond critique.

Consider, for example, the way we might explore the mental consequences of urban life. The differences, and the potential linkages, between socio-logical and neurobiological approaches can be seen if we contrast the recent work of Meyer-Lindenberg's group on urbanicity, stress, and mental disorder with Georg Simmel's classic study of the metropolis and mental life. Meyer-Lindenberg and his group "show, using functional magnetic resonance imag-ing in three independent experiments, that urban upbringing and city living have dissociable impacts on social evaluative stress processing in humans. Current city living was associated with increased amygdala activity, whereas urban upbringing affected the perigenual anterior cingulate cortex, a key region for regulation of amygdala activity, negative affect and stress" (Led-erbogen, Kirsch, et al. 2011, 498). Simmel locates the psychological form of metropolitan individuality in the shift of humans from rural to urban living, in the space, texture, and temporality of the urban experience, in the constant encounter with differences in urban interactions, in the novel economic and occupational forms that enmesh those who live in cities, and in the constant tension between the equalization of all and the uniqueness of each that tra-verses the collective life of the metropolis (Simmel 1903). The neuroscientists argue that their work provides the basis for "a new empirical approach for integrating social sciences, neurosciences and public policy to respond to the major health challenge of urbanization" (Lederbogen, Kirsch, et al. 2011, 500). Yet they ignore a century of empirical research and conceptualization in sociology, social history, and cultural geography on these issues, on the ways in which they play out in the mental lives of the inhabitants of these environments, and on the policy difficulties that are entailed in transform-ing the experiences and consequences of contemporary forms of urban life. The challenge both to neuroscientists and to social scientists is to go beyond mutual critique—with each side accusing the other of gestural reference to the mental or the social—to a kind of conversation in which experimental reductionism does not imply the mindless empiricism so long criticized in the social sciences, and where experimental rigor does not become the enemy of social, historical, and cultural complexity.

Coda: The Human Sciences in a Neurobiological Age

For the human sciences, we suggest, there is nothing to fear in the rise to prominence of neurobiological attempts to understand and account for human behavior. It is important to point out the many weakness in the ex-perimental setups and procedures, for example, in the uses of animal models and in the interpretations of brain imaging data generated in the highly arti-ficial social situations of the laboratory. It is also important to challenge the attempts by neuroscientists and popular science writers to extrapolate wildly from small experiments to human practices, to attach the *neuro-* prefix to so many areas, and to overclaim the novelty of their findings and their impli-cations for everything from child rearing to marketing, and from criminal justice to political strategy. However, we should not be surprised to find, in contemporary neurosciences, all the features of inflated expectations, exag-gerated claims, hopeful anticipations, and unwise predictions that have been so well analyzed in other areas of contemporary biotechnologies (Brown, Rappert, and Webster 2000; Webster, Brown, and Rappert 2000; Brown 2003; Brown and Michael 2003; Brown and Webster 2004).

Historical and genealogical analysis can and should identify the conditions of possibility for the current prominence of neuroscientific styles of thought and the practices where these have gained traction. Science and technology studies can and should analyze the setups that have generated the truths of contemporary neuroscience and the impact of the sociopolitical shaping of contemporary science on these expectational and promissory landscapes. We need detailed social research to map out the interconnections between path-ways for the production of truth and hopes about the generation of profit that have shaped investment, research priorities, and the interpretation of evidence and its implications. We need to analyze the new practices for governing human conduct that are taking shape, the new powers of authority and expertise that deploy the *neuro-* prefix to legitimate themselves, the new ethical obligations associated with the valorization of the brain as a key sociopolitical resource, and the new modes of subjectivation that are being born. While we are not witnessing a wholesale revolution, in which the psychological domain brought into existence in the twentieth century is consigned to history by the rise of neuroscience, our conceptions of human personhood, of the topography of the human being, are undoubtedly being reshaped by these processes.

It is important constantly to remind neuroscientists that neither animal nor human brains exist in individualized isolation. That if evolution has acted on the brain, it has done so to the extent that brains are constitutively embodied in living creatures, dwelling in space and in time, interacting in small and large groups on which they depend for their existence, striving to survive, inhabiting and remaking their milieu across the course of their lives. That when it comes to humans, and perhaps other primates, cognitive capacities, affective flows, and the powers of volition and the will are not individuated, but distributed. That they are enabled by a supra-individual, material, symbolic, and cultural matrix—a matrix that the capacities of the human brain have themselves made possible, and that shapes the developing human brain (and the body within which it is situated) from the moment of conception to the moment of death. Whatever the virtues of reductionism as a research strategy, then, an account of the operation of the living brain in a living animal that does not recognize these transactions, that does not realize that the boundaries of skull and skin are not the boundaries of the processes that shape the irreal interior world in the brain, that does not recognize that human capacities and competencies emerge out of, and are possible only within, this wider milieu made and remade by living creatures, shaped by history, marked by culture in ways ranging from the design of space and material objects to the management of action and interaction and organization of time itself, will be scientifically flawed.

In the necessity for this criticism, there is also opportunity. There are many opportunities for a more positive role for the social and human sciences that engages directly with these truth claims, that seizes on the new openness provided by conceptions of the neuromolecular, plastic, and social brain to find some rapprochement. Opportunities exist to bring to bear the evidence from more than a century of empirical research on all those simplifications about sociality that currently inform even the most sophisticated accounts of the interactions between neurogenetics, brain development, and milieu, even the most subtle explanations of the affective shaping of human interaction, even most social theories of social neuroscience. The human sciences have nothing to fear from the argument that much of what makes us human occurs beneath the level of consciousness or from the endeavors of the new brain sciences to explore and describe these processes. If we take seriously the combined assault on human narcissism from historical and genealogical investigations, from cultural anthropology, from critical animal studies, and from contemporary neurobiology, we may indeed find the basis of a radical, and perhaps even progressive, way of moving beyond illusory notions of human beings as individualized, discrete, autonomous, coherent subjects who are, or should be, 'free to choose.'

We will undoubtedly find support for the core contention of the human sciences, that human societies are not formed by aggregations of such isolates, each bounded by the surfaces of its individual body. We will find much evidence that disproves the idea that the nature of humans is to seek to maximize self-interest, and hence to challenge the view that to govern in accord with human nature is to require each individual to bear the responsibilities and culpabilities of his or her selfish choices. Such unpredictable conversations between the social sciences and the neurosciences may, in short, enable us to begin to construct a very different idea of the human person, human societies, and human freedom. We have tried to show, in this book, how neuroscience has become what it is today; let us conclude with a simple hope for the future: that neuroscience should become a genuinely human science.

Appendix
. . . .
How We Wrote This Book

In this book we have drawn principally on arguments made by 'authorities,' that is to say, by those who claim to speak authoritatively about scientific research on the brains of human beings and other animals, and how humans should, in the light of that, be governed. We have examined the scientific literature as published in book and articles in the peer-reviewed journals, as well as popular books written by neuroscientists. We have looked at legal debates and law cases where neuroscientific evidence has been drawn on or referred to. We have researched the policy literature and policy proposals that make reference to neurobiology. We are aware, of course, that many of these matters have been the topic of public debate in the mass media and on the Internet, but, with some limited exceptions, we have not investigated these wider discussions and responses—while undoubtedly of great significance, they would have required a different kind of study.

We have, however, drawn extensively on a wider array of informal interactions with neuroscientists, lawyers, and regulators. Many of these have been made possible through the work of the European Neuroscience and Society Network (ENSN) funded by the European Science Foundation (ESF), which has brought neuroscientists, psychiatrists, social scientists, historians, philosophers, regulators, and lawyers together in conferences, workshops, symposia, and neuroschools over the past five years. In addition to the many opportunities to explore the key issues in intense yet informal dialogue, we have also been fortunate to be able to spend time with researchers in a number of laboratories, notably Cornelius Gross's laboratory at the European Molecular Biology Laboratory facility, Monterotondo, outside Rome; the imaging facilities at the University of Vienna; Klaus-Peter Lesch's Laboratory of Translational Neuroscience in the Department of Psychiatry, Psychosomatics and Psychotherapy at the University of Wurzburg; and Kenneth Hugdahl's Bergen fMRI Group, at the Department of Biological and Medical Psychology of the University of Bergen.

Our research on the legal issues raised by the new brain sciences was informed by a special interdisciplinary conference on Law and Neuroscience, also funded by the ESF, which we helped organize and which took place in Italy in October 2009, and by the involvement of one of us in the law and neuroscience module of the U.K. Royal Society's Brain Waves project. Our research on animal models was informed by our involvement in some heated discussions at a major international conference of animal modelers at the European Molecular Biology Laboratory facilities in Heidelberg, and by the involvement of one of us in the U.K. Academy of Medical Sciences working group on research with animals containing human materials, which involved quite extensive discussion of animal modeling in neuroscience. And one of us has also been involved in many discussions of these issues with regulators and lawyers, through membership of the U.K.'s Nuffield Council on Bioethics and other advisory groups. While we do not directly refer to these and other interactions, in Europe, North America, China, and Latin America, they have engaged us within a community of interlocutors, who have helped us form, test, and revise the arguments we make in this book. None of our colleagues in any of these encounters bears any responsibility for the arguments we make here, but their openness and friendship have shaped the tone of our argument, our belief in the possibilities of critical and affirmative dialogue, and our wish to avoid the many simplistic stereotypes that can be found in much critical writing on neuroscience from the side of the social sciences.

Notes

· · · · · · · · · · ·

Introduction

1. Our focus in this book is on "advanced liberal" societies (Rose 1999), but we should note that these developments are by no means confined to Europe, the Americas, and Australasia, and indeed that we have seen many novel attempts at rapprochement, via the brain, with practices of meditation, especially those from the Buddhist tradition (see for example, the activities of the Mind and Life Institute: http://www.mindandlife.org). Further, there is another whole pathway of research that we barely touch on in the account that follows—cognitive science and artificial intelligence. While there are many links between those developments and the ones we analyze in this book, to address them in any detail would overburden an already lengthy account. A good starting point for interested readers would be Elizabeth Wilson's study of Alan Turing (Wilson 2010). See also the discussion of the work of Warren McCulloch by Lily Kay (Kay 2002).

2. There is a growing body of work in the human sciences that shares this view, recognizing that human beings are, after all, living creatures, and seeking a more affirmative relation with biology. We can see this in the rise of 'corporeal' feminism, in 'affect studies,' and in projects to create a new materialism. While we do not discuss these endeavors here, we are sympathetic to them, although we are critical of the way in which some authors in this vein borrow very selectively from work in the life sciences to support pre-formed conceptual, political, or ethical commitments. In the main, however, we have not engaged with this literature in the present study.

3. Quoted from Lewis (1981, 215). We have drawn extensively on Lewis's account of this debate in this paragraph. Although it is undoubtedly true that the question dates back to Aristotle and was posed again by Descartes, the idea that it was answerable by a positive knowledge, rather than by philosophy, dates back to the start of the nineteenth century.

4. The phrase "the physical basis of mind" has a long history—for example, it is the title of a nineteenth-century book by George Henry Lewes (Lewes 1877). Contributors to the BBC series included Edgar Adrian, who was Sherrington's pupil and shared the Nobel Prize with him in 1932 for his work on the function of neurons; Elliot Slater, who was a central figure in psychiatric genetics in the postwar years; Gilbert Ryle, who rehearsed his critique from *The Concept of Mind* written the year before (Ryle 1949) that there was no "ghost in the machine"; and Wilder Penfield, pioneer of neurosurgery in epilepsy, who argued that there was a coordinating center in the upper brain stem that, together with areas of the cortex, was the

seat of consciousness and that this was "the physical basis of mind"—not mind itself, but the place through which "the mind connected to the brain" (a view he was to elaborate in subsequent years; cf. Penfield 1975).

5. The set of essays was later published as a book simply titled *The Brain* (Edelman and Changeux 2001).

6. As we discuss in chapter 1, Francis O. Schmitt of the Massachusetts Institute of Technology used it in 1962 to designate his Neurosciences Research Program, and the term *neuroscientist* was first used soon after (Swazey 1975).

7. Data from http://www.sfn.org/static/amstats/amstatsgraph.html; consulted December 2010. By the end of the decade, annual attendance at this event was more than thirty thousand scientists, and around four thousand nonscientists, including many staffing industry or pharmaceutical displays.

8. Data from LSE's Brain, Self, and Society project, 2007–10 (http://www.esrc.ac.uk/my-esrc/grants/RES-051-27-0194/read).

9. The British Library catalog lists 433 books with the terms *mind* and *brain* in the title, the earliest, by Spurzheim and Gall, dating from 1815. It contains 62 books with the words *mind* and *brain* in their titles published in the thirty-five years between 1945 and 1980. A further 68 were published in the decade between 1981 and 1990, 109 between 1991 and 2000, and 122 between 2001 and 2009. By 1984, John Searle thought he had resolved this problem (Searle 1984, 1992), as did Jean-Pierre Changeux, whose *Neuronal Man* was published in English a year after Searle's lectures (Changeux 1985), although many, including Paul Ricoeur, were not convinced (Changeux and Ricoeur 2002). Francis Crick, whose reductionist text *The Astonishing Hypothesis* was published in 1994 argued for a physical basis for consciousness—the "claustrum" in his last paper written with Christof Koch (Crick and Koch 2005). We will come across many versions of these debates and many words used to characterize the mind-brain relation in the course of this book, and we will return to this issue in our conclusion.

10. There have been earlier uses of the prefix. For example in his biography of Wilder Penfield cited above, Lewis notes that, in the 1940s, whenever a new researcher was recruited to the Montreal Neurology Institute, Penfield "would add the prefix 'neuro' to his or her speciality": thus when a chemist was recruited his speciality would become Neurochemistry, photography would become Neurophotography, and there was Neuropsychology, Neurocytology, Neuropathology and Neuroanatomy—the prefix was intended to draw all together towards a common purpose (Lewis 1981, 192).

11. In 1996, Peter Shizgal and Kent Conover sought to describe the neurobiological substrate for choice in rats using a normative economic theory, followed in 1999 by Michael Platt and Paul Glimcher's publication in *Nature* of "Neural Correlates of Decision Variables in Parietal Cortex"; by the early years of this century, there was a steady trickle of publications on these issues (Montague 2003; Sanfey, Rilling, Aronson, et al. 2003; Schultz 2003). A Society for Neuroeconomics was established in 2005; neuroeconomics programs and laboratories were set up at a number of U.S. universities (e.g., Caltech, UC Berkeley, UCLA, Stanford, Duke University); several handbooks and textbooks have now been published under titles such as *Decision Making and the Brain* (Glimcher 2003, 2009); and popular magazines began to report excitedly on the new science of decision making (Adler 2004).

12. For Johns Hopkins, see http://education.jhu.edu/nei; for developments at the University of London on "educational neuroscience," see http://www.educationalneuroscience.org.uk.

13. http://www.informaworld.com/smpp/title~db=all~content=t741771143; consulted April 2010.

14. The report is titled *MINDSPACE: Influencing Behavior through Public Policy* and can be downloaded at http://www.instituteforgovernment.org.uk/content/133/mindspace-influencing-behavior-through-public-policy; consulted January 2011.

15. http://royalsociety.org/policy/projects/brain-waves.

16. The program is called "Neurosciences et politiques publiques" (http://www.strategie.gouv.fr/article.php3?id_article=1029; consulted January 2011).

17. Although the term was apparently coined by Yuri Olesha, it is conventionally attributed to Josef Stalin, who used it in a speech to Soviet writers at the house of Maxim Gorky in 1932: "The production of souls is more important than the production of tanks. . . . And therefore I raise my glass to you, writers, the engineers of the human soul" (for some details, see http://www.newworldencyclopedia.org/entry/Yury_Olesha).

18. We have omitted most references here, as each of the points that we make in this overview is discussed in detail in the chapters that follow.

19. We use the term *complex* to bring to mind the interconnected resonances of this term. The *Oxford English Dictionary Online* includes the following: "A whole comprehending in its compass a number of parts, *esp.* (in later use) of interconnected parts or involved particulars, . . . An interweaving, contexture. . . . A group of emotionally charged ideas or mental factors, unconsciously associated by the individual with a particular subject . . . hence *colloq.*, in vague use, a fixed mental tendency or obsession."

20. The critical literature here is large, and we can only cite some representative samples. David Healy has been the most significant chronicler of the history of the links that we outline here, and the most persistent critic of the overclaiming made by the pharmaceutical companies (Healy 1996, 1997, 1998, 2000, 2004). Some of these points were tellingly made by Lauren Mosher, in his letter of resignation from the American Psychiatric Association after thirty-five years—he said it should now be known as "the American Psychopharmacological Association"; see http://www.successfulschizophrenia.org/stories/mosher.html. Recently, Marcia Angell, former editor of the *New England Journal of Medicine*, has also become an outspoken critic (Relman and Angell 2002; Angell 2004); see also her 2011 review of a number of polemical books on the "epidemic of mental illness" (http://www.nybooks.com/articles/archives/2011/jun/23/epidemic-mental-illness-why/?pagination=false). The most persistent, and perceptive, critic of "medicalization" in psychiatry has been Peter Conrad (Conrad 1976, 2005; Conrad and Schneider 1992). Our own view of the limits of the medicalization thesis and our alternative approach has been spelled out in a number of places, and we will not repeat it here (Rose 1994, 2006a, 2007a).

21. These transformations in genomics are discussed in much more detail in Rose (2007c). We can note here that in this new vision, it was also at this molecular level that evolutionary pressures operated, and that evolved species' differences in behavior were to be understood. We discuss this further in relation to animal models in chapter 3, and also in relation to the 'evolution of the social brain' in chapter 5.

22. The intriguing title of the U.K. House of Commons Public Administration Select Committee's Second Report for the Session 2006–7 was indeed "Governing the Future"; see http://www.publications.parliament.uk/pa/cm200607/cmselect/cmpubadm/123/123i.pdf; consulted July 1, 2011.

23. In the case of the NRP, funding was chiefly drawn from federal agencies—not only from the NIH and later the NIMH, but also NASA, NSF, and even the Office of Naval Research (ONR). This was complemented by funds from the Rockefeller Foundation and a private charity fund related to the NRP: the NRF, or Neurosciences Research Foundation (Schmitt 1990).

24. This is the widespread belief in biomedicine that Charles Rosenberg calls "reductionist means to achieve necessarily holistic ends" (Rosenberg 2006).

25. For example, the World Alzheimer Reports. The one for 2009 contained estimates of future prevalence (http://www.alz.co.uk/research/files/WorldAlzheimerReport.pdf). That for 2010 focused on the economic cost (http://www.alz.co.uk/research/files/WorldAlzheimerReport2010ExecutiveSummary.pdf).

26. As the reputable and influential commentator Polly Toynbee put it in the *Guardian* newspaper, December 4, 2010, "the brain hardens" by the age of three.

27. The quote is from a presentation by Viding and McCrory to a 2007 conference commissioned by the United Kingdom's Department of Health; the Department for Children, Schools and Families; Youth Justice Board; and the Cabinet Office that followed the publication of the Social Exclusion Action Plan (2006) and the Care Matters white paper (2007). See http://www.personalitydisorder.org.uk/resources/emerging-pd; consulted January 4, 2011.

28. See, for example, Beddington, Cooper, Field, et al. (2008), framed recently in the form of a "Grand Challenge of Global Mental Health" (Collins, Patel, Joestl, et al. 2011).

29. According to the *NIH Almanac* (http://www.nih.gov/about/almanac), funding for the NIMH remained more or less constant in the five years from 2005 to 2010, at between $1.4 billion and $1.5 billion, despite inflation at around 4% over this period. However, this was a significant increase from 1995, where the figure was just over $0.5 billion, and indeed from 1965, where it was around $186 million (data from http://www.nih.gov/about/almanac/appropriations/index.htm#one; consulted July 18, 2011). An indication of the complexity of deriving meaningful data in this area is that this covers all grants from the NIMH, not only for neuroscience, and that, as Dorsey et al. point out "NIMH separated from NIH in 1967 and was raised to bureau status in PHS, became a component of PHS's Health Services and Mental Health Administration (HSMHA), later became a component of ADAMHA (successor organization of HSMHA), and rejoined the NIH in 1993." Another estimate shows significant variations in NIH funding for research in different mental disorders over the period from 2005 to 2009 (http://report.nih.gov/rcdc/categories); while NIH funding for some disorders—notably depression and autism—had increased by between 20% and 25% in the five years from 2005 to 2009, that for other conditions such as schizophrenia had decreased by the same amount, and funding for research on Parkinson's, Alzheimer's, and ADHD had actually decreased by between 30% and 40%—http://brainposts.blogspot.com/2009/08/nih-trends-in-clinical-neuroscience.html; consulted July 18, 2011. Of course, these figures represent only a very partial glimpse of public funding for neuroscience and do not take account of inflation, which was between 3% and 4% over the first decade of the twenty-first century.

30. These data are from a presentation given by Zack Lynch in 2010 at the fifth annual Neurotech Investing and Partnering Conference; we would like to thanks Zack Lynch for making this presentation available to us and allowing us to quote from it.

31. In 2011, the European College of Neuropsychopharmacology (ECNP) expressed concern that pharmaceutical companies were withdrawing their own research funding in this area, despite the growing burden of brain diseases (see http://www .guardian.co.uk/science/2011/jun/13/research-brain-disorders-under-threat; consulted July 1, 2011).

32. See http://www.bbsrc.ac.uk/funding/news/2011/110211-bbsrc-neuroscience-funding .aspx; consulted July 1, 2011.

33. For IMS Health ("revealing the insights within the most comprehensive market intelligence available"), see http://www.imshealth.com/portal/site/imshealth. For NeuroInsights ("track market dynamics, develop investment strategies, identify partnering opportunities, and analyze comparables across this $143 billion global industry), see http://www.neuroinsights.com.

34. Sir David Cooksey, who reviewed U.K. health research funding at the request of the then government, concluded that the United Kingdom was at risk of failing to reap the full economic, health, and social benefits that its public investment in health research should generate, and argued that there were "two key gaps in the translation of health research: translating ideas from basic and clinical research into the development of new products and approaches to treatment of disease and illness; and—implementing those new products and approaches into clinical practice" (Cooksey 2006, 3). It is these translational gaps that led to attempts to reform the whole research funding system to overcome professional, financial, and institutional barriers to translation.

35. These issues are discussed in more detail in *Translating Neurobiological Research*, a 2009 report produced by the LSE's BIOS Centre for the Medical Research Council.

36. We use the term *translational platforms* for zones of knowledge exchange and hybridization of techniques, practices, and styles of thought whose driving rationale is the translation of potential products and artifacts produced in the lab or the clinic into meaningful and useful clinical and social applications of clinical, commercial, or social value. They can be thought of as a special case of what Peter Keating and Albert Cambrosio call "biomedical platforms" (Keating and Cambrosio 2003). The idea of a translational platform derived from our mapping of seventy thousand peer-reviewed articles published in 2008 in almost three hundred high-impact journals; see Abi-Rached (2008); Abi-Rached, Rose, and Mogoutov (2010).

37. This is especially problematic in the much-criticized links between psychiatrists and the pharmaceutical companies. One particularly notorious example was the promotion of the diagnosis of pediatric bipolar disorder in the United States by Dr. Joseph Biederman, leading to a fortyfold increase in the diagnosis over the decade from 1994 to 2003 and the off-label use of antipsychotics, notably Risperdal, manufactured by Janssen, a division of Johnson and Johnson, who funded the research center that he directed. Some details of this case are given in a press report in the *New York Times* in November 2008 (http://www.nytimes .com/2008/11/25/health/25psych.html?bl&ex=1227762000&en=ab700f6adb9c70e5 &ei=5087%0A; consulted April 4, 2011).

38. http://www.lrb.co.uk/v30/n06/shapo1_.html; consulted April 4, 2008.
39. The following passage derives more or less verbatim from a paper given by NR, "Commerce versus the Common Good," at the LSE Asia Forum in Singapore in April 2008, and subsequently published as Rose (2009). Note that in theory the federal government retains the right to reclaim intellectual property and reallocate it elsewhere—the so-called march-in provision. The march-in provision of the act, 35 U.S.C. § 203, implemented by 37 C.F.R. § 401.6, authorizes the government, in certain specified circumstances, to require the funding recipient or its exclusive licensee to license a federally-funded invention to a responsible applicant or applicants on reasonable terms, or to grant such a license itself.
40. For example, according to the U.S. Association of University Technology Managers Report in 2006, $45 billion in R&D expenditures were received by U.S. academic centers in that year alone; 697 new products were introduced into the market; and as a result of these relations, a total of 4,350 new products were introduced from Financial Year 98 through Financial Year 06; 12,672 licenses and options were managed, each yielding active income to a university, hospital, or research center; and 5,724 new spinouts were launched from 1980 to 2006 (Association of University Technology Managers, 2007); http://www.autm.net/Surveys.htm; consulted July 1, 2011.
41. Quoted from its press release of August 29, 2002 (http://royalsociety.org/news.asp?id=2507). The report published in 2003 concluded: "Intellectual property rights (IPRs) can stimulate innovation by protecting creative work and investment, and by encouraging the ordered exploitation of scientific discoveries for the good of society. . . . A narrow focus on research most likely to lead directly to IPRs would damage the health of science in the longer term. . . . Patents can provide valuable, although sometimes expensive, protection for inventions. They therefore encourage invention and exploitation, but usually limit competition. They can make it impracticable for others to pursue scientific research within the areas claimed, and because inventions cannot be patented if they are already public knowledge, they can encourage a climate of secrecy. This is anathema to many scientists who feel that a free flow of ideas and information is vital for productive research. Additionally, research by others may be constrained by patents being granted that are inordinately broad in scope—a particular risk in the early stages of development of a field. This is bad for science and bad for society" (Royal Society of London 2003, v).
42. For example, in the celebrated Berkeley-Novartis deal, a pharmaceutical company, Novartis, gave UC Berkeley $25 million over five years, in exchange for the first rights to license any discovery that was made on the basis of research that was supported by Novartis funds; critics of such arrangements within the university were reported to have suffered various forms of disadvantage, including in relation to tenure processes (Triggle 2004, 144).
43. It is notable that bodies such as the NIH have recently begun to address this issue of conflicts of interest—see its 2004 "Conflict of Interest" report (http://www.nih.gov/about/ethics_COI_panelreport.pdf). The scientific journals are also increasingly preoccupied with this issue.
44. While some have argued that neuroscience will challenge ideas of free will and responsibility in the criminal justice system, given evidence on the nonconscious initiation of apparently willed acts (Greene and Cohen 2004), this does not seem

to be the case. As we discuss in chapter 6, while neurobiological evidence from CT and MRI scans showing structural damage or anomalies in the brain has begun to play a role, for example, in claims for compensation after injury, this has not fundamentally challenged the logics of personhood within those systems. Indeed, to the extent that individual' responsibility does come to be understood as somehow compromised by their neurobiology, as in the case of psychopathy, the response of the authorities of control, in a sociopolitical context that emphasizes precaution and preemption, is likely to be more, rather than less, severe—as in the rising numbers of individuals who are already detained, without reference to neuroscientific evidence, on the grounds that they pose a significant and continuing risk to others. This issue is discussed further in the "Neuroscience and Law" module of the U.K. Royal Society's Brain Waves project, in which one of us was involved (Royal Society 2011).

45. On inhibition, see R. Smith (1992).
46. We can see similar arguments coming from the Social Brain project in the United Kingdom (Rowson 2011).
47. These self-technologies have been the subject of research by Francisco Ortega, who uses the term *neuroascesis* to describe them (http://www.neuroculture.org/brain_gym.pdf; consulted July 1, 2011).

Chapter One. The Neuromolecular Brain

1. Schmitt is quoted from Worden, Swazey, Adelman, et al. (1975, 529–30).
2. This chapter draws on earlier analyses published as Abi-Rached and Rose (2010). Some of the information is presented in more detail in three working papers for the Brain, Self, and Society project written by JAR (http://eprints.lse.ac.uk/view/lseauthor/Abi-Rached,_Joelle_M=2E.html). Some parts are also derived from a series of talks given by NR at the University of California in February and March 2007. Thanks to Sara Lochlan Jain for the invitation to Stanford, to Charis Thompson for the invitation to UC Berkeley, to Jenny Reardon for the invitation to UC Santa Cruz, to Norton Wise for the invitation to UCLA, and to all those who participated in the discussions. A version of the paper was given by NR as the Bochner Lecture delivered at the Scientia Institute of Rice University in Houston, Texas on March 9, 2009. NR wishes to thank his hosts, especially Susan Macintosh, and those at Rice and at Baylor College of Medicine who commented on the presentation. We have also drawn on an unpublished research report, "The Age of Serotonin," funded by the Wellcome Trust Biomedical Ethics Program, and jointly written by NR, Mariam Fraser, and Angelique Pratt.
3. Schmitt uses "wet, dry, and moist" to refer to three different kinds of biophysics. "Wet biophysics" refers to the biochemical study of macromolecules and other cellular elements in their "normal aqueous environment." "Dry biophysics" is concerned with the study of cellular constituents or organisms as "systems" in mathematical or natural models. Finally, "moist biophysics" is the study of properties of the central nervous system through "bioelectric studies." See Swazey (1975, 532); Schmitt (1990, 199).
4. The citation to the first usage is given in the *Oxford English Dictionary Online*.
5. SfN was founded in 1969 (see http://web.sfn.org/home.aspx; consulted April 2010). It is the world's largest association for neuroscience, with more than forty

thousand members by 2010. For a historical account of the SfN, see Doty (1987). Six autobiographical volumes from the SfN are available at http://www.sfn.org/ index.cfm?pagename=HistoryofNeuroscience_autobiographies; consulted April 2010. Similarly, interviews with prominent British neuroscientists have been conducted in a project at UCL titled "Today's Neuroscience, Tomorrow's History" and supported by a Wellcome Trust Public Engagement grant in the History of Medicine (http://www.ucl.ac.uk/histmed/audio/neuroscience; consulted April 2010).

6. There is a plethora of books and articles on brain scientists and their particular contributions to the neurosciences. For example, Changeux's *Neuronal Man* (1985, 1997), Gross's *Brain, Vision, Memory* (1999), or Finger's *Origins of Neuroscience* (1994). Some other examples include a recent book by leading French scholars on the contribution of prominent French scientists to the neurosciences (Debru, Barbara, and Cherici 2008); numerous references on the contribution of British neuroscientists (Gardner-Thorpe 1999; Rose 1999; C.U.M Smith 1999, 2001; Parkin and Hunkin 2001; Wade and Bruce 2001; Laporte 2006); a special double issue of the *Journal of the History of the Neurosciences* 16 (1–2) (January–June 2007) on the contribution of Russian neuroscientists; and numerous references on the contribution of this or that Nobel laureate to our understanding of the brain and the nervous system (Sourkes 2006; Langmoen and Apuzzo 2007).

7. In this account we have drawn on Rose (1996, 42–43).

8. Cajal's lecture was titled "The Structure and Connexions of Neurons" (http:// nobelprize.org/nobel_prizes/medicine/laureates/1906/cajal-lecture.html), and Golgi's bore the title "The Neuron Doctrine—Theory and Facts" (http:// nobelprize.org/nobel_prizes/medicine/laureates/1906/golgi-lecture.html).

9. For more on the beginnings of neurochemistry, see McIlwain (1985, 1988); Bachelard (1988); Agranoff (2001).

10. In the early 1960s, he named his laboratory "Groupe de Neuropsychologie et de Neurolinguistique" (Boller 1999).

11. An overview of its historical development can be found in Boller (1999).

12. Of course, one could mention various other trajectories, for example, those of neural networks, cybernetics, communication theory, computing, and artificial intelligence, the latter of which was, of course, bound up with hopes for its usefulness for military and intelligence applications, which provided funding and support. Warren McCulloch also thought that these developments had significant implications for psychiatry, though his argument seems not to have convinced others. Kay argues that "the particular configurations of these tools and images bore the unmistakable marks of a new, postindustrial episteme: an emergent military technoculture of communication, control, and simulations. Within its regimes of signification, life and society were recast as relays of signals and as information systems . . . exemplifying the new episteme of the information age" (Kay 2002, 610). And we could also consider the borderlands with psychology, as it disciplined itself over the first half of the twentieth century, especially the intersections with psychophysiology. But this is not the place to write this history.

13. For more, see Bigl and Singer (1991); Marshall, Rosenblith, Gloor, et al. (1996).

14. J. Richter (2000) argues that the IBRO was the successor of the Brain Commission founded in 1903 in London. Although the commission was dissolved at the outbreak of the First World War, it was the first international organization whose

aim was the creation of national brain research institutes and the promotion of international cooperation. One notable example is the Russian psychiatrist, Bechterev, who served on the Brain Commission and who pioneered long before neuroanatomist David McKenzie Rioch (mentioned by Cowan, Harter, and Kandel 2000) the physiological underpinnings of psychiatric illness and mental behavior. Bechterev was actually the founder of the first institute that looked at psychiatry from a physiological perspective, the so-called Psycho-Neurological Institute, founded in 1907 in St. Petersburg (Richter 2000).

15. In organizational terms, the United Kingdom seems to have lagged behind: the Brain Research Association (BRA) was founded in 1968. In 1996, the association decided to rebrand itself by changing the name to the British Neuroscience Association (Abi-Rached 2012).

16. We discuss Moniz's techniques of cerebral angiography using X-rays to visualize intracranial tumors, vascular abnormalities, and aneurysms in a subsequent chapter.

17. Moniz shared the Nobel Prize with Walter Hess, who developed an early brain mapping technique focused on the diencephalon (http://nobelprize.org/nobel_prizes/medicine/laureates/1949/hess-lecture.html). He was, of course, not the first to intervene on the brain. Those who have written on this history, notably Valenstein (1986) and Swayze (1995), point out that Burckhardt, the superintendent of a Swiss mental hospital, conceived neurosurgical approaches based on early studies in cerebral localization. He reported his results in 1891. However, his procedure was considered reckless, since one of his patients died soon after the surgery and two others developed seizures. There were no further attempts until 1910: Puusepp, a Russian neurosurgeon, resected fibers in the parietal and frontal lobes of patients suffering from manic-depressive psychosis. But results were published after the advent of prefrontal leucotomy, and in any case the procedure had limited success. In 1936, the Geneva neurosurgeon Ody resected the right prefrontal lobe of a patient suffering from "childhood-onset schizophrenia." The result was published before Moniz's first publication and was inspired by Cushing's belief that mental symptoms could be reversed by resection of cerebral (frontal lobe) tumors. In 1935, Moniz attended the Second International Neurologic Congress in London where Charles Jacobsen presented work that he and John Fulton had done in bilaterally ablating the prefrontal cortex of two chimpanzees. Swayze comments: "Although in his memoirs Moniz downplayed the influence of this report, stating that he had conceived the idea of a neurosurgical approach 2 years before, most observers, including Fulton and Freeman, believed that the animal results were the primary impetus for his taking such a bold step in November 1935" (Swayze 1995).

18. See Cobb's obituary (http://www.harvardsquarelibrary.org/unitarians/cobb.html).

19. Note that there are more interconnections than our "three paths" might imply: Richter was one of the pioneers in neurochemistry and was instrumental in establishing the IBRO under the aegis of UNESCO in 1961. He also helped establish—though the extent of his involvement is disputed—the Brain Research Association (BRA) in 1968, which was renamed in 1997 as the British Neuroscience Association (BNA). For more on the history of the BRA, see Abi-Rached (2012).

20. A further related element, which we will not discuss here, is the history of the medical uses of LSD (Dyck 2008).

21. See his 2001 lecture on acceptance of the Nobel Prize in Physiology or Medicine, reprinted as Carlsson (2001).

22. Some very useful papers on these issues are contained in a special issue of *Science in Context* 14 (4) (2001), edited by Michael Hagner and Cornelius Borck. There is much more sociological work on the history and anthropology of psychiatry, some of which touches on biological psychiatry, usually in the context of reductionism and eugenics. And there is a growing body of ethnographic work that touches on disputes over biological psychiatry. A few empirical investigations have been carried out by social scientists. Historians of science have charted changing ideas of mind and brain (Harrington 1987, 2008). There are also studies of changing treatment of depression in clinical practice (Healy 1997, 2004; Kramer 1997; Elliott 2003); narratives of users of antidepressants (Stepnisky 2007); ethnographic accounts of transformations in psychiatric training and clinical practice (Lakoff 2005); and investigations of specific disorders: for example, post traumatic stress disorder (A. Young 1995), bipolar affective disorder (E. Martin 2007), and autism spectrum disorder (Silverman 2004). Others have studied the cultural impact of brain imaging (Dumit 1997, 2003; Beaulieu 2000b; Joyce 2008).

23. Susan Leigh Star was at the forefront of reflections on infrastructure from the perspective of ethnography and science studies (Star 1999). We share her emphasis on the importance of this, but, for present purposes, not the inventory of issues that were relevant for her ethnographic attention to the minutiae of space and interactions.

24. Horace Magoun has pointed to the key role of such research institutes in the twentieth-century history of neurology and brain research—for example, of Ranson's Institute of Neurology at Northwestern University, with its work on the role of the hypothalamus and the lower brainstem in the 1930s—pointing to the way that this can stabilize the lives of young scientists by providing them with space, equipment, mentors, and collaborators and by fostering a particular style of work (Magoun 1962, 1974).

25. Of course, those writing these histories tend to give the credit to those with whom they worked: Kandel worked in Kuffler's department, while Rioch is not mentioned in the 622 pages of *The Neurosciences: Paths of Discovery*, which celebrates Schmitt and the NRP (Worden, Swazey, Adelman, et al. 1975).

26. In his autobiographical account on his website (http://hubel.med.harvard.edu/bio.htm).

27. Purves's autobiographical account of his own movement, and his training at the hands of mentors including Eccles, Katz, Kuffler, and many others, provides a telling example of the ways in which these conceptual, methodological, and personal interconnections formed.

28. Among many honors, Kuffler was awarded the F. O. Schmitt Prize in Neuroscience in 1979.

29. Those approaching the issue from the perspective of investment opportunities and marketing divide the field differently; thus the entrepreneurial market analyst Zack Lynch maps the "neurotechnology industry" into three main sectors: neuropharmaceutical, neurodevices, and neurodiagnositic (Lynch and Lynch 2007).

30. See, for example, Aaron Panofsky's analysis of the diverse and rivalrous research groups working on the genetics of behavior, which we draw upon below (see note 32). Thanks to Aaron Panofsky for discussing this with NR and for letting him see

three chapters of his doctoral thesis, which discuss these issues in great detail, as well as considering some of the issues of scientific fields, which we also consider here (Panofsky 2006).

31. See the acclaimed biography of Benzer, *Time, Love, Memory: A Great Biologist and His Quest for the Origins of Behavior* (Weiner 1999).

32. Aaron Panofsky has analyzed differentiation of those who work on the genetics of behavior into a number of separate yet overlapping groups and associations (Panofsky 2006).

33. Of course, others had questioned this division before this time, for example, Sir Archibald Garrod in his idea of "chemical individuality" (Garrod 1902; Childs 1970).

34. This distinction was not merely a semantic or institutional one. For example, in the United Kingdom, this conception of the difference between treatable states of illness and untreatable variations in personality meant that those problematic individuals with "personality disorder" could not be legally confined under various Mental Health Acts until 2007—confinement had to enable treatment, and a personality disorder, a "trait," was not treatable.

35. For the rising rates of the use of psychiatric drugs, see Rose (2003a, 2004, 2006a, 2006b) and chapter 7 of *The Politics of Life Itself* (Rose 2007c).

36. http://en.wikipedia.org/wiki/Neuroplasticity; consulted December 2010.

37. Wikipedia provides a good account of the developments that led up to the investigations of the brains of these monkeys in 1990, despite earlier decisions by NIH that no further research should be carried out on them, and that in 1990 indicated significant reorganization of the sensory cortex (http://en.wikipedia.org/wiki/Silver_Spring_monkeys; consulted December 2010).

38. For Taub's official biography at the University of Alabama, which makes a rather brief reference to the research on monkeys, see http://www.uab.edu/psychology/primary-faculty/11-primary-faculty/27-dr-edward-taub. For the therapy, see his statement at http://www.taubtherapy.com; consulted December 2010: "Providing The Most Effective Stroke Therapy In The World. Taub Therapy, widely recognized as the most innovative form of CI therapy, empowers people to improve the use of their limbs, no matter how long ago their stroke or traumatic brain injury (TBI) occurred. The most effective stroke rehabilitation program in the world, Taub Therapy has been proven to be over 95% successful in helping patients in the clinic regain significant movement. Through the one-on-one encouragement of a therapist, patients can relearn to use their affected limb by restricting the use of the unaffected one. By causing neurons to 'rewire' themselves, Taub Therapy not only changes the brain, it changes lives. . . . Taub Therapy gives patients hope that they can recapture the life they had before suffering a stroke or TBI."

39. For Scientific Learning ("Fit Brains Learn Better"), see http://www.scilearn.com/our-approach/our-scientists/merzenich; for Posit Science ("Proven in Labs and Lives: The Posit Science Brain Fitness Programs Dramatically Improve Cognitive Performance"), see http://www.positscience.com/science/global-science-team/merzenich.

40. This story has been told many times with different emphases (Gross 2000; Gage 2002; Rubin 2009); for our present purposes, this brief outline is sufficient.

41. In the wake of her publications on these issues, Gould was invited to speak at many debates on exactly this topic, for instance at the United Kingdom's

Royal Society of Arts in 2010 (http://www.thersa.org/events/video/archive/elizabeth-gould).

42. Some have suggested a link, perhaps even a causal one, between the popularity of these ideas of plasticity and a sociopolitical emphasis in "neo-liberalism" on flexibility in employment relations (Rubin 2009). Catherine Malabou has tried to distinguish this sociopolitical emphasis on flexibility from a more progressive idea of plasticity (Malabou 2008). The affinities are intriguing. However, we do not find it useful to invoke some generalized actor such as neo-liberalism in explaining either their origins or their implications. Rather, as we have suggested in this chapter, ideas of plasticity bring the brain sciences into line with the 'hopeful' ethos of the life sciences in contemporary societies—that we can, and should, seek to improve ourselves by using techniques based in biomedical knowledge to act on our biology. This opens multiple routes for authorities, and individuals themselves, to act on human conduct and capacities by means of, and in the name of, the brain.

43. http://www.best-personal-growth-resources.com/brain-plasticity.html.

Chapter Two. The Visible Invisible

1. Writing in the U.K. *Times Literary Supplement* on November 28, 2007 (http://entertainment.timesonline.co.uk/tol/arts_and_entertainment/the_tls/article2960112.ece).

2. Of course, much philosophical ink has been spent on debating this issue, and some philosophical critics claim it has merely been misleadingly reframed by attributing to brains and their components matters that can only properly be attributed to organisms, and in particular to humans (see Bennett and Hacker 2003).

3. As we shall see throughout this book, a variety of terms are used to express this relationship, all of them somewhat vague, and none of them specifying precisely how to overcome the explanatory gap between the activities of molecules within the human brain and the phenomenology of the experience of an inner world of thoughts, feelings, memories, and desires by a human being. We return to this question in our conclusion.

4. The English version of the title of chapter 9 of that book is "The Visible Invisible."

5. These were shown in a wonderful exhibition produced at the turn of the century titled "Spectacular Bodies" (Kemp and Wallace 2000).

6. Foucault's well-known words are worth repeating: "The gaze plunges into the space that it has given itself the task of traversing. . . . In the anatomo-clinical experience, the medical eye must see the illness spread before it . . . as it penetrates into the body, as it advances into its bulk, as it circumvents or lifts its masses, as it descends into its depths. Disease is no longer a bundle of characters disseminated here and there over the surface of the body and linked together by statistically observable concomitances and successions. . . . [I]t is the body itself that has become ill" (Foucault [1963] 1973, 136).

7. We repeat in this section some of the arguments that can be found in chapter 7 of Rose (2007c).

8. Of course, as many careful histories show, one should take this triumphalism with a pinch of salt—see, for example, Goldstein (2002).

9. The title page of the 1595 edition shows an image of a man, presumably Mercator, measuring a globe held in his hands; the image of Atlas replaced this in later editions: see, for example, that of 1633. There is a considerable historiographical literature on these issues, with some dispute about timing and rationale for the use of the term *atlas*, which we do not discuss here. The more familiar image of Atlas with the globe on his shoulders dates back at least to Roman times but is often wrongly attributed to Mercator's collections of maps.

10. Diamond was founder and president of the Royal Photographic Society of London, established in 1853, and of the *Photographic Journal*: he is often credited with taking the first photographs of asylum inmates in the Surrey County Asylum in 1851. On Diamond, see Didi-Huberman, cited below, and Packard (1962).

11. These are presented and discussed in chapter 3 of Didi-Huberman (1982) (see also Didi-Huberman and Charcot 2003), who also notes that a session of the French Medico-Psychological Society in Paris in 1867 was organized around the theme of the application of photography to the study of mental illness.

12. There is a close relation between photography in the asylum and the use of photography in the form of criminal anthropology by the police forces that were established in many countries at this time: we discuss some of this in a later chapter.

13. We have drawn on aspects of the analysis provided by Georges Didi-Huberman, cited above, although not its psychoanalytic interpretations. We have had to sidestep a large literature on the historical transformation of modes of scientific representation and the history of 'truth to nature' in the image, most tellingly analyzed by Lorraine Daston and Peter Galison in their account of how objectivity, as truth to nature, shifts in the nineteenth century from the image of the ideal form hidden within the empirical exemplars, to the image of the example itself, in all its anomalies and specificities (Daston and Galison 1992). We will return to some of these issues later in the chapter.

14. Finger credits Emanuel Swedenborg's work as an earlier example, though his treatises on the brain were only 'discovered' in 1868 (Finger 2000, 119).

15. The term was not used by Gall, and according to the *OED*, was probably coined by the American psychiatrist Benjamin Rush in 1811 to designate the science of the mind.

16. Supporters of Gall argued that such animal experiments were beside the point, but they did not prevail against their critics.

17. Many of these images collected together by John van Wyhe can be found at http://www.historyofphrenology.org.uk/overview.htm.

18. Stephen Jay Gould has discussed these attempts in detail, reanalyzed the available data, and pointed to the many ways in which those engaged in these brain-measuring activities fudged and distorted their data to fit their preconceptions concerning the differences in relative size of brain regions—for example, frontal versus posterior cranial regions—in "inferior" versus "superior" races, and between men and women (S. Gould 1981, see chap. 3).

19. Golgi and Cajal were jointly awarded the Nobel Prize in Physiology or Medicine in 1906 despite their bitter rivalry over this issue.

20. We will come across Vogt's work again in our investigation of "the antisocial brain."

21. We will not deal directly with brain banking here. Cathy Gere tells us that what happened was a remarkable shift not only in the techniques of preservation of

specimens (from chemical fixation to cryopreservation) but perhaps a more remarkable change in the rationale of their collection and the wider web of entanglements they found themselves part of. She argues: "In the 1960s, brain archiving underwent the greatest social, technological, and organizational transformations in its two hundred year history. Brains began to be archived prospectively, rather than for specific research projects, with tissue samples made available upon application—the advent of brain 'banking' " (Gere 2003, 408). To fulfill that prospective usage, brains had to be frozen rather than fixed, which also meant that the biochemical processes that were preserved could later be studied at the molecular level. The shift from brain collections to brain banks indicates another change. A bank does not merely refer to a store of things for future use, it also refers to a stock that has some value; in other words, it refers to some sort of capital. Because these vital processes could now be stored, brain banks became part of a broader "economy of vitality."

22. The use of thorium was unfortunate, as this substance often leaked into the brain, causing brain tumors.

23. We have drawn on the account given by Penfield's biographer (Lewis 1981). Penfield's work inaugurated a continuing technique for exploring the living brain using the opportunity presented by patients awaiting surgery for the removal of the localized brain lesions thought to be the focal points triggering seizures. For a recent example, see Abott (2010).

24. Another imaging technology was ultrasound. In 1937, Karl and Friedrich Dussik used ultrasound to obtain rather crude images of the brain. In 1953, Lars Leksell used ultrasound to diagnose a hematoma in an infant's brain—a technique he used in 1956 to identify the midline in the adult brain. However, although ultrasound was used as a diagnostic tool, it did not play a major role in brain visualization.

25. The history of MRI and the controversy over the respective contributions of Lauterbur and Damadian are discussed by Amit Prasad (Prasad 2005, 2007).

26. We discuss some subsequent uses of brain scans in the courtroom in chapter 6. See also chapter 8 of Rose (2007c).

27. The first symposium on the topic of molecular neuroimaging, with the theme "envisioning the future of neuroscience," was held in May 2010 at the National Institutes of Health in Bethesda, Maryland, to provide "an overview of the potential of molecular neuroimaging in improving our understanding and management of critical CNS pathophysiological processes, such as neurodegeneration, brain tumors, and psychiatric disease." The conference program is available at http://www.molecularimagingcenter.org/docs/Brain2010/MI_Brain_PrelimProgram.pdf; consulted July 2010.

28. Many new technologies are currently under development to increase the acuity of imaging and to reduce the need for cumbersome apparatus so that imaging may occur in more naturalistic settings. For example, there are developments in the use of the EEG to investigate connectivity between brain regions, and, more particularly, in the use of near infrared spectroscopy (NIRS).

29. In 2009, there were two widely cited indications of this disquiet. First was the widely reported fMRI experiment by Craig Bennett and colleagues that showed apparent variations in levels of brain activation in a dead salmon when exposed to various tasks. The poster by Craig Bennett, Abigail Baird, Michael Miller, and

George Walford was titled "Neural Correlates of Interspecies Perspective Taking in the Post-Mortem Atlantic Salmon" and pointed out the dangers of false positives and the need for correction for chance (http://prefrontal.org/files/posters/Bennett-Salmon-2009.pdf), but it received much wider attention when reported in *Wired* magazine (http://www.wired.com/wiredscience/2009/09/fmrisalmon). Second was the paper by Ed Vul and colleagues, widely circulated under its original title of "voodoo correlations," arguing that mistakes in statistical procedures in many cases had led to the "Puzzlingly High Correlations in fMRI Studies of Emotion, Personality, and Social Cognition" (Vul, Harris, et al. 2009).

30. Personal communication.

31. This, perhaps, suggests some caution in relation to recent attempts to image activity in single neurons and relate this to specific tasks or even specific thoughts.

32. Science and technology scholars have addressed similar issues in relation to digital images in other areas, for example, in Lynch's examination of digital image processing in astronomy (M. Lynch 1991). More recently, the manipulation of images in scientific publications, with the ready availability of software packages to manage such processes, has become a matter of controversy: see the discussion of this issue in *Nature* in 2009 (http://www.nature.com/news/2009/091009/full/news.2009.991.htm). Note that there are differences between different kinds of data manipulation. A technology such as fMRI requires choice of parameters such as those of threshold and color; it is good practice to note these explicitly in the publication. This differs from the silent use of techniques to enhance the visibility of an effect in an illustration in a scientific article. It is different again from the deliberate manipulation of data to mislead readers. However, the precise borderlines are sometimes difficult to define.

33. SPM is usefully discussed by researchers at the UCL Wellcome Trust Centre for Neuroimaging (http://www.fil.ion.ucl.ac.uk/spm).

34. See our discussion of the antisocial brain in chapter 6.

35. Thanks to Hauke Heekeren, now of the Max Planck Institute for Human Development, for his account of these problems at an ENSN workshop, which we have reproduced here.

36. *Molecules and Minds* was the title of a provocative collection of essays by the neurobiologist Steven Rose (1986), whose critique of reductionist explanations has many similarities to our own position.

Chapter Three. What's Wrong with Their Mice?

1. Voltaire, *The Philosophical Dictionary*, trans. H. I. Woolf (New York: Knopf, 1924), available at http://history.hanover.edu/texts/voltaire/volanima.html.

2. This chapter had its origin in discussions at the ENSN Neuroschool on Behavioral Genetics held in 2008 at the European Molecular Biology Laboratory (EMBL) outside Rome, and we benefited from reading the work in progress by one of those who attended that event. Nicole Nelson's doctoral dissertation is now completed ("Epistemic Scaffolds in Animal Behavior Genetics," PhD diss., Cornell University), and an abstract is available at http://nicolenelson.net/abstract.html. An earlier version of the chapter was presented as a keynote lecture at the EMBL conference on "Translating Behavior: Bridging Clinical and Animal Model Research," held in Heidelberg in November 2009, and we would like to thank

Cornelius Gross, Klaus-Peter Lesch, and Haldor Stefansson for their invitation to give "the social science lecture" at that event and the participants in the conference for their critical comments. The preparation of this paper was greatly facilitated by the comments of our colleague Carrie Friese on an early draft, and we also thank her for letting us see some of her own forthcoming work on this topic, especially a draft version of her paper "Models of Cloning, Models for the Zoo: Rethinking the Sociological Significance of Cloned Animals" (now published as Friese [2009]).

3. Thanks to Carrie Friese for pointing us to this ethnographic and sociological literature on models.

4. Although, as we shall discuss later, evolutionary theory has also been used by some to argue against animal experimentation, explicitly contesting the principle of conservation (LaFollette and Shanks 1996).

5. Note that Wittgenstein is not so clearly on the side of those who draw a bright line between humans and other animals. See also Vicki Hearne's impassioned argument that some animals also have desires to achieve goals and the capacity for moral understanding (Hearne 1986). Thanks to one of our reviewers for suggesting that we make this point more explicit.

6. The mereological fallacy is that of ascribing to the part that which is properly a property of the whole.

7. Nicole Nelson's ethnographic research provides us with several examples of these arguments and of the anxiety (if we may put it like that) expressed by the behavioral geneticists about their validity: see her dissertation, cited in note 2.

8. See, for example, Hogg (1996); Geyer and Markou (2000); Richter, Garner, and Würbel (2009).

9. These issues have certainly been addressed in the literature on animal models; for example, Geyer and Markou distinguish a number of different types of validity: predictive, constructive or convergent, discriminant, etiological, and face validity (Geyer and Markou 2000). This characterization problem is precisely the issue addressed by Jacqueline Crawley.

10. Of course, as we saw with brain imaging in the previous chapter, this question arises in different ways in all laboratory research in neuroscience.

11. Of course, this was a key argument in the work of Ludwik Fleck, who had a particular interest in the training in these ways of seeing (Fleck [1935] 1979).

12. Lynch provides us with some compelling ethnographic descriptions of the transformation of individual living creatures into data (M. Lynch 1989).

13. The *OED* gives the following: "1605 TIMME *Quersit.* III. 191 Wee commonly prouide that they bee prepared in our laboratorie. 1637 B. JONSON *Mercury Vind.* Induction, A Laboratory or Alchymist's workehouse."

14. We say "rhetoric of replication" because, of course, few have an interest in precisely replicating a published study, so replication is almost conducted through a series of displacements.

15. Ian Hacking is having fun by emphasizing the stability of the sciences against all the talk in the philosophy of science about scientific revolutions and so forth. So we are more than a little mischievous in using his arguments here.

16. Note that the term *model organism* is a relatively recent addition to the terms used to describe these experimental animals and animal models. As argued by Jean Gayon, the expression emerged in the 1980s, notably in relation to the funding priorities of the U.S. National Institutes of Health (Gayon 2006).

17. This quote is also used by Angela Creager, Elizabeth Lunbeck, and Norton Wise in the introduction to their interesting collection of essays on the roles of models, cases, and exemplary narratives in scientific reasoning (Creager, Lunbeck, and Wise 2007).

18. Note that Steven Rose, arguing that the "young chick is a powerful model system in which to study the biochemical and morphological processes underlying memory formation" titles his article "God's Organism?"—albeit with a question mark (Rose 2000).

19. Logan quotes Krogh: "[T]he route by which we can strive toward the ideal [of generality] is by a study of the vital functions in all their aspects *throughout the myriad of organisms*. We may find out, nay, we will find out before very long the essential mechanisms of mammalian kidney function, but the general problem of excretion can be solved only when excretory organs are studied wherever we find them and in *all their essential modifications*" (C. Logan 2002, 331; Logan's emphasis).

20. Creager et al. also make this point, citing the work of Jessica Bolker to point out that "some biologists argue that the highly canalised development of *Drosophila* that makes its development so easy to study in the laboratory also makes it a relatively poor evolutionary representative of its own phylum" (Creager, Lunbeck, and Wise 2007, 6; referring to Bolker [1995]). It is also relevant to note that the selection of the so-called wild type—against which the models are compared—is also a complex matter, as Rachel Ankeny points out in the same collection (Ankeny 2007).

21. There are, of course, many different classifications of types of model. One distinction, that we believe goes back to Clifford Geertz, is the distinction between "models of," which are representations or devices of things that are seemingly inaccessible to vision, and "models for," which can be thought of as designs that might aid in the later construction of an actual artifact (Geertz 1973). Some of these issues are helpfully discussed in McCarty (2003). For other classification schemes and discussions, see, for example, Black (1962); Suppes (1962); Achinstein (1968); Nouvel (2002); Harré (2009).

22. There are many different versions of the idea of a conceptual model, many of which are well discussed in Morrison and Morgan (1999).

23. As we suggested in earlier chapters, this model is almost certainly misleading, perhaps fundamentally mistaken. In the first decade of this century, researchers increasingly questioned the model proposed by Schildkraut, grounded in levels of neurotransmitter present in the synaptic cleft, and argued that that the principal mode of action of the selective serotonin class of psychopharmaceuticals is not an increase in the availability of neurotransmitter in the synapse, but effects on gene expression and neurogenesis (Santarelli, Saxe, Gross, et al. 2003; Sapolsky 2004). However, this model was the basis of one of the foundational myths of psychopharmacology.

24. For them, the other two dimensions are theories and experiments.

25. Besides animals (like rodents, frogs, dogs, rabbits, and fruit flies), model organisms also include plants (see Leonelli [2007]) and microorganisms such as *Escherichia coli* and yeasts—the latter are especially popular in molecular and cellular biology. For more on the evolution of model organisms and their phylogenetic diversity and relatedness see Hedges (2002).

26. Katz was awarded the Nobel Prize in Physiology or Medicine in 1970 for his dis-coveries, together with Ulf von Euler and Julius Axelrod (http://nobelprize.org/nobel_prizes/medicine/laureates/1970/katz-bio.html).

27. Dale Purves, who provides a very good account of this early work, argues that the entry of molecular biologists such as Sydney Brenner, Seymour Benzer, and others into neurobiology in the 1970s "rapidly changed the scene with respect to simple nervous systems. With surprising speed, the neurogenetics and behavior of rapidly reproducing invertebrates, such as the roundworm *Cenorhabitis elegans* and the fruit fly *Drosophila melanogaster*, supplanted leeches, lobsters, and sea slugs as the invertebrates of choice, with molecular genetics considered the best way to ultimately understand their nervous systems" (Purves 2010, 32).

28. Other models, for example, those in chemistry, can now include a temporal dimension through the use of computer animation and simulation (Francoeur and Segal 2004). Thanks to one of our reviewers for pointing to this work.

29. The question of the 'simplification' of the environment is now coming under scrutiny, with critics within the animal modeling community arguing for the replacement of the impoverished environment that has been conventional with an 'enriched' environment thought to confer more 'ecological validity' on the animals' behavior and their neural development: recent research has explored the behavioral and neurobiological changes that occur when animals are placed in such environments and has considered the implications for previous work using animal models of mental disorder (Laviola, Hannan, et al. 2008).

30. Friese, whose work (2009) has focused on the cloning of model animals in the preservation of endangered species in zoos, develops this term drawing on the terminology of Deleuze and Guattari ([1980] 1987). She suggests that the cloned animals function as models that validate in practice the theoretical assumptions of somatic cell nuclear transfer, especially interspecies transfer, as well as models for a particular way in which research on reproduction could be conducted in order to improve biomedical technologies and outcomes that might have impor-tant commercial applications—see also Sarah Franklin's work on the creation of "Dolly" by somatic cell nuclear transfer (Franklin 2007).

31. "Tout ce qui est vrai pour le Colibacille est vrai pour l'éléphant" (Jacob 1988). Note that Mark Pallen, from whom we have quoted the French original of this remark, has demonstrated that, as he puts it: "By contrast, in the postgenomic era, our bioinformatics and laboratory-based studies lead us to conclude that what is true of one strain of *E. coli* is not even true of another strain from within the same species!"—a finding that rather confirms our own view, and that of Canguilhem (and indeed, according to Pallen, that of Charles Darwin) (Pallen 2006, 116).

32. See chapters 5–6 in LaFollette and Shanks (1996) and Pallen (2006).

33. It is relevant to point out that Insel himself has long made use of voles as a model organism to study the adult human social brain, or at least with the aim of explor-ing some aspects of human behavior, and he makes frequent references to the probable implications for psychiatric disorders such as autism (e.g. Insel, O'Brien, and Leckman 1999), although he argues that the work with animals and other genomic research should only be the precursors to suggest hypotheses to be fol-lowed up in clinical studies in humans. Thanks to Zoltan Sarnyai for forcefully pointing this out to us at the event we cite in note 2 above.

34. The director of the West Riding Lunatic Asylum was James Crichton-Browne: as we saw in chapter 1, Ferrier and Crichton-Browne, together with Hughlings Jackson, founded the journal *Brain* in 1878. We discuss some further aspects of this work in chapter 6.

35. Harlow's work has been continued into the genomic era by Stephen Suomi. Suomi, working with Klaus-Peter Lesch, who is credited with discovering the effects of the short or long alleles of the promoter region of the serotonin transporter gene on risk for various psychiatric problems, now correlates the responses of infant monkeys to styles of mothering with single nucleotide polymorphism (SNP) variants affecting the activity of the serotonin transporter gene (Lesch, Bengel, Heils, et al. 1996; Bennett, Lesch, Heils, et al. 2002).

36. This injunction is the title of one of the manuscripts by Nicole Nelson referred to earlier, and she also points out, on the basis of ethnographic observation in labs, how difficult this is for those working in the area to achieve.

37. See also Scott (1958).

38. This from an author who, elsewhere, has given some serious and thoughtful consideration to problems in translation and in animal models (Miczek and de Wit 2008).

39. The authors conclude their paper by asserting: "Neuroethical concerns require discussion in open discourse among representatives of the criminal justice system, psychiatry, neuroscience, and the social sciences." (Miczek, de Almeida, Kravitz, et al. 2007, 11,805). Perhaps, however, it would be well for them to consider their own conceptual problems before suggesting that others need to consider the consequences of their tendentious extrapolations. We will come across Miczek and his group again in chapter 6, where we discuss the neurobiology of aggressive and impulsive conduct and its implications for the criminal justice system.

40. This despite explicit arguments to the contrary, for example, by John Paul Scott (Scott 1969). Of course, there is a huge literature here that we cannot discuss—for example, the careful argument by Tinbergen (1968). One could also argue that the issue of internal states could be eliminated altogether if it could be established or proved that aggressive behavior in humans (rape, fighting) is correlated with brain states and/or genomics, and that such similar states could be replicated in animals, but this nonetheless runs the risk of self-vindication. We return to some of these issues in chapter 6.

41. Note that Rodgers has pointed to some similar difficulties with the modeling of 'anxiety' in animals as if it were a unitary phenomenon (Rodgers 1997).

42. Chlorpromazine was marketed as Thorazine in the United States and as Largactil in Europe. Animal work actually began soon after the compound known as RP-4560 was first synthesized by the chemist Paul Charpentier in December 1950, and before it was made available for clinical use in 1951; however, the results were not published until after the clinical reports. See Lopez-Munoz, Alamo, Cuenca, et al. (2005, 117).

43. The elevated plus maze consists of a crosslike structure of four arms elevated above the floor, two of which are open, and two of which are enclosed at the sides (though open at the top), meeting at an open platform at the junction of the arms. The experimenters measure the amount of time the mouse spends in each arm

(together with other behavior, such as 'nose poking,' when the mouse peers out from an enclosed arm into the open platform).

44. Some suggest an interpretation that is precisely the reverse: that the more alert the mouse, the quicker it will recognize that escape is impossible in these circumstances, and will preserve its energy by ceasing futile struggle.

45. Fragile X is a genetic condition in humans that leads to mental impairments of a range of severity: it is commonly believed to arise from a triplet repeat (GCG) on the X chromosome that affects the proteins involved in neural development.

46. There is a long history of the creation of simulacra in biology that are then taken to exemplify real features of the phenomenon simulated. Evelyn Fox Keller gives us the example of Stéphane Leduc's attempts in the first decades of the twentieth century to overcome the distinction between life and non-life by creating "osmotic organisms" by introducing metallic salts and alkaline silicates into a solution of water glass, thus generating structures showing "a quite dramatic similarity to the growth and form of ordinary vegetable and marine life" (Keller 2002, 26).

47. These debates were central to the so called *methodenstreit* at the end of the nineteenth century about the distinct epistemological character of the human and social sciences, supposedly arising from their distinct ontologies, and has continued across the history of sociology and allied disciplines. The argument for a fundamentally distinct epistemology in the social sciences is particularly associated with German thought and the work of Max Weber, but it came to the boil again in the 1970s, for example, in the writings of Peter Winch and Alasdair MacIntyre (MacIntyre 1962; Winch 1990).

48. We explore these questions further in chapter 4.

49. As we were writing an early draft of this paper, *Le Monde* held a three-day public event under the title "Qui sont les animaux?" as part of the twenty-first annual forum "Le Monde Le Mans," November 13–16, 2009 (http://forumlemans .univ-lemans.fr/index.php?option=com_content&task=view&id=12&Itemid=27; consulted November 24, 2009).

50. Patton was writing in the context of a wave of recent philosophical and conceptual work on animals in the human sciences, notably by Vicki Hearne (1986). Much of this is discussed helpfully by Cary Wolfe (2003).

51. On the application of the idea of "forms of life" across species, see J. Hunter (1968).

52. Animal experiments in medical research, while contentious, are rigorously regulated in most countries, requiring specific ethical approval, licenses, assurances of good practice in the animal facilities, and the general principles of "the three R's"—replacement, reduction, and refinement—see, for example, http://www .understandinganimalresearch.org.uk/homepage.

53. A somewhat similar view has been expressed by Francesco Marincola, editor in chief of the *Journal of Translational Medicine*, in the editorial quoted earlier (Marincola 2003).

54. This is, of course, a proposal that has been made many times since Claude Bernard wrote: "Certains chimistes . . . raisonnent du laboratoire à l'organisme, tandis qu'il faut raisonner de l'organisme au laboratoire" (Bernard 1858–77, 242).

55. There is a parallel argument in contemporary genomics, where Caspi and Moffitt have suggested that one should not frame explanations in terms of genetic causes of particular types of behavior modified by environmental effects, but in terms

of environmental causes of disorders, and then seek the genomic variants that increase or decrease susceptibility or resilience to those environmental causes (Caspi and Moffitt 2006).

56. "Hopkins Team Develops First Mouse Model of Schizophrenia," July 30, 2007 (http://www.physorg.com/news105036711.html; consulted November 2009).

57. To complicate the picture, since the turn of the century, many have begun to argue that the comorbidity between these disorders arises because they are linked neurodevelopmental disorders, whose commonalities are masked by diagnostic and clinical practice (Rutter, Kim-Cohen, et al. 2006; Owen, O'Donovan, et al. 2011). We return to this issue in a later chapter, in relation to arguments for screening and early intervention.

58. The research on DISC 1 and neural development has, in the main, been carried out in rodents (for example, in Brandon, Handford, Schurov, et al. [2004], all the primary references cited are to work on the rodent brain).

59. Although three papers published in *Nature* in mid-2009 were optimistic in their claims (International Schizophrenia Consortium, 2009), most commentary on the papers was downbeat; for example, the discussion in the Schizophrenia Research Forum (dated July 3, 2009) pointed out: "There are no break-out candidate genes, though there is support for previous linkage findings, several new candidates, as well as statistical modeling that supports the notion of genetic overlap between schizophrenia and bipolar disorder. Perhaps surprisingly, none of the studies alone identified any genetic marker with significant association to the disease (by the commonly applied genome wide significance benchmark of $p < 5 \times 10^{-8}$ [i.e., $p < 5 \times 10^{-8}$])" (http://www.schizophreniaforum.org/new/detail.asp?id=1532; consulted November 20, 2009).

60. Some of these problems are alluded to in a discussion by John Medina (2008).

61. This is one use suggested for the synthetically engineered cells created by Craig Venter and his team (Gibson, Glass, Lartigue, et al. 2010); other work is currently under way with stem cell lines created by the induced pluripotent stem cell technique (iPS cells) from people diagnosed with neurodegenerative diseases, as discussed at http://www.timesonline.co.uk/tol/news/science/medicine/article7005401.ece; consulted June 2011).

62. This is one of the promises of the Human Brain Project (http://www.humanbrainproject.eu; consulted June 1, 2011).

63. For an example of the use of virtual reality–based experiments, see Baas, Nugent, Lissek, et al. (2004).

64. This way of putting things was suggested to us by Carrie Friese.

Chapter Four. All in the Brain?

1. Thanks to Peter Kinderman, who drew our attention to this quote.

2. Vol. 8 of *American Handbook of Psychiatry* (New York: Basic Books, 1986), xiv.

3. Our title for this chapter is an allusion to that well-known aphorism, that this or that experience is nothing to worry about—it is "all in the mind"—which is also the title of a program on mental health issues on BBC radio.

4. Some will remember this phrase from the title of a 1966 movie directed by Karel Reisz that raises the question of the borderlines of eccentricity and mental disorder.

5. For a review of the concept of neuropsychiatry, see Berrios and Marková (2002).

6. In the French medical literature, Dr. Lucien Lagriffe used the term *neuro-psychiatrie* in a 1910 article titled "A propos du diagnostic de la paralysie générale" in *Annales Médico-Psychologique* (June 27, 1910: 323): ". . . le diagnostic s'impose à quiconque a des notions, même seulement élémentaires, de neuro-psychiatrie." In the Anglo-American literature, a few articles published in the 1920s mention neuropsychiatry as a 'synthetic' specialty encompassing clinical as well as anatomical and physiological aspects of the nervous system and approaches to mental and psychological conditions (e.g., Schaller 1922). *Neuropsychiatry and the War*, a literature review on military neuropsychiatry, was published by the War Work Committee of the U.S. National Committee for Mental Hygiene in 1918. A neuropsychiatric branch of the U.S. Army Medical Department was founded in 1942. A year later it became a full-fledged division because of the increasing demand for 'neuropsychiatrists' during the Second World War; the division encompassed psychiatry, neurology, preventive psychiatry, and clinical psychology (Anderson, Bernucci, et al. 1966).

7. Thomas Willis, Sedleian Professor of Natural Philosophy at Oxford, was a neuroanatomist and the author of the very influential *Anatomy of the Brain*, which assigned to the striatum the function of linking brain with mind. Of course, it is possible to find earlier uses of the term. Gaspar Peucer (1525–1602), a German physician, mathematician, reformer, and scholar, wrote about doctors' "signs and diagnosis" in his popular *Commentarius de praecipuis divinationum generibus* published in Wittenberg in 1553, which was translated into French by Simon Goulard in 1584 as *Les devins ou commentaries des principales sortes de divinations* (Anvers 1584). Peucer tells us that the signs are themselves divided into "healthy, sick or neutral," that is, they reflect the bodily "constitutions." They get further divided into "diagnosis, prognosis and anamnestics." Diagnosis, Peucer writes, is a heuristic device, as it "helps discover the exact disease which is presently in the ailed body." This definition of *diagnosis* is repeated in Blankaart's *A Physical Dictionary* of 1684, and extended to include the cause of the disease and the affected bodily part besides the disease process itself: "Diagnosis is the knowledge of present signs or a knowledge whereby we understand the present condition of a Distemper; and it is three-fold, either a right instigation of the *part affected*, of the *disease itself*, or of its *cause*."

8. We have discussed some of these early attempts in Chapter 2, in connection with the question of visualizing mental states.

9. Allan Young has argued that those who were responsible for drafting *DSM-III*—notably Robert Spitzer—"identified themselves, in a self-conscious way, with the nosological perspective of . . . Emil Kraepelin" (A. Young 1995, 95). Spitzer himself denies that he was a neo-Kraeplinian (Horwitz and Wakefield 2007).

10. Actually, much of the credit for these goes to Eli Robins, a biological psychiatrist, head of the Department of Psychiatry at Washington University School of Medicine (Kendler 2009): he was awarded the APA Foundations' Fund Prize in 1982 for his contributions.

11. While the 'molar' level is the tangible and visible body, the body of organs, of tissues and limbs, the 'molecular' body is, as we have seen in chapter 1, the one envisaged by the molecular gaze at the scale of neurotransmitters, of molecules and ionic pumps, of receptors and signals across neural networks. For more, see the chapter "Molecular Biopolitics" in Rose (2007c).

12. By this, we do not mean that it was uncontested—of course, the contemporary history of psychiatry is riven by conceptual and factual disputes, and this extends to the criteria for diagnosis and the utility of the *DSM*. But the idea of a single discipline, with an explanatory regime that should embrace all these conditions nonetheless is the premise for the contemporary discipline of psychiatry.

13. http://www.in-cites.com/papers/DrRonaldKessler.html; consulted June 2010. Lee Robins is the wife of Eli Robins, whose work on the Feigner criteria we discussed earlier.

14. There have been numerous other studies in the United States along these lines, such as the Epidemiological Catchment Area Study (Regier and Robins 1991). One of the authors (Darrel Regier) has had a key role in the task force developing *DSM-5*.

15. The EBC includes both the direct costs of medical and social care and the indirect costs of lost workdays and productivity: "Direct medical expenditures alone totalled €135 billion, comprising inpatient stays (€78 billion), outpatient visits (€45 billion) and drug costs (€13 billion). Attributable indirect costs resulting from lost workdays and productivity loss because of permanent disability caused by brain disorders and mortality were €179 billion, of which the mental disorders are the most prevalent. Direct non-medical costs (social services, informal care and other direct costs) totalled €72 billion. Mental disorders amounted to €240 billion and hence constitute 62% of the total cost (excluding dementia), followed by neurological diseases (excluding dementia) totalling €84 billion (22%). Neurosurgical diseases made up a smaller fraction of the total cost of brain disorders" (European Brain Council 2005, x).

16. http://www.in-cites.com/papers/DrRonaldKessler.html; consulted June 2010.

17. Note that some similar issues arose in earlier epidemiological studies, for example, the work of Hollinshead and Redlich (Hollinshead 1999) and the Midtown Manhattan Study (Srole, Langner, et al. 1962).

18. There is, of course, a tradition of critique of medicalization in the social sciences, most recently focusing on the role of the pharmaceutical industry in expanding the empire of conditions amenable to treatment with drugs. On "the new engines of medicalization," see Conrad (2005). We have discussed this kind of analysis in detail elsewhere and will not rehearse our views here (Rose 1994, 2006a): our focus in this chapter is specifically on its implications of, and for, neuroscience.

19. Note that Kessler and colleagues argue precisely the reverse—that it is important to include mild disorders in *DSM-5* because mild disorders may progress to major ones, and diagnosis at an early stage provides an opportunity for preventive intervention (Kessler, Merikangas, Berglund, et al. 2003).

20. This criterion was introduced into *DSM-IV* largely as a result of the leadership of Allen Frances.

21. Wakefield has argued repeatedly that the so-called harmful dysfunction criterion could enable a completely "factual" basis for determining that a mental condition is a medical one: "According to the HD analysis, a condition is a disorder if it is negatively valued ('harmful') and it is in fact due to a failure of some internal mechanism to perform a function for which it was biologically designed (i.e., naturally selected)" (Wakefield 2007, 149). This will, he believes, enable medical practitioners to distinguish disorders from negative mental conditions that should not be disorders: he is referring to conditions such as ignorance, lack of skill, lack

of talent, low intelligence, illiteracy, criminality, bad manners, foolishness, and moral weakness.

22. Indeed, this route seems to be implied by Wakefield's rather different appeal to biology as a means to protect contemporary *DSM*-based psychiatry against a repeat of the earlier backlash from antipsychiatry on the grounds that it is again, pathologizing and medicalizing difference.

23. According to the American Psychiatric Association's *DSM-5* website in March 2010, the process for revising the *DSM* began with a brief discussion between Steven Hyman, MD (then-director of the NIMH), Steven M. Mirin, MD (then-medical director of the APA), and David J. Kupfer, MD (then-chair of the American Psychiatric Association Committee on Psychiatric Diagnosis and Assessment) at the NIMH in 1999. The planning process included experts in family and twin studies, molecular genetics, basic and clinical neuroscience, cognitive and behavioral science, development throughout the life span, and disability. Planning work groups were created, including groups covering developmental issues, gaps in the current system, disability and impairment, neuroscience, nomenclature, and cross-cultural issues. Darrel A. Regier was recruited from the NIMH to serve as the research director for the APA and to coordinate the development of *DSM-5*. A series of five white papers were published together in a volume titled *A Research Agenda for DSM-5*, published in 2002; other papers were published; research was undertaken by the American Psychiatric Institute for Research and Education (APIRE), with Regier as the principal investigator; and thirteen conferences were held from 2004 to 2008 to help develop the research base for the *DSM-5* task force and work groups and for WHO as it develops revisions of the International Classification of Diseases. In 2006, Kupfer was appointed as chair and Regier as vice-chair of the task force to oversee the development of *DSM-5*. The work groups have developed draft *DSM-5* diagnostic criteria, with the release of the final, approved *DSM-5* planned for May 2013. See http://www.dsm5.org/about/Pages/DSMVOverview.aspx.

24. They added: "Ethnicity and culture represent important factors that should be included in all of these research endeavors," but little was said to indicate that they believed this suggestion had serious epistemological implications, rather than being merely a nod to political correctness.

25. Indeed, it has been argued by several critics that this very claim of specificity owed more to marketing than to either neurobiology or to the results of clinical trials.

26. Perhaps the most concerted attempt to recognize the complexity of the interactions that generate "phenotypes" of mental disorder is being undertaken by the Consortium for Neuropsychiatric Phenomics at UCLA (http://www.phenomics.ucla.edu).

27. http://www.dsm5.org/Newsroom/Pages/PressReleases.aspx.

28. http://www.corepsychblog.com/2010/02/psychiatric-diagnosis; consulted June 2010.

29. Van Os, a Dutch psychiatrist and epidemiologist, had earlier written an editorial for the *British Journal of Psychiatry*, based on epidemiological and related research, arguing that the diagnosis of schizophrenia should be abolished, in favor of "salience syndrome," on the basis that psychosis was best understood as aberrant salience regulation (van Os 2009).

30. Note that David Kingdon, Lars Hansen, and Douglas Turkington, writing in the same issue of *Psychosis* and also aware of the problem that the diagnosis "extends the breadth of conditions included as mental disorders and could lead to stigmatization and inappropriate treatment with associated side-effects for individuals, most of whom will never develop psychosis," think that nonetheless it "would focus attention on provision, and improve evaluation of, early intervention services. It might also lead to earlier availability of evidence based interventions, stimulate research and provide an agreed definition of psychosis risk" (Kingdon, Hansen, and Turkington 2010).

31. This despite the fact that in the article they reference, van Os and Kapur point to "key developments in biology, epidemiology, and pharmacology of schizophrenia" such as recent molecular genetic findings, demonstrable alterations in brain structure and changes in dopamine neurotransmission, and the effectiveness of pharmacological treatments that block the dopamine system on delusions and hallucinations. They argue that a "clear genetic susceptibility exists in schizophrenia; however, what one inherits is not the illness, but altered brain development, shared partly with developmental disorders, such as autism, and partly with affective disorders such as bipolar disorder." But the susceptibility only leads to the abnormal dopamine release that causes frank psychotic symptoms when this vulnerability is combined with environmental insults. Nevertheless, "100 years after being so named, research is beginning to understand the biological mechanisms underlying the symptoms of schizophrenia and the psychosocial factors that moderate their expression" (van Os and Kapur 2009, 641).

32. There is a proposal to rename this category "Attenuated Psychotic Symptoms Syndrome" (Cornblatt and Correll 2010).

33. AD is Alzheimer's disease; FDG-PET is fluorodeoxyglucose positron emission tomography; and CSF is cerebrospinal fluid.

34. The very diagnosis of Mild Cognitive Impairment is somewhat controversial (see Macher 2004; Rose 2010a). "Some claim that the diagnosis can be made with relative certainty, others say that the phenotype is too blurry to be of much clinical or predictive use. Estimates of prevalence in those aged over 65 range from 5.2% to 16.8% (Golomb, Kluger, Ferris, et al. 2004, 354). Some claim MCI is an early stage of Alzheimer's and will progress to full Alzheimer's; others dispute this inevitable progression. Some claim that recent advances in brain imaging enable the visualization of plaques and tangles at an early stage, and hence allow prediction of which among those who meet the behavioral criteria of MCI will progress to Alzheimer's, although the evidence from the Nun Study and elsewhere suggests we should be wary. Some say that the diagnosis enables early treatment, which will slow progression; others doubt that such treatment is available. There is much to be said, here, given the blurry character of Alzheimer's itself, of the apparent capacity of a diagnosis of MCI to shift an individual onto a social and experiential pathway to Alzheimer's. Nonetheless, since the naming of this phenomena in 1988 re-organised a complex and competing field of categories and definitions, citations have increased exponentially—according to the Institute for Scientific Information's Web of Science, there was one article on this topic in 1990, 158 in 2000, and 943 in 2007" (Rose 2010a, 77). A further 2,017 papers were published on this topic from 2008 to July 2011. For helpful discussions of the utility of this diagnosis, see Brayne (2007); Whitehouse and George (2008).

35. We discuss self-vindication in chapter 3.
36. This is also true of many common complex, nonpsychiatric conditions; see Rose (2007c).
37. As we discuss in a subsequent chapter, the logics of screen and intervene have recently come into question in relation both to breast cancer (Gøtzsche and Nielsen 2009) and to prostate cancer—in the latter case, Richard Ablin, the inventor of the prostate specific antigen test, widely used in the United States as a screening test, referred to the screening program as a "public health disaster" (http://www.nytimes.com/2010/03/10/opinion/10Ablin.html).

Chapter Five. The Social Brain

1. A first version of this chapter was given by Nikolas Rose at the "Global Minds" conference held at Aarhus University, Denmark, in November 2008. Thanks to Andreas Roepstorff and Nils Buband for organizing the event, and to those participants who gave helpful comments. In developing it for this chapter, we would like to thank John Cacioppo for responding to an e-mail out of the blue, and for sending us some very helpful papers prior to publication.
2. Book 2 of *The Second Sex* opens with, "One is not born, but rather becomes, a woman" (Beauvoir 1952).
3. Of course, wild children have historically been enrolled in many different arguments for different purposes, for example, by Locke, Condillac, and the sensationalist philosophers in relation to the issue of innate ideas, and more recently, in debates over "critical periods" in the development of capacities such as vision and language. Some of these issues are discussed in Rose (1985).
4. We discuss these split-brain experiments in chapter 7.
5. A rather different version of the modularity thesis was asserted a little earlier by Jerry Fodor, although his modules were delineated in terms of their functional properties, not their localizability in the brain (Fodor 1983). Of course, to some extent, modularity is both premise and outcome of the work on visualization discussed in chapter 2, and indeed, as we saw, Fodor—for whom modularity is a conceptual and philosophical issue rather than an anatomical one—is one of those who have cast doubt on the significance of that project of localization.
6. In 2009, Dunbar was the recipient of an award from the United Kingdom's British Academy for a project titled "From Lucy to Language: The Archaeology of the Social Brain," which "aimed to bring together archaeologists, evolutionary psychologists, social anthropologists, sociologists and linguists to reconstruct our ancestors' social lives and behavior from the archaeological evidence of bones and tools. New models developed for understanding primate behavior can now be applied to the hard evidence of our ancestors to help us understand how our brains have enlarged three-fold since early hominid Lucy, four million years ago" (http://www.britac.ac.uk/arp; consulted April 2011).
7. Dunbar gives a convenient account of this thesis in a lecture given at Gustavus Adolphus College in September 2009 (http://www.youtube.com/watch?v=i98XpBFWPrI). Dunbar also makes a great deal of the role of brain size for pair bonding, and the anthropomorphic and somewhat teleological style of thought here is characteristic.

8. "Mirror Neurons and the Brain in a Vat" (http://www.edge.org/3rd_culture/ ramachandran06/ramachandran06_index.html). See also Ramachandran's earlier essay, "Mirror Neurons and Imitation Learning as the Driving Force behind the 'Great Leap Forward' in Human Evolution" (http://www.edge.org/3rd_culture/ramachandran/ramachandran_p1.html; both consulted on November 12, 2008).

9. While the original observations of mirroring of actions seem to have occurred when monkeys observed human actions, there seems to be some ambiguity in the literature as to whether the mirroring that supposedly underpins empathy for the emotions of another is limited to cases where that other is of the same species, and whether it is more pronounced when that other has a prior emotional relationship with the subject.

10. http://www.edge.org/3rd_culture/ramachandran/ramachandran_p1.html. We would like to thank Daniel Margulies for sharing the research that he conducted in 2009 in the LSE's BIOS Centre on the rise and fall of mirror neuron theory.

11. See, for example, Olivier Morin's blog written in November 2008, http://www .cognitionandculture.net/Olivier-s-blog/do-we-have-mirror-neurons-at-all.html; consulted April 1, 2010).

12. See, for example, the dispute over a 2010 paper from Marco Iacoboni's group claiming for the first time to have made recordings from human mirror neurons (Mukamel, Ekstrom, et al. 2010), also at http://neurocritic.blogspot.com/2010/04/ mirror-neurons-join-marilyn-monroe.html.

13. Tania Singer, now at University College London, was a key early proponent of this line of work (e.g., Singer, Seymour, O'Doherty, et al. 2006). Another key figure is Ernst Fehr, who is based at the University of Zurich (http:// www.econ.uzh.ch/faculty/fehr.html). They are joined by a growing number of behavioral economists whose research usually entails various game playing scenarios.

14. Some anthropologists, notably Eve Danziger, have argued that capacities such as the ability to read the intentions of others are by no means universal, but are culturally specific (Danziger 2006, 2010).

15. This suggestion has troubled some, who wonder why, if the social brain arises in response to evolutionary pressures, such pathologies have persisted. Some suggest that they are the price that humans pay for other valuable attributes, such as language (Crow 2004); others hypothesize that they confer evolutionary advantages in certain settings (Bosman, Brunetti, and Aboitiz 2004); others believe that they persist because they share a common genetic basis with the evolving circuitry of the social brain (Burns 2004, 2006). This is not the place critically to evaluate these forms of argument.

16. They drew on their own earlier work (Oberman, Hubbard, McCleery, et al. 2005; Oberman and Ramachandran 2007) as well as the claims by Rizzolatti and Gallese cited above. See also Dapretto, Davies, Pfeifer, et al. (2005).

17. According to Essential Science Indicators, this paper had been cited 1,160 times by May 2007.

18. See the very helpful interview with Lesch (http://www.in-cites.com/papers/ KPLesch.html; consulted November 13, 2008).

19. See also Canli and Lesch (2007); Lesch (2008).

20. Thanks to John Cacioppo for letting us see a version of this history prior to publication.
21. Data derived from Web of Science citation index (November 2010) for articles with the topic "social neuroscience."
22. http://mitpress.mit.edu/catalog/item/default.asp?ttype=2&tid=10683; consulted November 18, 2010).
23. http://ccsn.uchicago.edu. The Social Neuroscience Lab at NYU, directed by a psychologist, David Amodio, was established at around the same time (http://www.psych.nyu.edu/amodiolab), and many other similar centers and labs followed, from Queensland, Australia, to Beijing, China. It is beyond the scope of this chapter to map out the networks of collaboration among a relatively small number of enthusiasts that established the basis for this new discipline.
24. "The social environment, therefore, is fundamentally involved in the sculpting and activation/inhibition of basic structures and processes in the human brain and biology. . . . [S]ocial isolation or perceived social isolation (loneliness) gets under the skin to affect social cognition and emotions, personality processes, brain, biology, and health." Both this quotation and the one in the text are from his website at the University of Chicago (http://psychology.uchicago.edu/people/faculty/cacioppo/index.shtml; consulted November 1, 2010).
25. Lieberman had hosted an early conference on social cognition at UCLA in 2001. On Lieberman, see http://www.scn.ucla.edu/people/pdf/APA07.pdf. Lieberman has published extensively with his wife, Naomi Eisenberger, on the neural basis of social rejection, notably "Why Rejection Hurts" (Eisenberger, Lieberman, and Williams 2003), which we discuss later, and which is seen by many neuroimagers as a textbook example of the flawed reasoning known as "reverse inference."
26. http://www.psy.pku.edu.cn/LABS/CSCN_lab/research.html; consulted November 26, 2010.
27. http://s4sn.org/drupal; consulted November 26, 2010.
28. http://s4sn.org/drupal/?q=node/3; consulted November 30, 2010.
29. They refer here to two histories: Cacioppo, Berntson, and Decety (2011); Matusall, Kaufmann, and Christen (2011).
30. http://grants.nih.gov/grants/guide/rfa-files/RFA-DA-06-004.html; consulted November 15, 2008.
31. http://www.esrc.ac.uk/ESRCInfoCentre/about/CI/CP/research_publications/index27.aspx?ComponentId=6545&SourcePageId=6557; consulted November 15, 2008.
32. It called for applications focused on "[u]nderstanding normal and abnormal cognition, social and economic behavior of individuals and groups, and the neuropsychological basis of such behavior (including, associated disorders), particularly utilising innovative tools and methods" and on "[s]ocial, economic and occupational factors affecting mental health, mental wellbeing and mental illness" (ESRC website; consulted November 15, 2008; page no longer active).
33. Posted on http://neuroanthropology.net/2008/03/04/neuroanthropology-session-at-the-aaa-conference; consulted November 15, 2008.
34. http://www.nyas.org/valconf; consulted November 15, 2008.
35. This is at the heart of Alain Ehrenberg's critique of the social brain—that the social is reduced to the interaction between two naturalized human individuals, whereas humans as social actors operate in a shared world of cultures, institu-

tions, meanings, and values, and human affects such as empathy arise only in certain kinds of societies and in relation to those persons to whom those societies deem it appropriate to empathize—they are not natural but social (Ehrenberg 2008). However, presumably, social neuroscientists would retort that features of the brain provide the *capacity* for empathy, but that different societies and cultures shape it in particular historically contingent ways.

36. http://psychology.uchicago.edu/people/faculty/cacioppo/index.shtml; consulted November 2010.

37. Ibid.

38. Note that they refer to this as a neurosociologic theory, not a neurosociology.

39. Franks also wrote the entry on neurosociology under this name for the online *Blackwell Encyclopedia of Sociology* and, in an interesting case of conceptual entrepreneurship, he also seems to have been responsible for the Wikipedia entry—comments by those editing that entry argue that the term is now obsolete and is synonymous with *social neuroscience*: see http://en.wikipedia.org/wiki/Neurosociology.

40. One important social science discipline that has shown an openness to these new ways of thinking is economics. A number of economists have used versions of social neuroscience to explore decision making in consumption, trading, and other areas of economic activity, contesting images of rational choice, most often using the game-playing methodologies that have become habitual, and which are conveniently adaptable for use in brain scanners. While these endeavors have, to date, remained largely in the world of academia, another version, claiming to be based on an understanding of the brain, has achieved much popular exposure and has been picked up by governments in the United Kingdom and elsewhere. So-called 'nudge' techniques seek to make it easier for people to make behavioral change in desired directions, not by instructing them but by changing the incentive structures in their environment and the largely nonconscious influences on decisions (Thaler and Sunstein 2008). The claim of such arguments to rest on an understanding of the brain derived from neuroscience is hollow, however, and the suggestion that these indirect modes of behavior shaping are in some way novel is historically ignorant—governments and other authorities have utilized such techniques since at least the mid-nineteenth century (Rose and Miller 1992; Miller and Rose 2008).

41. http://www.thersa.org/events/vision/vision-videos/matthew-taylor.

42. Grist was the lead researcher on the Social Brain project, a role later taken over by Jonathan Rowson.

43. This phrase was also the title of an anthology of papers on ethology published in 1969 (Parkin 1969) and an introductory sociology book favoring an approach loosely based on social evolution (Runciman 1999).

44. Emily Martin, whose excellent work in this area we have already discussed, has consistently warned of the danger of neuroreductionism, suggesting that brain-based views of personhood were intrinsically reductionist, but that they would only become widely deployed in a culture in which a "brain-centered" view of the person made cultural sense (Martin 2000); however, we argue that a different path has been followed by many, in which brain-based views of personhood come more closely into alignment with what Martin refers to as "our everyday mental concepts"—that is to say, an ontology informed by a century of cultural assimilation of psychological conceptions of human beings.

Chapter Six. The Antisocial Brain

1. Significantly, this passage was quoted by Graham Allen in his *Early Intervention: The Next Steps; An Independent Report to Her Majesty's Government* (2011).
2. http://news.sky.com/home/world-news/article/15620258; consulted December 1, 2010.
3. Another individual who has made something of a media career out of these claims is James (Jim) Fallon of the University of California, Irvine—see for instance, his recent appearance on U.S. National Public Radio (NPR) (http://www .npr.org/templates/story/story.php?storyId=127888976).
4. http://www.abc.net.au/news/2010-04-27/carl-williams-brain-unlikely-to-reveal -secrets/412892; consulted December 1, 2010.
5. http://www.nature.com/news/2010/100317/full/464340a.html; consulted December 1, 2010.
6. This shift is discussed in more detail in chapter 8 of Rose (2007c).
7. Our account here depends on that given by Robert M. Young (1970, 39). Young points to the crucial difference between Gall's work and that of later localization researchers such as Ferrier—Gall's faculties were hypothesized, then localized, whereas the work of such researchers as Ferrier started from evidence derived from accidental or artificial brain lesions in humans or in animals (Gall was opposed to animal experiments on the grounds of their cruelty).
8. Note that Benedikt's argument was criticized at the time by the celebrated psychiatrist and neurologist Theodor Meynert, who in 1876 wrote: "[A]ttempts to localize complex phenomena of cerebral life, which only find expression in contact with social life, are not worthy of discussion. The scientific research towards the localization of cerebral powers has to restrict itself to simple functional energies, such as perception, motion, association, judgement or moods" (quoted in Verplaetse 2004, 314).
9. We discuss the Society for Mutual Autopsy in chapter 2. Note that this was just one of a number of similar societies that, unhappy with the fact that most brains for study were those obtained from asylums, prisons, or almshouses, sought to collect, study, and preserve the brains of eminent persons, and to examine them in an attempt to identify the physical basis of their prowess in the size of the brain or the convolutions of the cortex (Gere 2003, 403–6). We will not repeat here the now familiar criticisms of craniometry in general, and Broca in particular, for their attempts to identify the peculiarities of the brains of different and inferior races, or to find, in the morphology of the brain, the basis for the differences between men and women (Gould 1981).
10. In 1880, the year of Broca's own death, a large study showed an average increase of 11 grams in the brains of 119 assassins, murderers, and thieves when compared with brains from non-criminals.
11. We can also note the work of Lorenzo Tenchini, who founded the Museum of Criminal Anthropology in Parma, and, alongside his other work in medicine and psychiatry, engaged in a lifelong study of the brains of criminals and the relationship between brain anatomy and criminal behavior (Lorusso, Cristini, and Porro 2007).
12. We have drawn on David Garland's helpful discussion here (Garland 1988).
13. Note that, in his contribution to the discussion that followed this address, Thomas Clouston, himself a former president of the association, disagreed strongly: "No

doubt most of us who have looked through the books of Lombroso and Havelock Ellis and others are inclined to admit that it is a little overdone by some of our continental brethren, but to say that the mass of criminals in this country are merely criminals by want of opportunity of doing good, by want of education, and not by their organization, is absolutely contrary to the results of psychological investigation for the last fifty years. I once had occasion to carefully examine the inmates of the Edinburgh prison, and if there was one thing that impressed itself upon me it was that I had to do with a degenerate aggregation of human beings"(Clouston, in discussion following Nicolson 1895, 589). Other contributors to the discussion supported Clouston in this debate.

14. We have drawn on Hagner's excellent account in what follows.
15. Vogt, who had links with the Moscow Brain Institute, was called as a neurologist to consult on Lenin's illness; after his death he received Lenin's brain for histological study: the story is told by Paul Gregory (2008). Vogt had been Forel's pupil and then his collaborator—they cofounded the *Journal für Psychologie und Neurologie*, later renamed the *Journal für Hinforschung* (Journal of Brain Research). As for Tonnies, he came from Vogt's hometown and had been his mentor and friend (see "Vogt, Oskar Georg Dieckmann," *Complete Dictionary of Scientific Biography* (http://www.encyclopedia.com/doc/1G2-2830906176.html; consulted July 29, 2011).
16. Volkow, later to become director of NIDA, is Leon Trotsky's great-granddaughter; she grew up in the house in Mexico City where Trotsky was killed by a blow from an ice axe to his brain, though not from a repetitively violent offender, but from an undercover NKVD agent.
17. Then at the Forensic Neuropsychology Laboratory, Kirby Forensic Psychiatric Center, and Nathan S. Kline Institute for Psychiatric Research, New York University School of Medicine.
18. See, for example, presentations from Kent Kiehl and Adrian Raine at the March 2011 symposium on neuroscience and the law organized by the U.K. Royal Society's Brain Waves project and the U.S. Sackler Foundation. Transcripts of their talks (and talks from some of the other key figures in this debate such as Hank Greely and Stephen Morse) are available at http://sites.nationalacademies.org/PGA/stl/PGA_062477; consulted July 10, 2011.
19. We will not discuss developments in other legal areas, for example, in civil law and in claims for damages relating to brain injury (where the term *neurolaw* was originally coined), where things are very different; see our earlier note on this.
20. Jurisdictions vary in how they make such exceptions in relation to age, gender, medical conditions, and the like, some of which we discuss below. Of course, as noted above, there are differences between different domains of law—tax law, shipping law, the law relating to specific types of activity such as driving a motorized vehicle, and so forth. Our discussion here focuses on the criminal justice system in jurisdictions with common law and adversarial systems, where the court acts as a supposedly impartial overseer of a combat between prosecution and defense, rather than, as in inquisitorial systems such as those in most countries in continental Europe, where the court is actively involved in the investigation of the offense. It is a regrettable fact that most published debate on the matters under discussion in this chapter also relates to such systems, and there is little published information on how these issues are being worked out in other jurisdictions.

21. Probabilities do, of course, figure in deliberations in civil cases, where the standard of proof is not 'beyond reasonable doubt' but 'the balance of probabilities.' As one reviewer of an earlier draft pointed out, there have been discussions of the use of Bayesian statistics in legal proceedings, but so far with little evidence of success (Fienberg and Kadane 1983; Taroni, Bozza, et al. 2010).

22. This issue is tellingly illustrated in the attempts by Dr. Richard J. Konkol to explain what was shown on his SPECT scans, when he gave evidence at the 1999 trial of Kip Kinkel, who was accused of the Oregon school shootings. In this case, the State chose not to cross-examine Konkol. The transcript is available at http://www.pbs.org/wgbh/pages/frontline/shows/kinkel/trial/konkol.html; consulted June 1, 2010.

23. There are many cases where brain scan evidence of structural abnormalities has been introduced, where this is relevant to questions of brain damage, as in cases where compensation is at stake. At the time of writing, December 2011, these are currently under research by Susan Wolf in the United States (http://www.law.umn.edu/facultyprofiles/wolfs.html#I4thmUSlx6Uir73faZ-K_w) and by Lisa Claydon in Europe (http://law.uwe.ac.uk/staff-directory/dol-staff/lisa-claydon.aspx); results are not yet published. In criminal trials in the United States, there was, of course, the famous case of John Hinckley Jr., found not guilty by reason of insanity in his 1982 trial after shooting Ronald Reagan: a CAT scan was introduced as part of the psychiatric evidence for the defense, although there is no indication that this was decisive. After this verdict, many U.S. states changed their legal codes to abolish the verdict of not guilty by reason of insanity, which mandates a psychiatric disposal, and replaced it with the verdict of guilty but insane, which allows the courts to choose among the full range of options for sentence.

24. We confine ourselves here to questions of violent or antisocial conduct in the criminal justice system and crime control policy. We will not discuss the many other areas where neurobiological evidence may be considered relevant, for example, in witness credibility, capacity to stand trial, etc.

25. There is no evidence that we know of, as yet, of neurobiological arguments being brought to bear in cases of infanticide—one key area where the English law, since the 1920s, has established a partial defense in murder cases, as juries were reluctant to convict mothers who killed their newborn children if this meant a mandatory death sentence. The Law Commission of England and Wales reviewed this legislation in 2006, but it remains in force at the time of writing.

26. Jay Aronson has provided some helpful discussions of these issues (J. Aronson 2007, 2009).

27. For example, in 2002, the Supreme Court ruled by a majority of six to three, in the case of *Atkins v. Virginia*, that executing individuals who are mentally retarded or mentally handicapped violated the Eighth Amendment ban on cruel and unusual punishment; the threshold for such a ban was left to individual states, but is usually deemed to come into effect with an IQ below 70. This case engendered much discussion. Wikipedia provides an accessible account of the case (http://en.wikipedia.org/wiki/Atkins_v._Virginia). Despite this Supreme Court ruling, and the appeals to it by his lawyers, in 2009 Bobby Wayne Woods, with an IQ below the threshold, was put to death by lethal injection for the rape and murder of the eleven-year-old daughter of his girlfriend (http://www.guardian.co.uk/world/2009/dec/04/texas-execution-mental-disability-iq; both websites were consulted in July 2011).

28. Note that, in England, this issue is under investigation by the Law Commission as we write (http://www.justice.gov.uk/lawcommission/areas/unfitness-to-plead .htm; consulted July 10, 2011). The commission published a consultation paper in October 2010, which does discuss issues of brain damage in relation to unfitness to plead, but apart from recommending the continued involvement of psychiatrists in the assessment, makes no reference to evidence from neuroscience. Some of these issues are discussed in Peay (2010).

29. Again note that our focus here is on the antisocial brain; we will therefore not discuss the use of neuroscientific arguments and neurotechnologies in other legal areas, for example, in relation to 'deception'—as in the invention of neural lie detector technology. These issues are under ongoing investigation by Andrew Balmer of the University of Sheffield and have been discussed in a number of helpful conference papers (http://sheffield.academia.edu/AndrewBalmer/ CurriculumVitae). See also Moriarty (2009). However, there is one emerging possibility here that is relevant—the potential use of neural lie detection technologies in risk assessment of certain groups of offenders, for example, sexual offenders.

30. For more detail see Rose (2007c).

31. The online edition of the MIT newspaper, the *Tech*, of October 22, 1993, reported it thus: "More than three decades ago, a Dutch schoolteacher, troubled by a pattern of violence among his male relations, traced the pattern's origin to a couple who married in 1780. He concluded his kin must be suffering from an inherited mental disability. Pretending to be a dispassionate outsider, he then wrote up his notes under the title 'A Curious Case.' The teacher has long since died. But Friday, his 'curious case' earns a page in the annals of science, as a team of researchers from the Netherlands and the United States report that some men in his family harbor a mutant gene that predisposes them to aggressive behavior" (http://tech .mit.edu/V113/N51/agression.51w.html; consulted June 10, 2010).

32. Van Praag was one of the group of researchers in the 1960s, influenced by the emergence of the first antidepressants and antipsychotic drugs, as well as the evidence from the use of LSD, who argued that psychiatric disorders had a biological basis. Initially working on the biological basis of endogenous depression and its links with serotonin, he later argued that genetically based variations in serotonin levels were linked to aspects of personality such as impulsivity and irritability. According to David Healy and others, his lectures on biological psychiatry in the Netherlands were picketed by students and he received death threats from antipsychiatry activists (Healy 1997). Further details of his work can be found in an interview conducted with him when he was elected to become an honorary member of the European College of Neuropsychopharmacology in 2007: (http:// matters.ecnp.nl/number13/interview3.shtml; consulted December 1, 2010).

33. 5-hydroxyindoleacetic acid (5-HIAA), the main metabolite of serotonin in the human body, is used in urine samples to determine the body's levels of serotonin.

34. For further details, including the Mobley trial, see Rose (2007c).

35. As of June 2010, Brunner was professor of medical genetics and head of the department of human genetics at the Radboud University Nijmegen Medical Centre, focusing his work on the genetics of disease (http://www.ncmls.eu/ NCMLS/MenuStructures/PI/theme3/HanBrunner.asp; consulted June 10, 2010).

36. For Miczek's biography, see http://ase.tufts.edu/psychology/peopleMiczek.htm; consulted December 1, 2010.

37. The work of this panel is discussed in chapter 8 of Rose (2007c).
38. The members of this cohort were assessed on a range of measures at age three, and then at ages five, seven, nine, eleven, thirteen, fifteen, eighteen, twenty-one, and twenty-six, with future assessments scheduled for ages thirty-two, thirty-eight, forty-four, and fifty. Retention rates at age twenty-six were over 95%.
39. We discuss this discovery by Klaus Peter Lesch in chapter 5.
40. Much of the text in this section is derived verbatim from Rose (2010a).
41. Interestingly, Lynam's early research was carried out with Terrie Moffitt.
42. http://www.dshs.state.tx.us/mhprograms/tmapover.shtm; consulted February 24, 2009.
43. http://www.pharmalot.com/2008/08/texas-suspends-psych-drug-program-for-kids; consulted December 1, 2010.
44. http://www.teenscreen.org/about. For the controversy, see, for example, http://www.teenscreentruth.com; both websites consulted January 4, 2011. Note that many of the critics are widely believed to be organized by the Citizens Commission on Human Rights (CCHR), which is funded by Scientology.
45. The problems surrounding the use of biomarkers as predictors in psychiatry are discussed in detail in Singh and Rose (2009).
46. Those who work in this area distinguish between a relatively large group of children with 'conduct disorder' (around 7%), a smaller group within this with 'antisocial personality' (around 3%), and an even smaller group with 'psychopathy' (1%), and argue that each has its own characteristic pattern of neural anomalies. Some suggest that research is needed to establish which particular interventions are most appropriate for which group. For our purposes, these distinctions are not relevant.
47. This work is based on the Twins Early Development Study (TEDS) (http://www.iop.kcl.ac.uk/departments/?locator=336): "a large-scale longitudinal study of twins from early childhood through adolescence. The twins were assessed longitudinally at 2, 3, 4, 7, 9, 10, 12, 14 and currently 16 years of age in order to investigate genetic and environmental contributions to change and continuity in language, cognitive and academic abilities and behavior problems from multivariate quantitative and molecular genetic perspectives. The twins were identified from birth records of twins born in the UK in 1994–96. More than 15,000 pairs of twins have been enrolled in TEDS and the participating families are representative of the UK population. TEDS data indicate that both genetic and environmental influences are important in nearly all areas of behavioral development. . . . The TEDS dataset is proving valuable in genome-wide association research that tries to identify some of the many genes responsible for the ubiquitous heritability of behavior." See also http://www.teds.ac.uk; consulted January 4, 2011.
48. The conference followed the publication of the Social Exclusion Action Plan (2006) and the "Care Matters" white paper (2007). See http://www.personalitydisorder.org.uk/resources/emerging-pd; consulted January 4, 2011.
49. These are the aspects emphasized in the presentation on "Multi-Systemic Therapy" by Brigitte Squire at the 2007 conference cited above (http://www.personalitydisorder.org.uk/news/wp-content/uploads/brigitte-squire.pdf; consulted January 4, 2011).
50. The phrase *incredible years* derives from the well-known training series, described as follows on the organization's website: "The Incredible Years, our

award-winning parent training, teacher training, and child social skills training approaches have been selected by the U.S. Office of Juvenile Justice and Delinquency Prevention as an 'exemplary' best practice program and as a 'Blueprints' program. The program was selected as a 'Model' program by the Center for Substance Abuse Prevention (CSAP). . . . The Incredible Years Parents, Teachers, and Children Training Series has two long-range goals. The first goal is to develop comprehensive treatment programs for young children with early onset conduct problems. The second goal is the development of cost-effective, community-based, universal prevention programs that all families and teachers of young children can use to promote social competence and to prevent children from developing conduct problems in the first place. The purpose of the series is to prevent delinquency, drug abuse, and violence" (http://incredibleyears.com; consulted January 4, 2011).

51. The details for the United States that follow in this paragraph are derived from the Prevention Action website (http://www.preventionaction.org; consulted January 4, 2011).

52. http://www.preventionaction.org/putting-brain-science-back-streets-los-angeles.

53. http://www.ChildTrauma.org; consulted January 4, 2011.

54. According to the *Guardian* on April 9, 2010, in a speech recorded by a member of the audience, Duncan Smith said: "We keep going back, and as you track back you begin to realise that actually, for far too many people in society crime began before they were born. And that is a really strongly held belief of mine now, and more and more work was done in both social science, but even particularly now backed up by neuroscience demonstrates that the damage that we start children with, is damage that they keep, and that damage becomes more and more difficult as they go through. . . . We now know that we can pretty much figure out where an 18-year-old will be at the time that they are two and a half or three years old." He added the inability of a child to have "imbibed the concept of empathy" from its parent could have profound impacts on its later life. Perry said that these remarks were "an oversimplification" that "greatly misrepresents the way we would explain the impact of neglect or trauma on the developing brain." He added: "to oversimplify this way is, essentially, to distort." "I do believe that overstating and misunderstanding the neurobiology can lead to confusion, anger, distortion and potentially to bad policy," he said, adding that the claims appeared to be "a terrible distraction from the important issues related to the need to create family friendly, and developmentally informed policy that is aware and informed about the importance of early childhood and brain development" (http://www .guardian.co.uk/politics/2010/apr/09/iain-duncan-smith-childrens-brains; consulted January 4, 2010).

55. There were a number of studies of the developmental consequences of the extreme neglect suffered by Romanian orphans (Rutter, Beckett, Castle, et al. 2007; Rutter, Sonuga Barke, and Castle 2010). Brain images from the Romanian orphans, and stories of their neglect and recovery, gained considerable popular attention. For an example of the public reporting, see http://americanradioworks .publicradio.org/features/romania/b1.html; consulted 4 January 4, 2011.

56. http://www.kidspeaceofmind.org; consulted January 4, 2011.

57. For other evaluations of parenting programs in the United Kingdom, including those using the Incredible Years approach, see Edwards, Céilleachair, Bywater,

et al. (2007); Hutchings, Bywater, Daley, et al. (2007); Jones, Daley, Hutchings, et al. (2008).

58. http://www.kidspeaceofmind.org; consulted January 4, 2011.

Chapter Seven. Personhood in a Neurobiological Age

1. Earlier versions of this chapter were presented at various places, including the University of California Los Angeles, the Centre for Research on Socio-Cultural Change, and the Universities of Durham and Exeter. Thanks to all for comments. We also return here to some themes discussed in a number of papers dating back to the 1980s, some of which were brought together under the title *Inventing Ourselves: Psychology, Power, and Personhood* (Rose 1996).
2. For a very small selection of neuroscience books on these issues written for a non-specialist audience, in addition to the work of Francis Crick, Joseph LeDoux, Chris Frith, Thomas Metzinger, Julian Baggini, and the Churchlands cited below, see Gazzaniga (2008a); Greenfield (2008); Damasio (2010); Baggini (2011).
3. We should note that one of the present authors suggested some time ago that we had witnessed the birth of what he termed "the neurochemical self" (Rose 2003a, 2003b, 2004)—a term that was meant to imply, not a wholesale mutation in personhood, but the availability of a neurochemical register within which individuals could describe, judge, and seek to modulate their mental states and ailments.
4. http://www.brainhood.net; consulted June 2009.
5. We use this compendious German term for the sciences of the spirit, incorporating the humanities, the social sciences and many other disciplines, as opposed to the natural and mathematical sciences, the *Naturwissenschaften*.
6. Of course, discussion of what Mauss really meant depends a great deal on questions of translation. We return to this question later in this chapter. See also note 18, below.
7. By the phrase *positive science* we do not mean to imply approval of their epistemological bases, let alone of their normative claims—merely that these endeavors justify their truth claims by reference to empirical evidence in a characteristic manner.
8. We should note that each author finds his or her own way of making sense of these; for example, Feinberg chooses a quasi-Freudian language when he asserts: "The neuropathologies of the self are a group of highly complex and multi-determined syndromes in which brain dysfunction causes a profound and specific disturbances [*sic*] of ego boundaries and ego functions (ego disequilibrium). This disturbance facilitates the emergence of primitive ego functions and primitive psychological defenses that are essential for full creation of the NPS. The neuropathology in the majority of these cases involves the right hemisphere with a predilection for frontal involvement especially within right medial heteromodal association cortices. This latter region is hypothesized to play a critical role in the establishment of ego boundaries and to mediate the relationship between self and world. According to this hypothesis, the preservation and activation of the largely verbal defenses, such as verbal denial, projection, splitting, and fantasy may be the result of the remaining, and presumably relatively intact left verbal hemisphere. The evidence suggests that given the fact these immature defenses and fantasies are preserved and even activated in the presence of right hemisphere damage, the

emergence of the mature defenses in childhood, and their functioning in adulthood, critically depends up [*sic*] right hemisphere functions" (Feinberg 2011, 79).

9. We have discussed some aspects of these experiments, and of Gazzaniga's conclusions from them, in previous chapters.

10. The references that Gazzaniga draws on to reach this conclusion include Turk,
Heatherton, Kelley, et al. (2002); Cooney and Gazzaniga (2003); Turk, Heatherton, Macrae, et al. (2003).

11. Oliver Sacks recounts a particularly striking case of this phenomenon, one he
terms "internal amputation" in *A Leg to Stand On*. However, his experience
comes, not from brain damage but from the immobilization of his badly broken
leg in a plaster cast for several months—suggesting that the relation between the
limb and the brain can equally work in the reverse direction (Sacks 1984).

12. The philosopher Daniel Dennett has been a persistent critic of Libet's work and
his conclusions (Dennett 2003a, 2003b).

13. The title of this section is taken from a conference held by the New York Academy of Sciences on September 26–28, 2002, and a subsequent volume containing
essays by many of the participants (LeDoux, Debiec, and Moss 2003).

14. This has been an argument particularly developed by Jaak Panksepp, who has
suggested that humans share with many other animals what he terms a "core self"
that is capable of integrating information about the animals' (or humans') own
internal body states and their goal orientations, generating the affects or emotions that promote the behaviors required for survival; see, for example, Panksepp
(2005); Northoff and Panksepp (2008).

15. Theory of mind is discussed in chapter 5.

16. Others follow a similar route in suggesting that it is humans' access to their
own experiences of self that underpins insight into the mental states of others
considered as similar selves. Thus those committed to the significance of mirror
neurons, discussed in chapter 5, also draw on evidence from those diagnosed with
autism spectrum disorder, claiming this is linked to a disruption in the capability
of the mirror neuron system to represent the action goals of others, and to link
those to the self's own action goals—for it is this that "critically contributes to
shaping our experience of ourselves and of other selves, providing us with a mul-
tilayered motor representation both of our own and of others' action possibilities
[and] paves the way for the higher-level forms of self- and other-awareness generally thought to be at the core of our full-fledged sense of self and sense of others"
(Sinigaglia and Rizzolatti 2011, 72).

17. Of course, not all of those who contribute to the emerging discipline of cultural
neuroscience frame their arguments in terms of neuroplasticity; others prefer
to think in terms of gene-environment interactions and consequential cultural
variation in patterns of gene expression in neural development and pathways
(Chiao 2009). This is not the place to review the very extensive neurobiologically
inflected literature on culture and cognition.

18. Many have sought to clarify Mauss's argument. When he titled his talk "Une caté-
gorie de l'esprit humaine: La notion de personne, celle de 'moi'," what exactly
did he mean by these terms? (Carrithers, Collins, and Lukes 1985). As we have
mentioned earlier, the issue is somewhat bedeviled by translation. The original
text, in French, reads: "Pas plus que de linguistique, je ne vous parlerai de psychologie. Je laisserai de côté tout ce qui concerne le 'moi,' la personnalité consci-

ente comme telle. Je dirai seulement: il est évident, surtout pour nous, qu'il n'y a jamais eu d'être humain qui n'ait eu le sens, non seulement de son corps, mais aussi de son individualité spirituelle et corporelle à la fois. . . . Mon sujet est tout autre, et est indépendant. C'est un sujet d'histoire sociale. Comment, au cours des siècles, à travers de nombreuses sociétés, s'est lentement élaboré, non pas le sens du 'moi,' mais la notion, le concept que les hommes des divers temps s'en sont créés? Ce que je veux vous montrer, c'est la série des formes que ce concept a revêtues dans la vie des hommes des sociétés, d'après leurs droits, leurs religions, leurs coutumes, leurs structures sociales et leurs mentalités" (http://classiques .uqac.ca/classiques/mauss_marcel/socio_et_anthropo/5_Une_categorie/Une_ categorie.html; consulted July 1, 2011). It is often argued that Mauss was trying to 'sociologize' Kant's universal categories, but as Steven Lukes points out, for Kant "the self, or thinking subject (*das Ich*), is *not* a category, but 'the vehicle of all concepts' and itself transcendental but not knowable," presupposed by all experience but not amenable to knowledge (Carrithers, Collins, and Lukes 1985, 283). And when Mauss spoke of ideas, was he concerned with specialized, articulated knowledge, with more or less formalized discourses, or was he concerned with the implicit presuppositions of lay beliefs or indigenous psychologies? These and other issues are debated in the volume cited above.

19. As Mauss refers to Ribot, we might note here that Théodule Ribot was one of many in the late nineteenth century (including Sigmund Freud) who argued that consciousness is but a small part of a much larger domain of mental life—for example, asserting in his *Maladies de la personalité* of 1885 that "what emerges of the *moi* into full consciousness is only a small part of what remains buried although active. The conscious personality is never more than a small part of physical personality. . . . [I]t is by studying how the *moi* unravels [*se defait*] that we can understand how it was originally constructed" (quoted in Nicolas and Murray 1999, 290).

20. The idea of "self-fashioning" was most articulately developed by Stephen Greenblatt in his study *Renaissance Self-Fashioning: From More to Shakespeare* (1980). Self-fashioning, for Greenblatt, is the self-conscious process of fashioning character or identity, which almost always occurs in relation to one or more authorities, often in competition, and usually in order to exclude or minimize some other form of the self that is perceived as alien, strange, or in some way threatening. We differ in our usage here, placing less emphasis on fashioning that concentrates on the outward person, and in our recognition that self-fashioning is not a process primarily undertaken in the element of language.

21. We know a great deal about the history of the use of these drugs and the rise in consumption of the different classes of psychiatric drugs in different countries and geographical regions over the last two decades of the twentieth century, especially of those newer drugs claiming to be based on neurobiological research and to target precise neural pathways that underpin disorders such as depression, anxiety, or attention deficit hyperactivity disorder. See chapter 7 of Rose (2007c) and also Rose (2006b).

22. There have been many fictional versions of such a fantasy, and its costs, most recently in Alan Glynn's *Limitless* (2011).

23. http://www.dana.org/brainweek; consulted April 1, 2011.

24. http://ibreaus.org/site/why-brain-education/; consulted April 1, 2011. The IBREA also enunciated the "Brain Declaration": "World-renowned scholars, thinkers, and social activists gathered in Seoul, South Korea on June 15, 2001, to attend the New Millennium World Peace Humanity Conference. It was the first Humanity Conference held to highlight the values of Humanity, the Earth, and the Brain. . . . The adoption of the Declaration of Humanity reflected a fundamental shift in consciousness, capable of resolving the problems facing the human race. The Declaration became the philosophical foundation of Brain Education and was later adapted and renamed the Brain Declaration. Essentially, the Brain Declaration is a set of statements that affirm the power of the human brain: I declare that I am the master of my brain; I declare that my brain has infinite possibilities and creative potential; I declare that my brain has the right to accept or refuse any information or knowledge that it is offered; I declare that my brain loves humanity and the earth; I declare that my brain desires peace" (http://www.ibreaus.org/newdesign2011/?page_id=13).
25. http://www.powerbrainedu.com; consulted April 1, 2011.
26. http://www.braingym.org.uk/about/about.htm; consulted April 1, 2011.
27. See, for example, SharpBrains (http://www.sharpbrains.com; consulted April 1, 2011).
28. http://www.neurobics.com/explain.html; consulted April 1, 2011—one of many such programs that appears to be a commercial spin-off developed by a research neurobiologist.
29. For the Mind and Life Institute, see http://www.mindandlife.org; consulted April 1, 2011. The work of Francisco Varela was one important element in this relationship, and he is remembered in the institute's annual Varela Awards. Varela was trained in medicine and biology and did his doctorate on information processing in the retina with Torsten Wiesel. He trained as a Buddhist monk in Tibet in the 1970s (Hayward and Varela 1992; Varela, Thompson, and Rosch 1993).
30. For example, Ilchi Lee (http://www.ilchi.com; consulted April 1, 2011): "Ilchi Lee is a pioneering Brain Philosopher and Educator. He has developed brain training programs that are widely used in many organizations around the world. In his homeland, South Korea, his programs have been adopted as mind and body training methods in the Ministry of Education, Samsung Corporation, and the Korean Army. . . . Ilchi Lee has spent several decades investigating ways to develop the potential of the human brain. Through his life-long pursuit of brain-centered training methods and programs, hundreds of thousands of people around the world have achieved the benefits of healthier bodies, improved learning, business success, and personal empowerment."

Conclusion. Managing Brains, Minds, and Selves

1. To aid the readability of our conclusion, we have chosen, in the main, not to include bibliographic references where we are summarizing arguments made in previous chapters.
2. *Neuronal Man* is the title of a much-reprinted book by the eminent French neuroscientist Jean-Pierre Changeux (1985, 1997).
3. *GxE* is the conventional notation used to indicate that there are "interactions" between genes—or a specific genomic sequence—and environmental factors.

4. Woese distinguishes what he terms "empirical reductionism"—the methodological analysis of constituent parts and their relations—from "fundamentalist
   reductionism"—by which he means a metaphysical belief that living systems can
   be completely understood in terms of the specific properties of their constituent
   parts—a view that came to prominence with the rise of molecular genetics, and
   that, in his view, denies what classically trained biologists took for granted: the
   existence of emergent properties.
5. This issue is perceptively discussed in two papers by Cathy Gere (2004a, 2004b).
6. The theorist who has pursued this line of thought most rigorously is Elizabeth
   Wilson, and there are many affinities between her project of "gut feminism" and
   the issues that we raise here. For some early products of this work, see Wilson
   (2004a, 2004b).

References

Abi-Rached, J. M. (2008). "Mapping the field of the new brain sciences: Methodologi-
    cal, conceptual and technical dimensions." Brain, Self and Society Working
    Paper, no. 3. London: BIOS Centre, London School of Economics and Political
    Science. http://eprints.lse.ac.uk/27942.
——— (2012). "From brain to neuro: The Brain Research Association and the making
    of British neuroscience, 1965–1996." Journal of the History of the Neurosciences
    21:1–25.
Abi-Rached, J. M., and Rose, N. (2010). "The birth of the neuromolecular gaze." History
    of the Human Sciences 23 (1):1–26.
Abi-Rached, J. M., Rose, N., and Mogoutov, A. (2010). " Mapping the rise of the new
    brain sciences." Brain, Self and Society Working Paper, no. 4. London: BIOS
    Centre, London School of Economics and Political Science.
Abott, A. (2010). " 'Marilyn Monroe' neuron aids mind control." Nature. doi: 10.1038/
    news.2010.568.
Abramson, L. Y., and Seligman, M.E.P. (1977). "Modeling psychopathology in the labo-
    ratory: History and rationale." In J. M. Maser and M.E.P. Seligman, eds., Psy-
    chopathology: Experimental Models. San Francisco: W. H. Freeman.
Achinstein, P. (1968). Concepts of science. Baltimore: Johns Hopkins University Press.
Adler, J. (2004). "Mind reading: The new science of decision making." Newsweek.
    http://www.thedailybeast.com/newsweek/2004/07/04/mind-reading.html.
Adolphs, R. (2003). "Cognitive neuroscience of human social behaviour." Nature Re-
    views Neuroscience 4 (3):165–78.
Adrian, E., and Matthews, B.H.C. (1934). "The Berger rhythm: Potential changes from
    the occipital lobes in man." Brain 57 (4):355–85.
Agranoff, B. W. (2001). History of Neurochemistry. New York: Wiley.
Albin, R. (2002). "Sham surgery controls: Intracerebral grafting of fetal tissue for Par-
    kinson's disease and proposed criteria for use of sham surgery controls." British
    Medical Journal 28 (5):322.
Allen, C. (2004). "The Dunbar number as a limit to group sizes." Life with Alacrity
    blog, http://www.lifewithalacrity.com/2004/03/the_dunbar_numb.html.
Allen, G. (2011). Early intervention: The next steps; an independent report to Her Maj-
    esty's Government. London: Cabinet Office.
Allen, G., and Duncan Smith, I. (2008). Early intervention: Good parents, great kids,
    better citizens. London: Centre for Social Justice / Smith Institute.
Allport, G. W. (1954). "The historical background of modern social psychology."
    In G. Linzey, ed., Handbook of social psychology, 1–80. Cambridge, MA:
    Addison-Wesley.

Alonso, W., and Starr, P., eds. (1987). *The politics of numbers.* New York: Russell Sage Foundation.

Althusser, L. (1971). "Ideology and ideological state apparatuses (notes towards an investigation)." In *Lenin and philosophy, and other essays.* London: New Left Books.

Amara, S. G., Grillner, S., Insel, T., et al. (2011). "Neuroscience in recession?" *Nature Reviews Neuroscience* 12 (5):297–302.

American Psychiatric Association. (1952). *Diagnostic and statistical manual for mental disorders.* Washington, DC: American Psychiatric Association.

—— (1968). *Diagnostic and statistical manual for mental disorders: DSM-II* Washington, DC: American Psychiatric Association.

—— (1980). *Diagnostic and statistical manual for mental disorders: DSM-III.* Washington, DC: American Psychiatric Association.

—— (1987). *Diagnostic and statistical manual for mental disorders: DSM-III-R.* Washington, DC: American Psychiatric Association.

—— (1994). *Diagnostic and statistical manual for mental disorders: DSM-IV.* Washington, DC: American Psychiatric Association.

Anderson, M. J. (1988). "The American census: A social history." New Haven, CT: Yale University Press.

Anderson, R. S., Bernucci, R. J., et al. (1966). *Neuropsychiatry in World War II.* Washington, DC: Office of the Surgeon General, Department of the Army.

Andreasen, N. C. (2001). *Brave new brain: Conquering mental illness in the era of the genome.* Oxford: Oxford University Press.

Andriezen, W. L. (1894). "On some of the newer aspects of the pathology of insanity." *Brain* 17 (4):548–692.

Angell, M. (2004). *The truth about the drug companies: How they deceive us and what to do about it.* New York: Random House.

Ankeny, R. (2007). "Wormy logic: Model organisms as case-based reasoning." In A.N.H. Creager, E. Lunbeck, and M. N. Wise, eds. *Science without laws: Model systems, cases, exemplary narratives,* 46–58. Durham, NC: Duke University Press.

Arbib, M. A., and Mundhenk, T. N. (2005). "Schizophrenia and the mirror system: An essay." *Neuropsychologia* 43 (2):268–80.

Arden, J. (2010). *Rewire your brain: Think your way to a better life.* New York: Wiley.

Ardrey, R. (1966). *The territorial imperative: A personal inquiry into the animal origins of property and nations.* New York: Atheneum.

Arnold, D. (1986). *Police power and colonial rule: Madras, 1859–1947.* Delhi: Oxford University Press.

—— (2000). *Science, technology and medicine in colonial India.* Cambridge: Cambridge University Press.

Aronson, E. (1972). *The social animal.* San Francisco: W. H. Freeman.

Aronson, J. D. (2007). "Brain imaging, culpability and the juvenile death penalty." *Psychology, Public Policy, and Law* 13 (2):115.

—— (2009). "Neuroscience and juvenile justice." *Akron Law Review* 42:917.

Association for Research in Nervous and Mental Disease (1948). *The frontal lobes: Proceedings of the association, December 12 and 13, 1947, New York.* New York: Hafner Publishing.

Association of University Technology Managers (2007). "US University Activity Survey: FY 2006." Association of University Technology Managers. http://www .autm.net/FY_2006_Licensing_Activity_Survey.htm.

Atran, S. (2009). *Values, empathy, and fairness across social barriers*. Hoboken, NJ: New York Academy of Sciences.

Baas, J. M., Nugent, M., Lissek, S., et al. (2004). "Fear conditioning in virtual reality contexts: A new tool for the study of anxiety." *Biological Psychiatry* 55 (11):1056–60.

Bachelard, H. S. (1988). "A brief history of neurochemistry in Britain and of the Neurochemical Group of the British Biochemical Society." *Journal of Neurochemistry* 50 (3):992–95.

Bach-y-Rita, P. (1967). "Sensory plasticity." *Acta Neurologica Scandinavica* 43 (4):417–26.

Baggini, J. (2011). *The ego trick: What does it mean to be you?* London: Granta.

Bahls, C., Weitzman, J., and Gallagher, R. (2003). "Biology's models." *Scientist* 17 (suppl. 1):5.

Bailey, K. R., Rustay, N. R., and Crawley, J. N. (2006). "Behavioral phenotyping of transgenic and knockout mice: Practical concerns and potential pitfalls." *ILAR Journal* 47 (2):124.

Baker, T., and Simon, J. (2002). *Embracing risk: The changing culture of insurance and responsibility*. Chicago: University of Chicago Press.

Bakker, C., Gardner, R., Koliatsos, V., et al. (2002). "The social brain: A unifying foundation for psychiatry." *Academic Psychiatry* 26 (3):219.

Baron-Cohen, S., and Belmonte, M. (2005). "Autism: A window onto the development of the social and the analytic brain." *Annual Review of Neuroscience* 28:109–26.

Baron-Cohen, S., Leslie, A. M., and Frith, U. (1985). "Does the autistic child have a theory of mind?" *Cognition* 21 (1):37–46.

Baron-Cohen, S., Tager-Flusberg, H., and Cohen, D. J. (2000). *Understanding other minds: Perspectives from developmental cognitive neuroscience*. Oxford: Oxford University Press.

Bateson, G. (1972). *Steps to an ecology of mind*. New York: Ballantine Books.

Bateson, P.P.G., and Hinde, R. A. (1976). *Growing points in ethology: Based on a conference sponsored by St John's College and King's College, Cambridge*. Cambridge: Cambridge University Press.

Bayer, R. (1981). *Homosexuality and American psychiatry: The politics of diagnosis*. New York: Basic Books.

Bayer, R., and Spitzer, R. L. (1985). "Neurosis, psychodynamics, and *DSM-III*: A history of the controversy." *Archives of General Psychiatry* 42 (2):187–96.

Bear, M. F., and Cooper, L. N. (1998). "From molecules to mental states." *Daedalus* 127 (2):131–44.

Beaulieu, A. (2000a). "The brain at the end of the rainbow: The promise of brain scans in the research field and in the media." In J. Marchessault and K. Sawchuk, eds., *Wild science: Reading feminism, medicine and the media*, 39–54. London: Routledge.

—— (2000b). "The space inside the skull: Digital representations, brain mapping and cognitive neuroscience in the decade of the brain." Ph.D. diss., University of Amsterdam.

Beaulieu, A(2002). "Images are not the (only) truth: Brain mapping, visual knowledge and iconoclasm." *Science, Technology and Human Values* 27 (1):53–86.

—— (2004). "From brainbank to database: The informational turn in the study of the brain." *Studies in History and Philosophy of Science Part C: Studies in History and Philosophy of Biological and Biomedical Sciences* 35 (2):367–90.

Beauvoir, S. de (1952). *The second sex*. New York: Vintage Books.

Beddington, J., Cooper, C. L., Field, J., et al. (2008). "The mental wealth of nations." *Nature* 455 (7216):1057–60.

Begley, S. (2007). "How thinking can change the brain." *Wall Stret Journal*, January 19, 2007. http://online.wsj.com/article/SB116915058061980596.

—— (2009). *The plastic mind: New science reveals our extraordinary potential to transform ourselves*. London: Constable.

Bell, C. S. (1806). *Essays on the Anatomy of Expression in Painting*. London: Longmans.

Belliveau, J. W., Kennedy, D. N., Jr., McKinstry, R. C., et al. (1991). "Functional mapping of the human visual cortex by magnetic resonance imaging." *Science* 254 (5032):716–19.

Bender, E. (2004). "Personality Disorder Prevalence Surprises Researchers." *Psychiatric News* 39 (17):12–40.

Benes, F. M. (2001). "Carlsson and the discovery of dopamine." *Trends in Pharmacological Science* 22 (1):46–47.

Bennett, A. J., Lesch, K. P., Heils, A., et al. (2002). "Early experience and serotonin transporter gene variation interact to influence primate CNS function." *Molecular Psychiatry* 7 (1):118–22.

Bennett, M. R., and Hacker, P.M.S. (2003). *Philosophical foundations of neuroscience*. Malden, MA: Blackwell.

Bennett, T. (2004). *Pasts beyond memory: Evolution, museums, colonialism*. London: Routledge.

Berlucchi, G., and Buchtel, H. (2009). "Neuronal plasticity: Historical roots and evolution of meaning." *Experimental Brain Research* 192 (3):307–19.

Bernard, C. ([1865] 1957). *An introduction to the study of experimental medicine*, with an introduction by L. J. Henderson and a foreword by I. B. Cohen. Reprint, New York: Dover.

——. (1858–77). *Principes de médecine expérimentale, ou de l'expérimentation appliquée à la physiologie, à la pathologie et à la thérapeutique*. Paris.

Berrios, G. E., and Marková, I. S. (2002). "The concept of neuropsychiatry: A historical overview." *Journal of Psychosomatic Research* 53 (2):629–38.

Bigl, V., and Singer, W. (1991). "Neuroscience in the former GDR." *Trends in Neurosciences* 14 (7): 278–81.

Bilder, R. M., and LeFever, F. F. (1998). *Neuroscience of the mind on the centennial of Freud's Project for a scientific psychology*. New York: New York Academy of Sciences.

Black, M. (1962). *Models and metaphors: Studies in language and philosophy*. Ithaca, NY: Cornell University Press.

Blair, K. S., Newman, C., Mitchell, D.G.V., et al. (2006). "Differentiating among prefrontal substrates in psychopathy: Neuropsychological test findings." *Neuropsychology* 20 (2):153–65.

Blair, R.J.R. (2004). "The roles of orbital frontal cortex in the modulation of antisocial behavior." *Brain and Cognition* 55 (1):198–208.

——— (2005). "Applying a cognitive neuroscience perspective to the disorder of psychopathy." *Development and Psychopathology* 17 (3):865–91.

Blakemore, S. J. (2008). "The social brain in adolescence." *Nature Reviews Neuroscience* 9 (4):267–77.

——— (2010). "The Developing Social Brain: Implications for Education." *Neuron* 65 (6):744–47.

Blank, R. H. (1999). *Brain policy: How the new neuroscience will change our lives and our politics.* Washington, DC: Georgetown University Press.

Bloom, F. E. (1997). "Francis O. Schmitt." *Proceedings of the American Philosophical Society* 141 (4):505–8.

Bockamp, E., Maringer, M., Spangenberg, C., et al. (2002). "Of mice and models: Improved animal models for biomedical research." *Physiological Genomics* 11 (3):115–32.

Bogen, J. E., DeZure, R., TenHouten, W. D., et al. (1972). "The other side of the brain. IV. The A-P ratio." *Bulletin of the Los Angeles Neurological Society* 37 (2):49–61.

Bolker, J. A. (1995). "Model systems in developmental biology." *Bioessays* 17 (5):451–55.

Boller, F. (1999). "History of the International Neuropsychological Symposium: A reflection of the evolution of a discipline." *Neuropsychologia* 37 (1):17–26.

Borg, E. (2007). "If mirror neurons are the answer, what was the question?" *Journal of Consciousness Studies* 14 (8):5–19.

Bosanac, P., Patton, G. C., and Castle, D. J. (2010). "Early intervention in psychotic disorders: Faith before facts?" *Psychological Medicine* 40 (3):353–58.

Bosman, C., Brunetti, E., and Aboitiz, F. (2004). "Schizophrenia is a disease of general connectivity more than a specifically "social brain" network." *Behavioral and Brain Sciences* 27 (6):856.

Botvinick, M., and Cohen, J. (1998). "Rubber hands 'feel' touch that eyes see." *Nature* 391 (6669):756.

Brandon, N. J., Handford, E. J., Schurov, I., et al. (2004). "Disrupted in Schizophrenia 1 and Nudel form a neurodevelopmentally regulated protein complex: Implications for schizophrenia and other major neurological disorders." *Molecular and Cellular Neuroscience* 25 (1):42–55.

Brayne, C. (2007). "The elephant in the room—healthy brains in later life, epidemiology and public health." *Nature Reviews: Neuroscience* 8 (3):233–39.

Brazier, M. (1984). *A history of neurophysiology in the 17th and 18th centuries: From concept to experiment.* New York: Raven Press.

——— (1988). *A history of neurophysiology in the 19th century.* New York: Raven Press.

Bremner, J. D., and Vermetten, E. (2001). "Stress and development: Behavioral and biological consequences." *Development and Psychopathology* 13 (3):473–89.

Brenner, S. (2008). "An interview with . . . Sydney Brenner. Interview by Errol C. Friedberg." *Nature Reviews Molecular Cell Biology* 9 (1):8–9.

Broadhurst, P. (1960). "Abnormal animal behaviour." In H. Eysenck, ed., *Handbook of abnormal psychology: An experimental approach,* 726–62. London: Pitman.

Broca, P. (1861). "Perte de la parole, ramollissement chronique et destruction partielle du lobe antérieur gauche du cerveau." *Bulletin de la Société Anthropologique* 2:235–38.

Brodie, B. B., Pletscher, A., and Shore, P. A. (1955). "Evidence that serotonin has a role in brain function." *Science* 122:968–69.

Brooks, D. (2011). *The social animal: How success happens.* New York: Short Books.

Brothers, L. (1990). "The social brain: A project for integrating primate behavior and neurophysiology in a new domain." *Concepts in Neuroscience* 1:27–61.

———(1997). *Friday's footprint: How society shapes the human mind.* New York: Oxford University Press.

———(2001). *Mistaken identity: The mind-brain problem reconsidered.* Albany, NY: State University of New York Press.

———(2002). "The social brain: A project for integrating primate behavior and neurophysiology in a new domain." Reprinted in J. T. Cacioppo, G. G. Berntson, R. Adolphs, et al., eds., *Foundations in social neuroscience,* 367–87. Cambridge, MA: MIT Press.

Brower, M. C., and Price, B. H. (2001). "Neuropsychiatry of frontal lobe dysfunction in violent and criminal behaviour: A critical review." *Journal of Neurology, Neurosurgery and Psychiatry* 71 (6):720.

Brown, N. (2003). "Hope against hype—accountability in biopasts, presents and futures." *Science Studies* 16 (2):3–21.

Brown, N., and Michael, M. (2003). "A sociology of expectations: Retrospecting prospects and prospecting retrospects." *Technology Analysis and Strategic Management* 15 (1):3–18.

Brown, N., Rappert, B., and Webster, A. (2000). *Contested futures: a sociology of prospective techno-science.* Aldershot, UK: Ashgate.

Brown, N., and Webster, A. (2004). *New medical technologies and society: Reordering life.* Cambridge: Polity.

Brunet-Gouet, E., and Decety, J. (2006). "Social brain dysfunctions in schizophrenia: A review of neuroimaging studies." *Psychiatry Research* 148 (2–3):75–92.

Brunner, H. G. (1996). "MAOA deficiency and abnormal behavior: Perspectives on an association." In CIBA Foundation, *Genetics of criminal and antisocial behavior,* Symposium no. 194, 155–64. New York: Wiley.

Brunner, H. G., Nelen, M., Breakefield, X. O., et al. (1993). "Abnormal-behavior associated with a point mutation in the structural gene for monoamine oxidase A." *Science* 262 (5133):578–80.

Buckholtz, J. W., and Meyer-Lindenberg, A. (2008). "MAOA and the neurogenetic architecture of human aggression." *Trends in Neurosciences* 31 (3):120–29.

Bucknill, J. C., and Tuke, D. H. (1874). *A manual of psychological medicine.* London: X. J. and A. Churchill.

Burleigh, M. (1994). *Death and deliverance: 'Euthanasia' in Germany c.1900–1945.* Cambridge: Cambridge University Press.

Burns, J. K. (2004). "Elaborating the social brain hypothesis of schizophrenia." *Behavioral and Brain Sciences* 27 (6):868–85.

———(2006). "Psychosis: A costly by-product of social brain evolution in *Homo sapiens.*" *Progress in Neuro-Psychopharmacology and Biological Psychiatry* 30 (5):797–814.

Burri, R., and Dumit, J. (2007). "Social Studies of Scientific Imaging and Visualization." in E. J. Hackett, O. Amsterdamska, M. Lynch, and J. Wajcman, eds., *Handbook of science and technology studies,* 297–317. Cambridge, MA: MIT Press.

Bush, V. (1945). "Science—the endless frontier. A report to the president." Office of Scientific Research and Development. Washington, DC: United States Government Printing Office.

Cacioppo, J. T. (1994). "Social neuroscience—autonomic, neuroendocrine, and immune-responses to stress." *Psychophysiology* 31 (2):113–28.

Cacioppo, J. T., and Berntson, G. G. (1992). "Social psychological contributions to the decade of the brain: Doctrine of multilevel analysis." *American Psychologist* 47 (8):1019–28.

———, eds. (2004). *Essays in social neuroscience.* Cambridge, MA.: MIT Press.

Cacioppo, J. T., Berntson, G. G., and Decety, J. (2011). "A brief history of social neuroscience." In A. Kruglanski and W. Stroebe, eds., *Handbook of the history of social psychology.* New York: Psychology Press.

Cacioppo, J. T., Berntson, G. G., Lorig, T. S., et al. (2003). "Just because you're imaging the brain doesn't mean you can stop using your head: A primer and set of first principles." *Journal of Personality and Social Psychology* 85 (4):650–61.

Cacioppo, J. T., Bernston, G. G., Neil, J. S., et al. (2001). "Social Neuroscience." In N. J. Smelser and P. B. Baltes, eds., *International Encyclopedia of the Social and Behavioral Sciences,* 14,388–91. Oxford: Pergamon.

Cacioppo, J. T., and Decety, J. (2011). "Challenges and Opportunities in Social Neuroscience." *Annals of the New York Academy of Sciences* 1224 (1):162–73.

Cacioppo, J. T., and Patrick, W. (2008). *Loneliness: Human nature and the need for social connection.* New York: W. W. Norton.

Cacioppo, J., and Tassinary, L. (1990). *Principles of psychophysiology: Physical, social, and inferential elements*: Cambridge: Cambridge University Press.

Cameron, H. A., and Gould, E. (1994). "Adult neurogenesis is regulated by adrenal-steroids in the dentate gyrus." *Neuroscience* 61 (2):203–9.

Cameron, H. A., McEwen, B. S., and Gould, E. (1995). "Regulation of adult neurogenesis by excitatory input and NMDA receptor activation in the dentate gyrus." *Journal of Neuroscience* 15 (6):4687–92.

Cameron, J., Duncan, G. J., Fox, N. A., et al. (2007). "The timing and quality of early experiences combine to shape brain architecture." Boston: Harvard University Center on the Developing Child.

Canguilhem, G. ([1965] 2008). *Knowledge of Life.* New York: Fordham University Press.

——— (1968). *Etudes d'histoire et de philosophie des sciences.* Paris: Vrin.

——— (1977). *La formation du concept de réflexe aux XVIIe et XVIIIe siècles.* Paris: Vrin.

——— (1994). *A vital rationalist: Selected writings from Georges Canguilhem.* Edited by François Delaporte. Translated by Arthur Goldhammer. New York: Zone Books.

Canli, T., and Lesch, K. P. (2007). "Long story short: The serotonin transporter in emotion regulation and social cognition." *Nature Neuroscience* 10 (9):1103–9.

Carlsson, A. (1959). "The occurence, distribution and physiological role of catecholamines in the nervous system." *Pharmacological Reviews* 11:490–93.

——— (1990). "Early psychopharmacology: The rise of modern brain research." *Journal of Psychopharmacology* 4:120–27.

——— (2001). "A half-century of neurotransmitter research: Impact on neurology and psychiatry (Nobel lecture)." *Chembiochem* 2:484–93.

Carlsson, A., and Lindqvist, M. (1963). "Effect of chlorpromazine or haloperidol on formation of 3-methoxytyramine and normetanephrine in mouse brain." *Acta Pharmacologica et Toxicologica* 20 (2):140–44.

Carlsson, A., Lindqvist, M., Magnusson, T., et al. (1958). "Presence of 3-hydroxytyramine in brain." *Science* 127 (3296):471.

Carpenter, M. (1851). *Reformatory schools for the children of the perishing and dangerous classes, and for juvenile offenders*. London: N.p.

Carrithers, M., Collins, S., and Lukes, S., eds. (1985). *The category of the person: Anthropology, philosophy, history*. Cambridge: Cambridge University Press.

Carruthers, P. (2009). "How we know our own minds: The relationship between mind-reading and metacognition." *Behavioral and Brain Sciences* 32 (2):121–38.

Carter, R. (1998). *Mapping the mind*. London: Weidenfeld and Nicholson.

Cartwright, L. (1995). *Screening the body: Tracing medicine's visual culture*. Minneapolis: University of Minnesota Press.

Cartwright, N. (1983). *How the laws of physics lie*. Oxford: Oxford University Press.

—— (1997). "Models: The blueprints for laws." *Philosophy of Science* 64:S292–S303.

Cases, O., Seif, I., Grimsby, J., et al. (1995). "Aggressive-behavior and altered amounts of brain-serotonin and norepinephrine in mice lacking MAOA." *Science* 268 (5218):1763–66.

Casey, B., Jones, R. M., and Hare, T. A. (2008). "The adolescent brain." *Annals of the New York Academy of Sciences* 1124 (1):111–26.

Caspi, A., McClay, J., Moffitt, T. E., et al. (2002). "Role of genotype in the cycle of violence in maltreated children." *Science* 297 (5582):851–54.

Caspi, A., and Moffitt, T. E. (2006). "Gene-environment interactions in psychiatry: Joining forces with neuroscience." *Nature Reviews Neuroscience* 7:583–90.

Caspi, A., Moffitt, T. E., Cannon, M., et al. (2005). "Moderation of the effect of adolescent-onset cannabis use on adult psychosis by a functional polymorphism in the catechol-O-methyltransferase gene: Longitudinal evidence of a gene X environment interaction." *Biological Psychiatry* 57 (10):1117–27.

Caspi, A., Sugden, K., Moffitt, T. E., et al. (2003). "Influence of life stress on depression: Moderation by a polymorphism in the 5-HTT gene." *Science* 301 (5631):386–89.

Castel, R. (1976). *L'ordre psychiatrique: L'âge d'or de l'aliénisme*. Paris: Editions de Minuit.

Castel, R., Castel, F., and Lovell, A. (1982). *The psychiatric society*. Translated by Arthur Goldhammer. New York: Columbia University Press.

Center for Mental Health Services and National Institute of Mental Health (1999). *Mental health: A report of the surgeon general*. Rockville, MD: U.S. Department of Health and Human Services.

Champagne, F. A., Chretien, P., Stevenson, C. W., et al. (2004). "Variations in nucleus accumbens dopamine associated with individual differences in maternal behavior in the rat." *Journal of Neuroscience* 24 (17):4113–23.

Champagne, F. A., and Meaney, M. J. (2006). "Stress during gestation alters postpartum maternal care and the development of the offspring in a rodent model." *Biological Psychiatry* 59 (12):1227–35.

Changeux, J.-P. (1985). *Neuronal man: The biology of mind*. New York: Pantheon Books.

—— (1997). *Neuronal man: The biology of mind*. 2nd ed. Princeton, NJ: Princeton University Press.

Changeux, J.-P., and Ricoeur, P. (2002). *What makes us think?: A neuroscientist and a philosopher argue about ethics, human nature, and the brain*. Princeton, NJ: Princeton University Press.

Charney, D. S., Barlow, D. H., Botteron, K., et al. (2002). "Neuroscience research agenda to guide development of a pathophysiologically based classification system." In D. J. Kupfer, M. B. First, and D. A. Regier, eds., *A research agenda for DSM-V*, 31–84. Washington, DC: American Psychiatric Association.

Chiao, J. Y. (2009). "Cultural neuroscience: A once and future discipline." in Y. C. Joan, ed., *Progress in brain research*, 287–304. Amsterdam: Elsevier.

Childs, B. (1970). "Sir Archibald Garrod's conception of chemical individuality: A modern appreciation." *New England Journal of Medicine* 282 (2):71–77.

Chimonas, S., Frosch, Z., and Rothman, D. (2011). "From disclosure to transparency: The use of company payment data." *Archives of Internal Medicine* 171 (1):81–86.

Chou, I. H., and Chouard, T. (2008). "Neuropsychiatric disease." *Nature* 455 (7215):889.

Churchland, P. M. (1995). *The engine of reason, the seat of the soul: A philosophical journey into the brain.* Cambridge, MA: MIT Press.

Churchland, P. S. (1986). *Neurophilosophy: Toward a unified science of the mind-brain.* Cambridge, MA: MIT Press.

——— (2002). *Brain-wise: Studies in neurophilosophy.* Cambridge, MA: MIT.

——— (2003). "Self-representation in nervous systems." Special issue, "The self: From soul to brain," *Annals of the New York Academy of Sciences* 1001:31–38.

Cleckley, H. M. ([1941] 1976). *The mask of sanity: An attempt to clarify some issues about the so-called psychopathic personality.* London: Mosby.

Clouston, T. S. (1894). "The developmental aspects of criminal anthropology." *Journal of the Anthropological Institute of Great Britain and Ireland* 23:215–25.

Cohen, M. (1998). "The monetary value of saving a high-risk youth." *Journal of Quantitative Criminology* 14 (1):5–33.

Cohn, S. (2008). "Petty cash and the neuroscientific mapping of pleasure." *BioSocieties* 3 (2):151–63.

Collins, P. Y., Patel, V., Joestl, S. S., et al. (2011). "Grand challenges in global mental health." *Nature* 475 (7354):27–30.

Congrès International de Psychiatrie, ed. (1950). *Anatamo-physiologie cérébrale à la lumière des lobotomies et topectomies.* Paris: Hermann and Cie.

Conrad, L. I., Neve, M., Nutton, V., et al. (1995). *The western medical tradition 800 BC to AD 1800.* Cambridge: Cambridge University Press.

Conrad, P. (1976). *Identifying hyperactive children: The medicalization of deviant behavior.* Lexington, MA: D. C. Heath.

——— (2005). "The shifting engines of medicalization." *Journal of Health and Social Behavior* 46 (1):3–14.

Conrad, P., and Schneider, J. W. (1992). *Deviance and medicalization: From badness to sickness.* Philadelphia: Temple Universtiy Press.

Cooksey, D. (2006). "A review of UK health research funding." London: Stationery Office.

Cooney, J. W., and Gazzaniga, M. S. (2003). "Neurological disorders and the structure of human consciousness." *Trends in Cognitive Sciences* 7 (4):161–65.

Cooper, R. (2004). "What is wrong with the *DSM*?" *History of Psychiatry* 15 (57):5–25.

Corcoran, R., Mercer, G., and Frith, C. (1995). "Schizophrenia, symptomatology and social inference: Investigating 'theory of mind' in people with schizophrenia." *Schizophrenia Research* 17 (1):5–13.

Cornblatt, B. A., and Correll, C. U. (2010). "A new diagnostic entity in *DSM-5*?" Medscape Today, http://www.medscape.com/viewarticle/727682.

Couldwell, W. T., Feindel, W., and Rovit, R. L. (2004). "William Osler at McGill University: the baby professor and his early contributions to neurosurgery." *Journal of Neurosurgery: Pediatrics* 101 (4):705–13.

Cowan, W. M., Harter, D. H., and Kandel, E. R. (2000). "The emergence of modern neuroscience: Some implications for neurology and psychiatry." *Annual Review of Neuroscience* 23:343–91.

Crane, D. (1972). *Invisible colleges: Diffusion of knowledge in scientific communities.* Chicago: University of Chicago Press.

Crawley, J. N. (2007). *What's wrong with my mouse? Behavioral phenotyping of transgenic and knockout mice.* Hoboken, NJ: Wiley.

Creager, A.N.H., Lunbeck, E., and Wise, M. N., eds. (2007). *Science without laws: Model systems, cases, exemplary narratives.* Durham, NC: Duke University Press.

Crick, F. (1994). *The astonishing hypothesis: The scientific search for the soul.* New York: Charles Scribner's Sons.

—— (1999). "The impact of molecular biology on neuroscience." *Proceedings of the Royal Society B: Biological Sciences* 354 (1392):2021.

Crick, F. C., and Koch, C. (2005). "What is the function of the claustrum?" *Proceedings of the Royal Society B: Biological Sciences* 360 (1458):1271.

Crow, T. J. (2004). "Language and asymmetry versus the social brain—where are the testable predictions?" *Behavioral and Brain Sciences* 27 (6):857.

Crowe, S. L., and Blair, R.J.R. (2008). "The development of antisocial behavior: What can we learn from functional neuroimaging studies?" *Development and Psychopathology* 20 (4):1145–59.

Cryan, J. F., and Holmes, A. (2005). "The ascent of mouse: Advances in modelling human depression and anxiety." *Nature Reviews: Drug Discovery* 4 (9):775–90.

Cunningham, W. A., Johnson, M. K., Gatenby, J. C., et al. (2003). "Neural components of social evaluation." *Journal of Personality and Social Psychology* 85 (4):639–49.

Dadds, M. R., Fraser, J., Frost, A., et al. (2005). "Disentangling the underlying dimensions of psychopathy and conduct problems in childhood: A community study." *Journal of Consulting and Clinical Psychology* 73 (3):400–410.

Dahlstrom, A., and Carlsson, A. (1986). "Making visible the invisible." In M. J. Parnham and J. Bruinvels, eds., *Discoveries in pharmacology: Pharmacological methods, receptors and chemotherapy*, 97–125. Amsterdam: Elsevier.

Damasio, A. R. (1994). *Descartes' error: Emotion, reason, and the human brain.* New York: Putnam.

—— (1999). *The feeling of what happens: Body and emotion in the making of consciousness.* New York: Harcourt Brace.

—— (2010). *Self comes to mind: Constructing the conscious brain.* New York: Pantheon Books.

Damasio, A. R., Tranel, D., and Damasio, H. (1990). "Individuals with sociopathic behavior caused by frontal damage fail to respond autonomically to social-stimuli." *Behavioural Brain Research* 41 (2):81–94.

Danziger, E. (2006). "The thought that counts: Interactional consequences of variation in cultural theories of meaning." In N. J. Enfield and Stephen C. Levinson, eds., *Roots of human sociality: Culture, cognition, and interaction*, 259–78. Oxford: Berg.

—— (2010). "On trying and lying: Cultural configurations of Grice's maxim of quality." *Intercultural Pragmatics* 7 (2):199–219.

Danziger, K. (1990). *Constructing the subject: Historical origins of psychological research.* New York: Cambridge University Press.

Dapretto, M., Davies, M., Pfeifer, J., et al. (2005). "Understanding emotions in others: Mirror neuron dysfunction in children with autism spectrum disorders." *Nature Neuroscience* 9 (1):28–30.

Darwin, C. ([1872] 2009). *The expression of the emotions in man and animals.* Reprint, London: Penguin.

Daston, L., and Galison, P. (1992). "The image of objectivity." *Representation* 40 (Fall):135–56.

Davini, V. (1950). "Rapid action barbiturates in neuropsychiatry." *Sistema Nervoso* 2 (1):61–75.

Debru, C. (2006). "L'animal va-t-il céder la place aux biotechnologies?" In G. Gachelin, ed., *Les organismes modèles dans la recherche médicale,* 229–41. Paris: Presses Universitaires de France.

Debru, C., Barbara, J.-G., and Cherici, C., eds. (2008). *Neurosciences: Son essor de 1945 à 1975.* Paris: Hermann.

Decety, J., and Keenan, J. P. (2006). "Social neuroscience: A new journal." *Social Neuroscience* (1):1–4.

Decety, J., and Meyer, M. (2008). "From emotion resonance to empathic understanding: A social developmental neuroscience account." *Development and Psychopathology* 20 (4):1053–80.

de Chadarevian, S., and Hopwood, N., eds. (2004). *Models: The third dimension of science.* Stanford, CA: Stanford University Press.

Decker, H. S. (2007). "How Kraepelinian was Kraepelin? How Kraepelinian are the neo-Kraepelinians?—from Emil Kraepelin to *DSM-III.*" *History of Psychiatry* 18 (3):337–60.

Degos, L. (2006). "Jean Bernard: Medicine, science, humanity." *Haematologica* 91 (9):1163.

DeJong, R. N. (1982). *A history of American neurology.* New York: Raven Press.

De Koning, M. B., Bloemen, O.J.N., Van Amelsvoort, T.a.M.J., et al. (2009). "Early intervention in patients at ultra high risk of psychosis: Benefits and risks." *Acta Psychiatrica Scandinavica* 119 (6):426–42.

Delay, J. (1953). *Etudes de psychologie médicale.* Paris: Presses Universitaires de France.

Delay, J., and Buissoj, J. F. (1958). "Psychic action of isoniazid in the treatment of depressive states." *Journal of Clinical and Experimental Pyschopathology and Quarterly Review of Psychiatry and Neurology* 19 (2 [suppl.]):51–55.

Deleuze, G., and Guattari, F. ([1980] 1987). *A thousand plateaus: Capitalism and schizophrenia.* Minneapolis: University of Minnesota Press.

Dennett, D. C. (2003a). *Freedom evolves.* New York: Viking.

——— (2003b). "The self as a responding—and responsible—artifact." Special issue, "The self: From soul to brain," *Annals of the New York Academy of Sciences* 1001:39–50.

Deutsch, A. (1944). "The first US census of the insane (1840) and its use as pro-slavery propaganda." *Bulletin of the History of Medicine* 15:469–82.

Dickman, S. (1989). "Brain sections to be buried." *Nature* 339:498.

Didi-Huberman, G. (1982). *Invention de l'hystérie: Charcot et l'iconographie photographique de la Salpêtrière.* Paris: Macula.

Didi-Huberman, G. (2003). *The invention of hysteria: Charcot and the photographic iconography of the Salpêtrière.* Cambridge, MA: MIT Press.

Dinstein, I. (2008). "Human cortex: Reflections of mirror neurons." *Current Biology* 18 (20):956–59.

Dinstein, I., Thomas, C., Behrmann, M., et al. (2008). "A mirror up to nature." *Current Biology* 18 (1):R13–R18.

Di Pellegrino, G., Fadiga, L., Fogassi, L., et al. (1992). "Understanding motor events—a neurophysiological study." *Experimental Brain Research* 91 (1):176–80.

Dodds, E. R. (1968). *The Greeks and the irrational.* Berkeley: University of California Press.

Doidge, N. (2008). *The brain that changes itself.* New York: Viking.

Dolan, M. (1994). "Psychopathy—a neurobiological perspective." *British Journal of Psychiatry* 165:151–59.

Donzelot, J. (1979). *The policing of families.* New York: Pantheon Books.

Dorsey, E. R., De Roulet, J., Thompson, J. P., et al. (2010). "Funding of US Biomedical Research, 2003–2008." *JAMA: The Journal of the American Medical Association* 303 (2):137–43.

Dorsey, E. R., Vitticore, P., De Roulet, J., et al. (2006). "Financial anatomy of neuroscience research." *Annals of Neurology* 60 (6):652–59.

Doty, R. W. (1987). "Neuroscience." In John R. Brobeck, Orr E. Reynolds, and T. A. Appel, eds., *History of the American Physiological Society: The first century, 1887–1987,* 427–34. Bethesda, MD: American Physiological Society.

Dumit, J. (1997). "A digital image of the category of the person: PET scanning and objective self-fashioning." In G. L. Downey and J. Dumit, eds., *Cyborgs and citadels: Anthropological interventions in emerging sciences,* 83–102. Sante Fe, NM: School of American Research Press.

—— (1999). "Objective brains, prejudicial images." *Science in Context* 12 (1):173–201.

—— (2003). *Picturing personhood: Brain scans and biomedical identity.* Princeton, NJ: Princeton University Press.

Dunbar, R.I.M. (1993). "Coevolution of neocortical size, group size and language in humans." *Behavioral and Brain Sciences* 16 (4):681–93.

—— (1996). *Grooming, gossip and the evolution of language.* London: Faber and Faber.

—— (1998). "The social brain hypothesis." *Evolutionary Anthropology* 6 (5):178–90.

—— (2003a). "Evolution of the social brain." *Science* 302 (5648):1160–61.

—— (2003b). "The social brain: Mind, language, and society in evolutionary perspective." *Annual Review of Anthropology* 32:163–81.

Dunbar, R.I.M., and Shultz, S. (2007). "Evolution in the social brain." *Science* 317 (5843):1344–47.

Dupré, J. (1993). *The disorder of things: Metaphysical foundations of the disunity of science.* Cambridge, MA: Harvard University Press.

Dyck, E. (2008). *Psychedelic psychiatry: LSD from clinic to campus.* Baltimore: Johns Hopkins University Press.

Edelman, G. M., and Changeux, J. P., eds. (2001). *The brain.* New Brunswick, NJ: Transaction Publishers.

Edwards, R., Céilleachair, A., Bywater, T., et al. (2007). "Parenting programme for parents of children at risk of developing conduct disorder: Cost effectiveness analysis." *British Medical Journal* 334 (7595):682.

Ehrenberg, A. (2004). "Le sujet cérébral." *Esprit* 309:130–55.

—— (2008). "Le cerveau 'social': Chimère épistémologique et vérité sociologique." *Esprit* (January 2008):79–103.

Eisenberger, N. I., Lieberman, M. D., and Williams, K. D. (2003). "Does rejection hurt? An fMRI study of social exclusion." *Science* 302 (5643):290–92.

Eling, P. (2001). "Neuro-anniversary 2001." *Journal of the History of the Neurosciences* 10 (1):2–5.

Ellenberger, Henri F. (1970). *The discovery of the unconscious: The history and evolution of dynamic psychiatry.* New York: Basic Books.

Elliott, C. (2003). *Better than well: American medicine meets the American dream.* New York: W. W. Norton.

Ellis, H. (1890a). *The criminal.* 1st ed. London: Scott.

—— (1890b). "The study of the criminal." *Journal of Mental Science* 36 (152):1–15.

—— (1901). *The criminal.* 3rd ed., rev. and enlarged. London: Walter Scott.

English, D., and Annand, A. (2010). "Neural stem cell therapy." *Annals of Neurosciences* 17 (1):1.

Ericson, R. V., and Haggerty, K. D. (1997). *Policing the risk society.* Oxford: Clarendon Press.

Esman, A. H. (1999). "Henri Dagonet and the origins of psychiatric photography." *American Journal of Psychiatry* 156 (9):1439.

Esquirol, J.E.D. (1838). *Des maladies mentales considérées sous les rapports médical, hygiénique et médico-légal.* Paris: Baillière.

—— (1845). *Mental maladies: A treatise on insanity. Translated from the French with additions by E. K. Hunt.* Philadelphia: N.p.

—— ([1845] 1965). [*Des maladies mentales.*] *Mental maladies: A facsimile of the English edition of 1845.* With an introduction by Raymond de Saussure. New York: Hafner.

European Brain Council. (2005). "Cost of disorders of the brain in Europe." *European Journal of Neurology* 12 (Suppl. 1):x–xi.

Farah, M. J. (2007). "Social, legal, and ethical implications of cognitive neuroscience: "Neuroethics" for short." *Journal of Cognitive Neuroscience* 19 (3):363–64.

Feeley, M., and Simon, J. (1992). "The new penology: Notes on the emerging strategy of correction and its implications." *Criminology* 30 (4):449–74.

—— (1994). "Actuarial justice: Power/knowledge in contemporary criminal justice." In D. Nelken, ed., *The futures of criminology*, 173–201. London: Sage.

Feighner, J. P., Robins, E., Guze, S. B., et al. (1972). "Diagnostic criteria for use in psychiatric research." *Archives of General Psychiatry* 26 (1):57–63.

Feinberg, T. E. (2011). "Neuropathologies of the self: Clinical and anatomical features." *Consciousness and Cognition* 20 (1):75–81.

Feinberg, T. E., and Keenan, J. P. (2005). "Where in the brain is the self?" *Consciousness and Cognition* 14 (4):661–78.

Ferrier, D. (1876). *The Functions of the Brain.* London: Smith and Elder.

—— (1886). *The Functions of the Brain.* 2nd ed. London: Smith and Elder.

Fienberg, S. E., and Kadane, J. B. (1983). "The presentation of Bayesian statistical analyses in legal proceedings." *Statistician* 32:88–98.

Finger, S. (1994). *Origins of neuroscience: A history of explorations into brain function.* New York: Oxford University Press.

—— (2000). *Minds behind the brain: A history of the pioneers and their discoveries.* Oxford: Oxford University Press.

Fleck, L. ([1935] 1979). *Genesis and development of a scientific fact.* Edited by Thaddeus J. Trenn and Robert K. Merton. Translated by Fred Bradley and Thaddeus J. Trenn. Chicago: University of Chicago Press.

Flourens, M.J.P. (1824). *Recherches expérimentales sur les propriétés et les fonctions du système nerveux, dans les animaux vertébrés.* Paris: Crevot.

Fodor, J. A. (1983). *The modularity of mind.* Cambridge, MA: MIT Press.

——— (1999). "Let your brain alone." *London Review of Books* 21:19.

——— (2001). *The mind doesn't work that way: The scope and limits of computational psychology.* Cambridge, MA: MIT Press.

Forrester, J. (1996). "If *p*, then what? Thinking in cases." *History of the Human Sciences* 9 (3):1–25.

Foucault, M. ([1961] 2005). *History of madness.* Translated by Jonathan Murphy and Jean Khalfa. London: Routledge.

——— ([1963] 1973). *The birth of the clinic: An archaeology of medical perception.* London: Tavistock.

——— (1972). *The archaeology of knowledge.* London: Tavistock.

——— (1984). "What is enlightenment?" In P. Rabinow, ed., *The Foucault Reader.* London: Penguin.

Frackowiak, R.S.J. (2004). *Human brain function.* Amsterdam: Elsevier Academic.

Frances, A. (2010). "Alert to the research community—be prepared to weigh in on *DSM-V.*" *Psychiatric Times* 27 (2):1–5.

Frances, A., and Cooper, A. M. (1981). "Descriptive and dynamic psychiatry: A perspective on *DSM-III.*" *American Journal of Psychiatry* 138 (9):1198–1202.

Francis, D., and Meaney, M. J. (1999). "Maternal care and the development of stress responses." *Current Opinion in Neurobiology* 9 (1):128–34.

Francoeur, E., and Segal, J. (2004). From model kits to interactive computer graphics. In N. Hopwood and S. de Chadarevia, eds., *Models: The third dimension of science,* 402–29. Stanford, CA, Stanford University Press.

Franklin, S. (2007). *Dolly mixtures: The remaking of genealogy.* Durham, N.C.: Duke University Press.

Franks, D. (2010). *Neurosociology: The nexus between neuroscience and social psychology.* New York: Springer.

Franks, D., and Smith, T. (2001). *Mind, Brain and Society: Toward a Neurosociology of Emotions.* London: Emerald.

Freeman, W. J., and Watts, J. W. (1942). *Psychosurgery.* Springfield, IL: C. C. Thomas.

——— (1950). *Psychosurgery: On the treatment of mental disorders and intractable pain.* Oxford: Blackwell Scientific.

Freud, S. ([1893] 1962). *Charcot.* London: Hogarth Press.

———, ed. ([1916] 1953–74). *Introductory lectures on psychoanalysis.* London: Hogarth Press and Institute of Psychoanalysis.

Freygang, W. H., Jr., and Sokoloff, L. (1958). "Quantitative measurement of regional circulation in the central nervous system by the use of radioactive inert gas." *Advances in Biological and Medical Physics* 6:263.

Frick, P., and Hare, R. D. (2001). *The antisocial processes screening device.* Toronto: Multi-Health Systems.

Friese, C. (2009). "Models of cloning, models for the zoo: Rethinking the sociological significance of cloned animals." *BioSocieties* 4:367–90.

Frith, C. (2002). "Attention to action and awareness of other minds." *Consciousness and Cognition* 11 (4):481–87.

——— (2007a). *Making up the mind: How the brain creates our mental world.* Oxford: Blackwell.

——— (2007b). "The social brain?" *Proceedings of the Royal Society B: Biological Sciences* 362 (1480):671–78.

Frith, C., and Corcoran, R. (1996). "Exploring 'theory of mind' in people with schizophrenia." *Psychological Medicine* 26 (3):521–30.

Frith, U. E., and Hill, E.L.E. (2003). *Autism: Mind and brain:* London: Royal Society.

Fritsch, G., and Hitzig, E. (2009). "Electric excitability of the cerebrum (Uber die elektrische Erregbarkeit des Grosshirns)." *Epilepsy and Behavior* 15 (2):123–30.

Fulton, J. F. (1951). *Frontal lobotomy and affective behavior: A neuro-physiological analysis.* New York: W. W. Norton.

Gafoor, R., Nitsch, D., McCrone, P., et al. (2010). "Effect of early intervention on 5-year outcome in non-affective psychosis." *British Journal of Psychiatry* 196 (5):372–76.

Gage, F. (2002). "Neurogenesis in the adult brain." *Journal of Neuroscience* 22 (3):612.

Gale, C., and Howard, R. (2003). *Presumed curable: An illustrated casebook of Victorian psychiatric patients in Bethlem Hospital.* Petersfield, UK: Wrightson Biomedical.

Galison, P. L. (1997). *Image and logic: A material culture of microphysics.* Chicago: University of Chicago Press.

Galison, P., and Stump, D. J., eds. (1996). *The disunity of science: Boundaries, contexts, and power.* Stanford, CA: Stanford University Press.

Gall, F. J. ([1822] 1835). *On the functions of the brain and of each of its parts: With observations on the possibility of determining the instincts, propensities, and talents, or the moral and intellectual dispositions of men and animals, by the configuration of the brain and head.* Translated from the French by W. Lewis. 6 vols. Boston: Marsh, Capen and Lyon.

Gallese, V., Fadiga, L., Fogassi, L., et al. (1996). "Action recognition in the premotor cortex." *Brain* 119:593–609.

Gallese, V., and Goldman, A. (1998). "Mirror neurons and the simulation theory of mind-reading." *Trends in Cognitive Sciences* 2 (12):493–501.

Gallese, V., Keysers, C., and Rizzolatti, G. (2004). "A unifying view of the basis of social cognition." *Trends in Cognitive Sciences* 8 (9):396–403.

Gantt, W. H. (1936). "An experimental approach to psychiatry." *American Journal of Psychiatry* 92 (5):1007.

——— (1942). "The origin and development of nervous disturbances experimentally produced." *American Journal of Psychiatry* 98 (4):475.

——— (1953). "Principles of nervous breakdown—schizokinesis and autokinesis." *Annals of the New York Academy of Sciences* 56 (2):143–63.

Gardner-Thorpe, C. (1999). "The British contribution to the neurosciences." *Journal of the History of the Neurosciences* 8 (1):96–97.

Garfield, E. (1989). "A tribute to Eli and Lee Robins—citation superstars: A citationist perspective on biological psychiatry." *Current Contents* 12 (46):321–29.

Garland, B., ed. (2004). *Neuroscience and the law: Brain, mind and the scales of justice.* New York: Dana Press.

Garland, B., and Frankel, M. S. (2006). "Considering convergence: A policy dialogue about behavioral genetics, neuroscience and law." *Law and Contemporary Problems* 69:101–13.

Garland, D. (1985). *Punishment and welfare: A history of penal strategies*. Aldershot, UK: Gower.

—— (1988). "British criminology before 1935." *British Journal of Criminology* 28 (2):1–17.

Garrard, R. L. (1950). "Neuropsychiatry in private practice." *North Carolina Medical Journal* 11 (4):200–205.

Garrison, F. H., and McHenry, L. C. (1969). *History of neurology*. Springfield, IL: Charles C. Thomas.

Garrod, A. E. (1902). "The incidence of alkaptonuria: A study in chemical individuality." *Lancet* 2:1616–20.

Gaudillière, J.-P., and Löwy, I., eds. (1998). *The invisible industrialist: Manufactures and the production of scientific knowledge*. Basingstoke, UK: Macmillan Press.

Gawrylewski, A. (2007). "The trouble with animal models." *Scientist* 21 (7):45–51.

Gayon, J. (2006). "Les organismes modèles en biologie et en médecine." In G. Gachelin, ed., *Les organismes modèles dans la recherche médicale*, 9–43. Paris: Presses Universitaires de France.

Gazzaniga, M. S. (1985a). "The social brain." *Psychology Today* 19 (11):28.

—— (1985b). *The social brain: Discovering the networks of the mind*. New York: Basic Books.

—— (2005). "Forty-five years of split-brain research and still going strong." *Nature Reviews Neuroscience* 6 (8):653–59.

—— (2008a). *Human: The science behind what makes us unique*. New York: Ecco.

—— (2008b). "The law and neuroscience." *Neuron* 60 (3):412–15.

Geertz, C. (1973). *The interpretation of cultures*. New York: Basic Books.

—— (1974). "'From the native's point of view': On the nature of anthropological understanding." *Bulletin of the American Academy of Arts and Sciences* 28 (1):26–45.

Gere, C. (2003). "A brief history of brain archiving." *Journal of the History of the Neurosciences* 12 (4):396–410.

—— (2004a). "Brains-in-vats, giant brains and world brains: The brain as metaphor in digital culture." *Studies in History and Philosophy of Science Part C: Studies in History and Philosophy of Biological and Biomedical Sciences* 35 (2): 351–66.

—— (2004b). "Thought in a vat: Thinking through Annie Cattrell." *Studies in History and Philosophy of Science Part C: Studies in History and Philosophy of Biological and Biomedical Sciences* 35 (2): 415–36.

Geyer, M. A., and Markou, A. (2000). "Animal models of psychiatric disorders." In K. L. Davis, D. Charney, J. T. Coyle, et al., eds., *Neuropsychopharmacology: The fourth generation of progress*, 787–98. Philadelphia: Lippincott Williams and Wilkins; American College of Neuropsychopharmacology.

Gibson, D. G., Glass, J. I., Lartigue, C., et al. (2010). "Creation of a bacterial cell controlled by a chemically synthesized genome." *Science* 329 (5987):52–56.

Gigerenzer, G. (1991). "From tools to theories: A heuristic of discovery in cognitive psychology." *Psychological Review* 98:254–57.

Gillihan, S. J., and Farah, M. J. (2005). "Is self special? A critical review of evidence from experimental psychology and cognitive neuroscience." *Psychological Bulletin* 131 (1):76.

Gilman, S. (1982). *Seeing the insane*. New York: Wiley.

—— (1985). *Difference and pathology: Stereotypes of sexuality, race, and madness*. Ithaca, NY: Cornell University Press.

Giordano, S. D. (2010). *Exercise and eating disorders: An ethical and legal analysis*. London: Routledge.

Glimcher, P. W. (2003). *Decisions, uncertainty, and the brain: The science of neuroeconomics*. Cambridge, MA: MIT Press.

—— (2009). *Neuroeconomics: Decision making and the brain*. Amsterdam: Elsevier Academic Press.

Glynn, A. (2011). *Limitless*. New York: Picador.

Goffman, E. (1978). "Response cries." *Language* 54 (4):787–815.

Goldstein, J. E. (2002). *Console and classify: The French psychiatric profession in the nineteenth century*. Chicago: University of Chicago Press.

Golla, F. L., and Richter, D. (1950). *Perspectives in neuropsychiatry: Essays presented to Professor Frederick Lucien Golla by past pupils and associates*. Edited by Derek Richter. London: H. K. Lewis.

Golomb, J., Kluger, A. and Ferris, S. H. 2004 "Mild cognitive impairment: Historical development and summary of research." *Dialogues in Clinical Neuroscience: Mild Cognitive Impairment* 6(4): 351–67.

Goodman, N. (1978). *Ways of worldmaking*. Hassocks, UK: Harvester Press.

—— (1983). "Notes on the well-made world." *Erkenntnis* 19 (1):99–107.

Gøtzsche, P. C., and M. Nielsen (2006). "Screening for breast cancer with mammography." *Cochrane database of systematic reviews†18* (4):CD001877.

Gould, E., and McEwen, B. S. (1993). "Neuronal birth and death." *Current Opinion in Neurobiology* 3 (5):676–82.

Gould, E., Reeves, A. J., Graziano, M.S.A., et al. (1999). "Neurogenesis in the neocortex of adult primates." *Science* 286 (5439):548–52.

Gould, E., Tanapat, P., Hastings, N. B., et al. (1999). "Neurogenesis in adulthood: A possible role in learning." *Trends in Cognitive Sciences* 3 (5):186–92.

Gould, S. J. (1981). *The mismeasure of man*. New York: W. W. Norton.

Goyer, P. F., Cohen, R. M., Andreason, P. J., et al. (1994). "Positron-emission tomography and personality-disorders." *Neuropsychopharmacology* 10 (1):21–28.

Grant, B. F., Hasin, D. S., Stinson, F. S., et al. (2004). "Prevalence, correlates, and disability of personality disorders in the United States: Results from the National Epidemiologic Survey on Alcohol and Related Conditions." *Journal of Clinical Psychiatry* 65 (7):948–58.

Greenblatt, M., Arnot, R.M.D., and Solomon, H. C. (1951). *Studies in lobotomy*. London: William Heinemann Medical Books.

Greenblatt, S. (1980). *Renaissance self-fashioning: From More to Shakespeare*. Chicago: Chicago University Press.

Greene, J., and Cohen, J. (2004). "For the law, neuroscience changes nothing and everything." *Proceedings of the Royal Society B: Biological Sciences* 359 (1451):1775–85.

Greene, J. D., Sommerville, R. B., Nystrom, L. E., et al. (2001). "An fMRI investigation of emotional engagement in moral judgment." *Science* 293 (5537):2105–8.

Greenfield, S. (2008). *I.D.: The quest for meaning in the 21st century*. London: Hodder and Stoughton.

Gregory, P. R. (2008). *Lenin's brain and other tales from the secret Soviet archives.* Stanford, CA: Hoover Institution Press.

Grisso, T. (1986). *Evaluating competencies: Forensic assessments and instruments.* New York: Plenum.

Grist, M. (2009). *Changing the subject: How new ways of thinking about human behaviour may change politics, policy and practice.* London: Royal Society for the Encouragement of Arts, Manufactures and Commerce.

—— (2010). *Steer: Mastering our behaviour through instinct, environment and reason.* London: Royal Society for the Encouragement of Arts, Manufactures and Commerce.

Grob, G. N. (1991). "Origins of *DSM-I*: A study in appearance and reality." *American Journal of Psychiatry* 148 (4):421–31.

Gross, C. G. (1999). *Brain, vision, memory: Tales in the history of neuroscience.* Cambridge, MA: MIT Press.

—— (2000). "Neurogenesis in the adult brain: Death of a dogma." *Nature Reviews Neuroscience* 1 (1):67–73.

Haag, A. L. (2007). "Biomarkers trump behavior in mental illness diagnosis." *Nature Medicine* 13 (1):3–237.

Hacking, I. (1992a). "The self-vindication of the laboratory sciences." In A. Pickering, ed., *Science as practice and culture*, 29–64. Chicago: University of Chicago Press.

—— (1992b). "'Style' for historians and philosophers." *Studies in the History and Philosophy of Science* 23 (1):1–20.

—— (1995). *Rewriting the soul: Multiple personality and the sciences of memory.* Princeton, NJ: Princeton University Press.

—— (1996). "The disunities of the sciences." In P. Galison and D. J. Stump, eds., *The Disunity of science: Boundaries, contexts and power*, Stanford, CA: Stanford University Press.

—— (2009). *Scientific reason.* Taipei: National Taiwan University.

Haggard, P. (2008). "Human volition: Towards a neuroscience of will." *Nature Reviews: Neuroscience* 9 (12):934–46.

Hagner, M. (1997). *Homo cerebralis: Der Wandel vom Seelenorgan zum Gehirn.* Berlin, Berlin Verlag.

—— (2001). "Cultivating the cortex in German neuroanatorny." *Science in Context* 14 (4):541–63.

Hagner, M., and Borck, C. (2001). "Mindful practices: On the neurosciences in the twentieth century." *Science in Context* 14(4): 507–10.

Haider, A. (2005). "*Roper v. Simmons*: The role of the science brief." *Ohio State Journal of Criminal Law* 3:369.

Hall, E. F., and Brevoort, J. C. (1878). "Gerard Mercator: His life and works." *Journal of the American Geographical Society of New York* 10:163–96.

Halstead, W. C. (1947). *Brain and intelligence: A quantitative study of the frontal lobes.* Chicago: University of Chicago Press.

—— (1950). *Brain and behavior, a symposium.* Berkeley: University of California Press.

Haraway, D. J. (2003). *The companion species manifesto: Dogs, people, and significant otherness.* Chicago: Prickly Paradigm Press.

—— (2007). *When species meet.* Minneapolis: University of Minnesota Press.

Harcourt, B. E. (2007). *Against prediction: Profiling, policing, and punishing in an actuarial age*. Chicago: University of Chicago Press.

Hare, R. (1991). *The Hare psychopathy checklist—revised*. North Tonawanda, NY: Multi-Health Systems.

—— (1999). *Without conscience: The disturbing world of psychopaths among us*. New York: Guildford Press.

Hari, R. (2004). "Towards studies of the social brain: The human mirror-neuron system." *Brain and Cognition* 54 (2):128.

Hariharan, I. K., and Haber, D. A. (2003). "Yeast, flies, worms, and fish in the study of human disease." *New England Journal of Medicine* 348 (24):2457.

Harlow, H. F. (1958). "The nature of love." *American Psychologist* 13 (12):673–85.

Harmon-Jones, E., and Devine, P. G. (2003). "Introduction to the special section on social neuroscience: Promise and caveats." *Journal of Personality and Social Psychology* 85 (4):589–93.

Harré, R. (2009). *Pavlov's dogs and Schrödinger's cat: Scenes from the living laboratory*. Oxford: Oxford University Press.

Harrington, A. (1987). *Medicine, mind and the double brain: A study in nineteenth-century thought*. Princeton, NJ: Princeton University Press.

—— (2008). *The cure within: A history of mind-body medicine*. New York: W. W. Norton.

Harrington, R., and Bailey, S. (2003). "The scope for preventing antisocial personality disorder by intervening in adolescence." Manchester: NHS National Programme on Forensic Mental Health Research and Development.

Harrison, T. (2000). "Five Scientists at Johns Hopkins in the Modern Evolution of Neuroscience." *Journal of the History of the Neurosciences* 9 (2):165–79.

Hawes, D. J., and Dadds, M. R. (2005). "The treatment of conduct problems in children with callous-unemotional traits." *Journal of Consulting and Clinical Psychology* 73 (4):737–41.

—— (2007). "Stability and malleability of callous-unemotional traits during treatment for childhood conduct problems." *Journal of Clinical Child and Adolescent Psychology* 36 (3):347–55.

Hawkley, L., and Cacioppo, J. (2010). "Loneliness matters: A theoretical and empirical review of consequences and mechanisms." *Annals of Behavioral Medicine* 40:218–27.

Hayward, J. W., and Varela, F. J. (1992). *Gentle bridges: Conversations with the Dalai Lama on the sciences of mind*: Boston: Shambhala.

Hayward, R. (2002). "The tortoise and the love-machine: Grey Walter and the politics of electroencephalography." *Science in Context* 14 (4):615–41.

Healy, D. (1996). *The Psychopharmacologists*. London: Chapman and Hall.

—— (1997). *The Antidepressant Era*. Cambridge, MA: Harvard University Press.

—— (1998). *The Psychopharmacologists II*. London: Altman.

—— (2000). *The Psychopharmacologists III*. London: Arnold.

—— (2004). *Let them eat Prozac: The unhealthy relationship between the pharmaceutical industry and depression*. New York: New York University Press.

Hearne, V. (1986). *Adam's task: Calling animals by name*. New York: Knopf.

Heath, D., Rapp, R., and Taussig, K.-S. (2004). "Genetic citizenship." In D. Nugent and J. Vincent, eds., *Companion to the anthropology of politics*, 152–67. Oxford: Blackwell.

Heatherton, T. F. (2004). "Introduction to special issue on social cognitive neuroscience." *Journal of Cognitive Neuroscience* 16 (10):1681–82.

Heatherton, T. F., Macrae, C. N., and Kelley, W. M. (2004). "What the social brain sciences can tell us about the self." *Current Directions in Psychological Science* 13 (5):190–93.

Hebb, D. O. (1949). *The organization of behavior. A neuropsychological theory.* New York: Wiley.

Hécaen, H. (1950). "Le behaviorisme et la neuropsychiatrie" [Behaviorism and neuropsychiatry]. *Evolution Psychiatrique* (Paris) 1:23–58.

Hecht, J. M. (2003). *The end of the soul: Scientific modernity, atheism, and anthropology in France.* New York: Columbia University Press.

Hedges, S. B. (2002). "The origin and evolution of model organisms." *Nature Reviews: Genetics* 3 (11):838–49.

Hennah, W., Thomson, P., Peltonen, L., et al. (2006). "Genes and schizophrenia: Beyond schizophrenia; The role of DISC1 in major mental illness." *Schizophrenia Bulletin* 32 (3):409–16.

Herzig, R. M. (2005). *Suffering for science: Reason and sacrifice in modern America.* New Brunswick, NJ: Rutgers University Press.

Hikida, T., Jaaro-Peled, H., Seshadri, S., et al. (2007). "Dominant-negative DISC1 transgenic mice display schizophrenia-associated phenotypes detected by measures translatable to humans." *Proceedings of the National Academy of Sciences* 104 (36):14,501.

Hill, A. (1891). "Current nerve anatomy and physiology." *Brain*†14†(4):567–88.

Hill, E. L., and Frith, U. (2003). "Understanding autism: Insights from mind and brain." *Proceedings of the Royal Society B: Biological Sciences* 358 (1430):281–89.

Hill, J. (2003). "Early identification of individuals at risk for antisocial personality disorder." *British Journal of Psychiatry* 182 (44):s11–s14.

Hinde, R. A. (1967). "The nature of aggression." *New Society* 9:302–4.

——— (1970). *Animal behaviour: A synthesis of ethology and comparative psychology.* New York: McGraw-Hill.

Hirst, P. Q., and Woolley, P. (1982). *Social relations and human attributes.* London: Tavistock.

Hodgkin, A. L. (1992). *Chance and design: Reminiscences of science in peace and war.* Cambridge: Cambridge University Press.

Hogg, S. (1996). "A review of the validity and variability of the elevated plus-maze as an animal model of anxiety." *Pharmacology Biochemistry and Behavior* 54 (1):21–30.

Holland, G. N., Moore, W. S., and Hawkes, R. C. (1980). "Nuclear magnetic-resonance tomography of the brain." *Journal of Computer Assisted Tomography* 4 (1):1–3.

Holley, A. (1984). "Les neurosciences: Unité et diversité." *Courrier du CNRS* (55–56):12–15.

Hollinshead, K. (1999). "Surveillance of the worlds of tourism: Foucault and the eye-of-power." *Tourism Management* 20 (1):7–23.

Holman, B. L., Gibson, R. E., Hill, T. C., et al. (1985). "Muscarinic acetylcholine receptors in Alzheimer's disease: In vivo imaging with iodine 123-labeled 3-quinuclidinyl-4-iodobenzilate and emission tomography." *JAMA: The Journal of the American Medical Association* 254 (21):3063–66.

Horwitz, A. V. (2002). *Creating mental illness.* Chicago: University of Chicago Press.

Horwitz, A. V., and Wakefield, J. C. (2007). *The loss of sadness: How psychiatry transformed normal sorrow into depressive disorder*. Oxford: Oxford University Press.

Houts, A. C. (2000). "Fifty years of psychiatric nomenclature: Reflections on the 1943 War Department Technical Bulletin, Medical 203." *Journal of Clinical Psychology* 56 (7):935–67

Howes, O. D., Montgomery, A. J., Asselin, M., et al. (2007). "Molecular imaging studies of the striatal dopaminergic system in psychosis and predictions for the prodromal phase of psychosis." *British Journal of Psychiatry* 191 (51):s13–s18.

Hughes, V. (2010). " Science in court: Head case." *Nature* 464:340–42.

Humphrey, N. (1976). "The social function of intellect." In P. Bateson and R. A. Hinde, eds. *Growing points in ethology*, 303–17. Cambridge: Cambridge University Press.

Hunter, I. (1988). *Culture and government: The emergence of literary education*. Basingstoke, UK: Macmillan.

Hunter, J.F.M. (1968). " 'Forms of life' in Wittgenstein's philosophical investigations." *American Philosophical Quarterly* 5 (4): 233–43.

Hutchings, J., Bywater, T., Daley, D., et al. (2007). "Parenting intervention in Sure Start services for children at risk of developing conduct disorder: Pragmatic randomised controlled trial." *British Medical Journal* 334 (7595):678.

Hyman, S. E. (2003). "Diagnosing disorders." *Scientific American* 289 (3):96–103.

——— (2008). "A glimmer of light for neuropsychiatric disorders." *Nature* 455 (7215):890–93.

Iacoboni, M. (2003). "Monkey see, monkey do." *Nature Neuroscience* 6 (2):109.

——— (2005). "Neural mechanisms of imitation." *Current Opinion in Neurobiology* 15 (6):632–37.

——— (2008). *Mirroring people: The new science of how we connect with others*. New York: Farrar, Straus and Giroux.

Iacoboni, M., Molnar-Szakacs, I., Gallese, V., et al. (2005). "Grasping the intentions of others with one's own mirror neuron system." *PLoS Biology* 3 (3):e79.

Iacoboni, M., Woods, R. P., Brass, M., et al. (1999). "Cortical mechanisms of human imitation." *Science* 286 (5449):2526–28.

Illes, J. (2006). *Neuroethics: Defining the issues in theory, practice, and policy*. New York: Oxford University Press.

Insel, T. R. (2007). "From animal models to model animals." *Biological Psychiatry* 62 (12):1337–39.

Insel, T. R., and Fenton, W. S. (2005). "Psychiatric epidemiology: It's not just about counting anymore." *Archives of General Psychiatry* 62 (6):590.

Insel, T. R., and Fernald, R. D. (2004). "How the brain processes social information: Searching for the social brain." *Annual Review of Neuroscience* 27:697–722.

Insel, T. R., O'Brien, D. J., and Leckman, J. F. (1999). "Oxytocin, vasopressin, and autism: Is there a connection?" *Biological Psychiatry* 45 (2):145–57.

Insel, T., and Shapiro, L. (1992). "Oxytocin receptor distribution reflects social organization in monogamous and polygamous voles." *Proceedings of the National Academy of Sciences* 89 (13):5981.

Insel, T., and Young, L. (2000). "Neuropeptides and the evolution of social behavior." *Current Opinion in Neurobiology* 10 (6):784–89.

International Brain Research Organization. (1968). "IBRO survey of research facilities and manpower in the United States, supervised by the Committeee on Brain Sciences, Division of Medical Sciences, National Research Council." Washington, DC: National Academy of Sciences.

International Schizophrenia Consortium. (2009). "Common polygenic variation contributes to risk of schizophrenia and bipolar disorder." *Nature* 460 (7256):748–52.

Irino, Shozo. (2006). "A tribute to Professor Jean Bernard." *International Journal of Hematology* 84:385–86.

Ishizuka, K., Paek, M., Kamiya, A., et al. (2006). "A review of disrupted-in-schizophrenia-1 (DISC1): Neurodevelopment, cognition, and mental conditions." *Biological Psychiatry* 59 (12):1189–97.

Jackson, P. L., and Decety, J. (2004). "Motor cognition: A new paradigm to study self-other interactions." *Current Opinion in Neurobiology* 14 (2):259–63.

Jacob, F. (1988). *The statue within: An autobiography.* New York: Basic Books.

Jacob, P., and Jeannerod, M. (2005). "The motor theory of social cognition: A critique." *Trends in Cognitive Sciences* 9 (1):21–25.

Jenkins, W., Merzenich, M., Ochs, M., et al. (1990). "Functional reorganization of primary somatosensory cortex in adult owl monkeys after behaviorally controlled tactile stimulation." *Journal of Neurophysiology* 63 (1):82.

Jones, E. G., and Pons, T. P. (1998). "Thalamic and brainstem contributions to large-scale plasticity of primate somatosensory cortex." *Science* 282 (5391):1121–25.

Jones, K., Daley, D., Hutchings, J., et al. (2008). "Efficacy of the Incredible Years Programme as an early intervention for children with conduct problems and ADHD: Long term follow up." *Child: Care, Health and Development* 34 (3):380–90.

Jordan, B. (1974). *Poor parents: Social policy and the "cycle of deprivation."* London: Routledge and Kegan Paul.

Joseph, K. (1972). "The cycle of deprivation." *Midwife and Health Visitor* 8 (12):414.

Joyce, K. A. (2008). *Magnetic appeal: MRI and the myth of transparency.* Ithaca, NY: Cornell University Press.

Kahn, E. M. (1992). "Imaging of brain electrophysiologic activity—applications in psychiatry." *General Hospital Psychiatry* 14 (2):99–106.

Kandel, E. R. (1982). "The origins of modern neuroscience." *Annual Review of Neuroscience* 5:299–303.

——— (2006). *In search of memory: The emergence of a new science of mind.* New York: W. W. Norton.

Karp, D. A. (2006). *Is it me or my meds? Living with antidepressants.* Cambridge, MA: Harvard University Press.

Katz, B. (1982). "Stephen William Kuffler. 24 August 1913–11 October 1980." *Biographical Memoirs of Fellows of the Royal Society* 28:225–59.

Katz, M. M., Secunda, S. K., Hirschfeld, R., et al. (1979). "NIMH clinical research branch collaborative program on the psychobiology of depression." *Archives of General Psychiatry* 36 (7):765.

Kay, L. E. (1993). *The molecular vision of life: Caltech, the Rockefeller Foundation, and the rise of the new biology.* New York: Oxford University Press.

——— (2002). "From logical neurons to poetic embodiments of mind: Warren S. McCulloch's project in neuroscience." *Science in Context* 14 (4):591–614.

Kaymaz, N., and van Os, J. (2010). "*DSM-5* and the 'psychosis risk syndrome': Babylonic confusion." *Psychosis: Psychological, Social and Integrative Approaches* 2 (2):100–103.

Keating, P., and Cambrosio, A. (2003). *Biomedical platforms: Realigning the normal and the pathological in late-twentieth-century medicine.* Cambridge, MA: MIT Press.

Keller, E. F. (2000). "Models of and models for: Theory and practice in contemporary biology." *Philosophy of Science* 67 (suppl.):S72–S86.

——— (2002). *Making sense of life: Explaining biological development with models, metaphors, and machines.* Cambridge, MA: Harvard University Press.

Kemp, M., and Wallace, M. (2000). *Spectacular bodies: The art and science of the human body from Leonardo to now.* London: Hayward Gallery.

Kendell, R. E. (1975). *The role of diagnosis in psychiatry.* Oxford: Blackwell.

Kendler, K. S. (2009). "An historical framework for psychiatric nosology." *Psychological Medicine* 39 (12):1935–41.

Kendler, K. S., Munoz, R. A., and Murphy, G. (2010). "The development of the Feighner criteria: A historical perspective." *American Journal of Psychiatry* 167 (2):134–42.

Kennedy, D. (2001). "Enclosing the Research Commons." *Science* 294 (5550):2249.

——— (2004). "Neuroscience and neuroethics." *Science* 306 (5695):373.

Kessler, R. C., Berglund, P., Demler, O., et al. (2005). "Lifetime prevalence and age-of-onset distributions of *DSM-IV* disorders in the national comorbidity survey replication." *Archives of General Psychiatry* 62 (6):593–602.

Kessler, R. C., Chiu, W. T., Demler, O., et al. (2005). "Prevalence, severity, and comorbidity of 12-month *DSM-IV* disorders in the national comorbidity survey replication." *Archives of General Psychiatry* 62 (6):617–27.

Kessler, R. C., McGonagle, K. A., Zhao, S. Y., et al. (1994). "Lifetime and 12-month prevalence of *DSM-III-R* psychiatric-disorders in the United States—results from the national comorbidity survey." *Archives of General Psychiatry* 51 (1):8–19.

Kessler, R. C., Merikangas, K. R., Berglund, P., et al. (2003). "Mild disorders should not be eliminated from the *DSM-V*." *Archives of General Psychiatry* 60 (11):1117.

Kety, S. S. (1960). "Measurement of local blood flow by the exchange on an inert diffusible substance." *Methods in Medical Research* 8:228–36.

Kety, S. S., and Schmidt, C. F. (1948). "The effects of altered arterial tensions of carbon dioxide and oxygen on cerebral blood flow and cerebral oxygen consumption of normal young men." *Journal of Clinical Investigation* 27 (4):484–92.

Kevles, B. (1997). *Naked to the bone: Medical imaging in the twentieth century.* New Brunswick, NJ: Rutgers University Press.

Keynes, R. (2005). "J.Z. and the discovery of squid giant nerve fibres." *Journal of Experimental Biology* 208 (2): 179–80.

Kieser, D. G. (1855). *Elemente der Psychiatrik.* Bonn: Breslau.

Kilner, J., and Frith, C. (2008). "Action observation: Inferring intentions without mirror neurons." *Current Biology* 18 (1):R32–R33.

Kim-Cohen, J., Moffitt, T. E., Taylor, A., et al. (2005). "Maternal depression and children's antisocial behavior: Nature and nurture effects." *Archives of General Psychiatry* 62 (2):173–81.

Kinderman, P. (2005). "A psychological model of mental disorder." *Harvard Review of Psychiatry* 13 (4):206–17.

Kingdon, D., Hansen, L., and Turkington, D. (2010). "*DSM-5* and the 'psychosis risk syndrome': Would it be useful and where would it fit?" *Psychosis: Psychological, Social and Integrative Approaches* 2 (2):103–6.

Klerman, G. L. (1978). "The evolution of a scientific nosology." In J. C. Shershow, ed., *Schizophrenia: Science and practice*, 99–121. Cambridge, MA: Harvard University Press.

Kline, N. S. (1954). "Use of rauwolfina serpenthina benth. in neuropsychiatric conditions." *Annals of the New York Academy of Sciences* 59:107–32.

Knutson, B., Wolkowitz, O. M., Cole, S. W., et al. (1998). "Selective alteration of personality and social behavior by serotonergic intervention." *American Journal of Psychiatry* 155 (3):373–79.

Koenigs, M., Young, L., Adolphs, R., et al. (2007). "Damage to the prefrontal cortex increases utilitarian moral judgements." *Nature* 446 (7138):908–11.

Kohler, R. E. (1993). "*Drosophila*: A life in the laboratory." *Journal of the History of Biology* 26 (2):281–310.

——— (1994). *Lords of the fly: Drosophila genetics and the experimental life*. Chicago: University of Chicago Press.

Koyré, A. (1973). *The astronomical revolution: Copernicus, Kepler, Borelli*. Translated by R.E.W. Maddison. Ithaca, NY: Cornell University Press.

Kraepelin, E. ([1899] 1902). *Textbook of psychiatry: Abstracted and adapted from the 6th German edition of* Lehrbuch der psychiatrie. New York: Macmillan.

——— (1904). *Lectures on clinical psychiatry . . . authorized translation from the German*. Revised and edited by T. Johnstone. London: Baillière.

Kramer, P. D. (1992). *Listening to Prozac*. New York: Viking.

Krausse, J. (1998). *Information at a glance: On the history of the diagram*. Nijmegen, NL: OASE Sun.

Krebs, H. A. (1975). "The August Krogh principle: 'For many problems there is an animal on which it can be most conveniently studied.'" *Journal of Experimental Zoology* 194 (1):221–26.

Krishnan, V., and Nestler, E. J. (2008). "The molecular neurobiology of depression." *Nature* 455 (7215):894–902.

Kuhn, R. (1958). "The treatment of depressive states with G22355 (imipramine hydrochloride)." *American Journal of Psychiatry* 115:459–64.

Kupfer, D. J., First, M. B., and Regier, D. A., eds. (2002). *A research agenda for* DSM-V. Washington, DC: American Psychiatric Association.

Kutchins, H., and Kirk, S. A. (1997). *Making us crazy: The psychiatric bible and the creation of mental disorders*. New York: Free Press.

Lacan, J. ([1957] 1977). "The agency of the letter in the unconscious or reason since Freud." In *Ecrits*. Translated by Alan Sheridan. New York: W. W. Norton.

LaFollette, H., and Shanks, N. (1996). *Brute science: Dilemmas of animal experimentation*. New York: Routledge.

Lakoff, A. (2005). *Pharmaceutical reason: Knowledge and value in global psychiatry*. Cambridge: Cambridge University Press.

Lancet, The (2004). "Neuromarketing: Beyond branding." *Lancet Neurology* 3 (2):71.

Landau, W. M., Freygang, W. H., Jr., Roland, L. P., and Kety, S. S. (1955). "The local circulation of the living brain: Values in the unanesthetized and anesthetized cat." *Transactions of the American Neurology Association* (80th meeting):125–29.

Langmoen, I. A., and Apuzzo, M. L. (2007). "The brain on itself: Nobel laureates and the history of fundamental nervous system function." *Neurosurgery* 61 (5):891–907.

Laporte, Y. (2006). "Charles-Edouard Brown-Sequard: An eventful life and a significant contribution to the study of the nervous system." *Comptes Rendus Biologies* 329 (5–6):363–68.

Laslett, T.P.R. (1950). *The physical basis of mind: A series of broadcast talks.* Edited by P. Laslett. Oxford: Basil Blackwell.

Lassen, N.E.A. (1978). "Brain function and blood flow." *Scientific American* 239:62–71.

Latour, B. (1987). *Science in action: How to follow scientists and engineers through society.* Cambridge, MA: Harvard University Press.

——— (2002). "A well-articulated primatology—reflexions of a fellow traveller." In S. C. Strum and L. M. Fedigan, eds., *Primate encounters: Models of science, gender, and society.* Chicago: University of Chicago Press.

Lauterbur, P. (1973). "Image formation by induced local interactions: Examples employing nuclear magnetic-resonance." *Nature* 242 (5394):190–91.

Laviola, G., Hannan, A. J., Macrì, S., et al. (2008). "Effects of enriched environment on animal models of neurodegenerative diseases and psychiatric disorders." *Neurobiology of Disease* 31 (2): 159–68.

Lederbogen, F., Kirsch, P., Haddad, L., et al. (2011). "City living and urban upbringing affect neural social stress processing in humans." *Nature* 474(7352): 498–501.

LeDoux, J. E. (2002). *Synaptic self: How our brains become who we are.* New York: Viking.

——— (2003). "The self: Clues from the brain." *Annals of the New York Academy of Sciences* 1001:295.

LeDoux, J. E., Debiec, J., and Moss, H. (2003). *The self: From soul to brain.* New York: New York Academy of Sciences.

Lee, N., Broderick, A. J., and Chamberlain, L. (2007). "What is 'neuromarketing'? A discussion and agenda for future research." *International Journal of Psychophysiology* 63 (2):199–204.

Lee, T. S., Ng, B. Y., and Lee, W. L. (2008). "Neuropsychiatry—an emerging field." *Annals of the Academy of Medicine, Singapore* 37 (7):601–5.

Leidesdorf, M. (1865). *Lehrbuch der Psychischen Krankheiten.* Erlangen: F. Enke.

Lemov, R. M. (2005). *World as laboratory: Experiments with mice, mazes, and men.* New York: Hill and Wang.

Leonelli, S. (2007). "*Arabidopsis*, the botanical *Drosophila*: From mouse cress to model organism." *Endeavour* 31 (1):34–38.

Lesch, K. P. (2007). "Linking emotion to the social brain: The role of the serotonin transporter in human social behaviour." *EMBO Reports* 8:S24–S29.

——— (2008). "Serotonin transporter and depression: From the emotional to the social brain." *Actas Españolas de Psiquiatría* 36 (1):21–24.

Lesch, K. P., Bengel, D., Heils, A., et al. (1996). "Association of anxiety-related traits with a polymorphism in the serotonin transporter gene regulatory region." *Science* 274 (5292):1527–31.

Leuner, B., Glasper, E. R., and Gould, E. (2010). "Parenting and plasticity." *Trends in Neurosciences* 33 (10):465–73.

Levi-Montalcini, R. (1982). "Developmental neurobiology and the natural history of nerve growth factor." *Annual Review of Neuroscience* 5 (1):341–62.

Levy, N. (2007). *Neuroethics: Challenges for the 21st century*. Cambridge: Cambridge University Press.

Lewes, G. H. (1877). *Physical basis of mind*. London: Trubner.

Lewis, J. (1981). *Something hidden: A biography of Wilder Penfield*. Toronto: Doubleday Canada.

Libet, B., Freeman, A., and Sutherland, K., eds. (1999). *The volitional brain: Towards a neuroscience of free will*. Thorverton, UK: Imprint Academic.

Libet, B., Gleason, C. A., Wright, E. W., et al. (1983). "Time of conscious intention to act in relation to onset of cerebral activity (readiness-potential): The unconscious initiation of a freely voluntary act." *Brain* 106 (3):623.

Liddell, H. (1938). "The experimental neurosis and the problem of mental disorder." *American Journal of Psychiatry* 94 (5):1035.

—— (1947). "The experimental neurosis." *Annual Review of Physiology* 9 (1):569–80.

Liddell, H. S., Anderson, O. D., Kotyuka, E., et al. (1935). "Effect of extract of adrenal cortex on experimental neurosis in sheep." *Archives of Neurology and Psychiatry* 34 (5):973.

Lieberman, M. D. (2005). "Principles, processes, and puzzles of social cognition: An introduction for the special issue on social cognitive neuroscience." *NeuroImage* 28 (4):745–56.

—— (2006). "Social cognitive and affective neuroscience: When opposites attract." *Social Cognitive and Affective Neuroscience* 1 (1):1.

Lingnau, A., Gesierich, B., and Caramazza, A. (2009). "Asymmetric fMRI adaptation reveals no evidence for mirror neurons in humans." *Proceedings of the National Academy of Sciences* 106 (24):9925.

Logan, C. A. (2002). "Before there were standards: The role of test animals in the production of empirical generality in physiology." *Journal of the History of Biology* 35 (2):329–63.

Logan, P. M. (2008). "Imitations of insanity and Victorian medical aesthetics." *Romanticism and Victorianism on the Net (RaVoN)* 49. http://www.erudit.org/revue/ravon/2008/v/n49/017855ar.html.

Logothetis, N. K. (2008). "What we can do and what we cannot do with fMRI." *Nature* 453 (7197):869–78.

Lombardo, M. V., and Baron-Cohen, S. (2011). "The role of the self in mindblindness in autism." *Consciousness and Cognition* 20 (1):130–40.

Lombardo, M. V., Chakrabarti, B., Bullmore, E. T., et al. (2010). "Shared neural circuits for mentalizing about the self and others." *Journal of Cognitive Neuroscience* 22 (7):1623–35.

Lopez-Munoz, F., Alamo, C., Cuenca, E., et al. (2005). "History of the discovery and clinical introduction of chlorpromazine." *Annals of Clinical Psychiatry* 17 (3):113–35.

Lorenz, K. (1964). *On aggression*. London: Methuen.

Lorusso, L., Cristini, C., and Porro, A. (2007). "Lorenzo Tenchini (1852–1906): Neuroanatomy and criminal anthropology." *Medicina nei Secoli* 19 (2):353–60.

Löwy, I. (1992). "From guinea pigs to man: The development of Haffkine's anticholera vaccine." *Journal of the History of Medicine and Allied Science* 47:270–309.

Lucas, M. (2012). *Rewire your brain for love: Creating vibrant relationships using the science of mindfulness*. Carlsbad, CA: Hay House.

Luria, A. R. (1973). *The man with a shattered world: The history of a brain wound.* Translated by Lynn Solotaroff. London: Jonathan Cape.

Lutz, C., and White, G. (1986). "The anthropology of emotions." *Annual Review of Anthropology* 15:405–36.

Lynam, D. (1996). "Early identification of chronic offenders: Who is the fledgling psychopath?" *Psychological Bulletin* 120 (2):209–34.

——— (1998). "Early identification of the fledgling psychopath: Locating the psychopathic child in the current nomenclature." *Journal of Abnormal Psychology* 107 (4):566–75.

Lynam, D., and Gudonis, L. (2005). "The development of psychopathy." *Clinical Psychology* 1 (1): 381–407.

Lynch, C., and Lynch, Z. (2007). *The neurotechnology industry 2007 report: Drugs, devices and diagnostics for the brain and nervous system* San Francisco: NeuroInsights.

Lynch, M. (1989). "Sacrifice and the transformation of the animal body into a scientific object: Laboratory culture and ritual practice in the neurosciences." *Social Studies of Science* 18: 265–89.

——— (1991). "Laboratory space and the technological complex: An investigation of topical contextures." *Science in Context* 4(1): 51–78.

Lynch, Z., and Laursen, B. (2009). *The neuro revolution: How brain science is changing our world.* New York: St. Martin's Press.

Macher, J.-P. ed., (2004). Special issue, "Mild cognitive impairment." *Dialogues in Clinical Neuroscience* 6 (4).

Macintyre, A. (1962). "A mistake about causality in social science." In Peter Laslett and W. G. Runciman, eds., *Philosophy, Politics and Society*, 48–70. Oxford: Blackwell.

Macmillan, M. (2000). *An odd kind of fame: Stories of Phineas Gage.* Cambridge, MA: MIT Press.

Magoun, H. (1962). "Development of brain research institutes." In J. D. French, ed., *Frontiers in Brain Research*, 1–40. New York: Columbia University Press.

——— (1974). "The role of research institutes in the advancement of neuroscience: Ranson's Institute of Neurology, 1928–1942." In F. G. Worden, J. P. Swayze, and G. Adelman, eds., *The neurosciences: Paths of discovery*, 515–27. Cambridge, MA: MIT Press.

Malabou, C. (2008). *What should we do with our brain?* New York: Fordham University Press.

Marcus, S. J., ed. (2002). *Neuroethics: Mapping the field.* San Francisco: Dana Press.

Margo, J., and Stevens, A. (2008). *Make me a criminal: Preventing youth crime.* London: Institute for Public Policy Reearch.

Marincola, F. M. (2003). "Translational medicine: A two-way road." *Journal of Translational Medicine* 1 (1):1–3.

Markus, H. R., and Kitayama, S. (1991). "Culture and the self: Implications for cognition, emotion, and motivation." *Psychological Review* 98:224–53.

Maroney, T. (2009). "The false promise of adolescent brain science in juvenile justice." *Notre Dame Law Review* 85 (89):90–176.

——— (2011). "Adolescent brain science after *Graham v. Florida.*" *Notre Dame Law Review* 86 (2):765–93.

Marsella, A. J., De Vos, G. A., and Hsu, F.L.K. (1985). *Culture and self: Asian and western perspectives.* New York: Tavistock.

Marshall, L. H., Rosenblith, W. A., Gloor, P., et al. (1996). "Early history of IBRO: The birth of organized neuroscience." *Neuroscience* 72 (1):283–306.

Martell, D. A. (1992a). "Estimating the prevalence of organic brain-dysfunction in maximum-security forensic psychiatric-patients." *Journal of Forensic Sciences* 37 (3):878–93.

—— (1992b). "Forensic neuropsychology and the criminal law." *Law and Human Behavior* 16 (3):313–36.

Martin, E. (2000). "Mind body problems." *American Ethnologist* 27 (3):569–90.

—— (2004). "Talking back to neuro-reductionism." In H. Thomas and J. Ahmed, eds. *Cultural bodies: Ethnography and theory,* 190–212. Oxford: Blackwell.

—— (2007). *Bipolar expeditions: Mania and depression in American culture.* Princeton, NJ: Princeton University Press.

Martin, J. B. (2002). "The integration of neurology, psychiatry, and neuroscience in the 21st century." *American Journal of Psychiatry* 159 (5):695–704.

Marzi, C. A., and Sagvolden, T. (1997). "European brain and behaviour society." *European Psychologist* 2 (1):59–60.

Massoud, T. F., and Gambhir, S. S. (2003). "Molecular imaging in living subjects: Seeing fundamental biological processes in a new light." *Genes and Development* 17 (5):545.

Maturana, H. R., and Varela, F. J. (1987). *The tree of knowledge: The biological roots of human understanding:* Boston: Shambhala.

Matusall, S., Kaufmann, I. M., and Christen, M. (2011). "The emergence of social neuroscience as an academic discipline." In Jean Decety and John Cacioppo, eds., *The Oxford Handbook of Social Neuroscience,* 9–27. New York: Oxford University Press.

Maudsley, H. (1872). *Responsibility in Mental Disease.* New York: Appleton.

Mauss, M. ([1938] 1985). "A category of the human mind: The notion of person; the notion of self." In M. Carrithers, S. Colllins, and S. Lukes, eds., *The category of the person: Anthropology, philosophy, history,* 1–25. Cambridge: Cambridge University Press.

Mayes, R., and Horwitz, A. V. (2005). "*DSM-III* and the revolution in the classification of mental illness." *Journal of the History of the Behavioral Sciences* 41 (3):249–67.

Maynard, P. (1997). *The engine of visualization: Thinking through photography.* Ithaca, NY: Cornell University Press.

McCabe, D. P., and Castel, A. D. (2008). "Seeing is believing: The effect of brain images on judgments of scientific reasoning." *Cognition* 107 (1):343–52.

McCarty, W. (2003). " 'Knowing true things by what their mockeries be': Modelling in the Humanities." Computing in the Humanities and Social Sciences, *CH Working Papers.* http://projects.chass.utoronto.ca/chwp/CHC2003/McCarty2.htm.

McGorry, P. D., Nelson, B., Amminger, G. P., et al. (2009). "Intervention in individuals at ultra-high risk for psychosis: A review and future directions." *Journal of Clinical Psychiatry* 70 (9):1206–12.

McGowan, P. O., Sasaki, A., D'Alessio, A. C., et al. (2009). "Epigenetic regulation of the glucocorticoid receptor in human brain associates with childhood abuse." *Nature Neuroscience* 12 (3):342–48.

McIlwain, H. (1985). "In the beginning: To celebrate 20 years of the International Society for Neurochemistry (ISN)." *Journal of Neurochemistry* 45 (1):1–10.

—— (1988). "Neurochemistry and related terms—their introduction and acceptance." *Neurochemistry International* 12 (4):431–38.

Meaney, M. J., Aitken, D. H., Bodnoff, S. R., et al. (1985). "The effects of postnatal handling on the development of the glucocorticoid receptor systems and stress recovery in the rat." *Progress in Neuro-Psychopharmacology and Biological Psychiatry* 9 (5–6):731–34.

Meaney, M. J., and Ferguson-Smith, A. C. (2010). "Epigenetic regulation of the neural transcriptome: The meaning of the marks." *Nature Neuroscience* 13 (11):1313–8.

Meaney, M. J., and Stewart, J. (1979). "Environmental factors influencing the affiliative behavior of male and female rats (*Rattus norvegicus*)." *Animal Learning and Behavior* 7 (3):397–405.

Medina, J. (2008). "Schizophrenia, DISC1 and animal models." *Psychiatric Times* (April):22–23.

Meisenzahl, E., Koutsouleris, N., Gaser, C., et al. (2008). "Structural brain alterations in subjects at high-risk of psychosis: A voxel-based morphometric study." *Schizophrenia Research* 102 (1–3):150–62.

Menninger, K. A. (1965). *The vital balance: The life process in mental health and illness.* New York: Viking Press.

Menninger, W. C. ([1947] 1994). "The role of psychiatry in the world today: 1947 [classical article]." *American Journal of Psychiatry* 151 (6):75–81.

Merzenich, M., Recanzone, G., Jenkins, W., et al., eds. (1988). *Cortical representational plasticity.* New York: Wiley.

Mettler, F. A. (1949). *Selective partial ablation of the frontal cortex: A correlative study of its effects on human psychotic subjects by Columbia-Greystone associates.* New York: Paul B. Hoeber.

Metzinger, T. (2009). *The ego tunnel: The science of the mind and the myth of the self.* New York: Basic Books.

Metzinger, T., and Gallese, V. (2003). "The emergence of a shared action ontology: Building blocks for a theory." *Consciousness and Cognition* 12 (4):549–71.

Meulders, M. (2006). "La patte de grenouille: Pré-modèle de fonctions nerveuses." In G. Gachelin, ed., *Les organismes modèles dans la recherche médicale*, 95–109. Paris: Presses Universitaires de France.

Miczek, K. A., de Almeida, R.M.M., Kravitz, E. A., et al. (2007). "Neurobiology of escalated aggression and violence." *Journal of Neuroscience* 27 (44):11,803–6.

Miczek, K. A., and de Wit, H. (2008). "Challenges for translational psychopharmacology research—some basic principles." *Psychopharmacology* 199 (3):291–301.

Miczek, K. A., Faccidomo, S., de Almeida, R.M.M., et al. (2004). "Escalated aggressive behavior: New pharmacotherapeutic approaches and opportunities." *Youth Violence: Scientific Approaches to Prevention* 1036:336–55.

Miczek, K. A., Fish, E. W., DeBold, J. F., et al. (2002). "Social and neural determinants of aggressive behavior: Pharmacotherapeutic targets at serotonin, dopamine and gamma-aminobutyric acid systems." *Psychopharmacology* 163 (3–4):434–58.

Miczek, K. A., Weerts, E., Haney, M., et al. (1994). "Neurobiological mechanisms controlling aggression: Preclinical developments for pharmacotherapeutic interventions." *Neuroscience and Biobehavioral Reviews* 18 (1):97–110.

Millar, J. K., Wilson-Annan, J. C., Anderson, S., et al. (2000). "Disruption of two novel genes by a translocation co-segregating with schizophrenia." *Human Molecular Genetics* 9 (9):1415.

Miller, G. (2010). "Beyond *DSM*: Seeking a brain-based classification of mental illness." *Science* 327 (5972):1437.

Miller, P., and Rose, N. (1988). "The Tavistock program: The government of subjectivity and social life." *Sociology* 22 (2):171–92.

—— (1990). "Governing economic life." *Economy and Society* 19 (1):1–31.

—— (1994). "On therapeutic authority: Psychoanalytical expertise under advanced liberalism." *History of the Human Sciences* 7 (3):29–64.

—— (1995a). "Political thought and the limits of orthodoxy: A response to Curtis." *British Journal of Sociology* 46 (4):590–97.

—— (1995b). "Production, identity, and democracy." *Theory and Society* 24 (3):427–67.

—— (1997). "Mobilizing the consumer: Assembling the subject of consumption." *Theory, Culture and Society* 14 (1):1–36.

—— (2008). *Governing the present: Administering economic, social and personal life.* Cambridge: Polity Press.

Minsky, M. L. (1986). *The society of mind.* New York: Simon and Schuster.

Mirescu, C., Peters, J. D., and Gould, E. (2004). "Early life experience alters response of adult neurogenesis to stress." *Nature Neuroscience* 7 (8):841–46.

Moffitt, T. E., Caspi, A., Taylor, A., et al. (2010). "How common are common mental disorders? Evidence that lifetime prevalence rates are doubled by prospective versus retrospective ascertainment." *Psychological Medicine* 40 (6):899–909.

Moncrieff, J. (2008). *The myth of the chemical cure: A critique of psychiatric drug treatment.* Basingstoke, UK: Palgrave Macmillan.

Moncrieff, J., and Cohen, D. (2005). "Rethinking models of psychotropic drug action." *Psychotherapy and Psychosomatics* 74 (3):145–53.

Moniz, E. A. (1948). "How I came to perform prefrontal leucotomy." Lecture presented in English at the Congress of Psychosurgery, August. Lisbon: Ediçoes Atica.

—— (1954). "A leucotomia está em causa." Lecture presented at the Academy of Sciences of Lisbon, May 20. *Revista Filosófica de Coimbra* 10.

Montague, P. R. (2003). "Decisions, uncertainty, and the brain: The science of neuroeconomics." *Nature* 424 (6947):371–72.

Moran, P., Ford, T., Butler, G., et al. (2008). "Callous and unemotional traits in children and adolescents living in Great Britain." *British Journal of Psychiatry* 192 (1):65–66.

Morel, B. A. (1852–53). *Traité des maladies mentales.* 2 vols. Paris: Baillière.

—— (1857). *Traité des dégénérescences physiques, intellectuelles et morales de l'espèce humaine et des causes qui produisent ces variétés maladives.* Paris: Baillière.

—— (1860). "Traité des maladies mentales." *American Journal of Psychiatry* 17 (2):199–211.

Moreno, J. D. (2003). "Neuroethics: An agenda for neuroscience and society." *Nature Reviews Neuroscience* 4 (2):149–53.

Morgan, M. S., and Morrison, M. (1999). *Models as mediators: Perspectives on natural and social science.* Cambridge: Cambridge University Press.

Moriarty, J. (2009). "Visions of deception: Neuroimages and the search for truth." *Akron Law Review* 42:739.

Morris, J. A., Kandpal, G., Ma, L., et al. (2003). "DISC1 (disrupted-in-schizophrenia 1) is a centrosome-associated protein that interacts with MAP1A, MIPT3, ATF4/5 and NUDEL: Regulation and loss of interaction with mutation." *Human Molecular Genetics* 12 (13):1591–1608.

Morrison, W. D. (1891). *Crime and Its Causes.* London: Sonnenschein.

Morse, S. J. (2004). "New Neuroscience, Old Problems." In B. Garland, ed., *Neuroscience and the law: Brain, mind and the scales of justice,* 157–98. New York: Dana Press.

—— (2006). "Brain overclaim syndrome and criminal responsibility: A diagnostic note." Research paper no. 06-35. *University of Pennsylvania Law School* 3:397–412.

—— (2008). "Determinism and the death of folk psychology: Two challenges to responsibility from neuroscience." *Minnesota Journal of Law, Science and Technology* 9:1–36.

Morse, S. J., and Hoffman, M. B. (2007). "The uneasy entente between insanity and *mens rea*: Beyond *Clark v. Arizona.*" *Journal of Criminal Law and Behavior* 97 (4):1071–1149.

Morton, S. G. (1844). *Crania Ægyptiaca; or, Observations on Egyptian ethnography, derived from anatomy, history and the monuments.* From *Transactions of the American Philosophical Society,* vol. 9. Philadelphia: American Philosophical Society.

Moss, H. (2003). "Implicit selves." Special issue, "The self: From soul to brain," *Annals of the New York Academy of Sciences* 1001:1–30.

Mosso, A. (1880). "Sulla circolazione del sangue nel cervello dell'uomo." *Mem. Reale Accademia dei Lincei* 5:237–358.

Mountcastle, V. B. (1998). "Brain science at the century's ebb." *Daedalus* 127 (2):1–36.

Moynihan, R. (2002). "Alosetron: A case study in regulatory capture, or a victory for patients' rights?" *British Medical Journal* 325 (7364):592–95.

—— (2008). "Key opinion leaders: Independent experts or drug representatives in disguise?" *British Medical Journal* 336 (7658):1402–3.

Moynihan, R., and Cassels, A. (2005). *Selling sickness: How the world's biggest pharmaceutical companies are turning us all into patients.* New York: Nation Books.

Mukamel, R., Ekstrom, A. D., Kaplan, J., et al. (2010). "Single-neuron responses in humans during execution and observation of actions." *Current Biology* 20 (8):750–56.

Narrow, W. E., Kuhl, E. A., and Regier, D. A. (2009). "*DSM-V* perspectives on disentangling disability from clinical significance." *World Psychiatry* 8 (2):88–89.

Narrow, W. E., Rae, D. S., Robins, L. N., et al. (2002). "Revised prevalence estimates of mental disorders in the United States: Using a clinical significance criterion to reconcile 2 surveys' estimates." *Archives of General Psychiatry* 59 (2):115.

National Institute for Health Research. (2008). *National Institute for Health Research progress report 2006–2008.* London: National Institute for Health Research.

National Institute for Mental Health. (2001). *Borderline personality disorder: Raising questions, finding answers.* London: National Institute of Mental Health.

National Institute of Mental Health (1951). *Proceedings of the First [etc.] Research Conference on Psychosurgery . . . 1949 [etc.]:* Washington, DC: NIMH.

Nicolas, S., and Murray, D. J. (1999). "Théodule Ribot (1839–1916), founder of French psychology: A biographical introduction." *History of Psychology* 2 (4):277.

Nicolson, D. (1873a). "The morbid psychology of criminals." *Journal of Mental Science* 19 (86):222–32.

—— (1873b). "The morbid psychology of criminals." *Journal of Mental Science* 19 (87):398–409.

—— (1874). "The morbid psychology of criminals." *Journal of Mental Science* 20 (90):167–85.

—— (1875a). "The morbid psychology of criminals." *Journal of Mental Science* 21 (93):18–31.

—— (1875b). "The morbid psychology of criminals." *Journal of Mental Science* 21 (94):225–250.

—— (1895). "Presidential address delivered at the fifty-fourth annual meeting of the Medico-Psychological Association, held in London, 25th and 26th July, 1895." *Journal of Mental Science* 41 (175):567–91.

Norman, G., Cacioppo, J., and Berntson, G. (2010). "Social neuroscience." *Wiley Interdisciplinary Reviews: Cognitive Science* 1 (1):60–68.

Northoff, G., and Panksepp, J. (2008). "The trans-species concept of self and the subcortical-cortical midline system." *Trends in Cognitive Sciences* 12 (7):259–64.

Northoff, G., Qin, P., and Feinberg, T. E. (2011). "Brain imaging of the self: Conceptual, anatomical and methodological issues." *Consciousness and Cognition* 20 (1):52–63.

Nouvel, P. (2002). *Enquête sur le concept de modèle.* Paris: Presses Universitaires de France.

Novas, C. (2006). "The political economy of hope: Patients' organizations, science and biovalue." *BioSocieties* 1 (3):289–305.

Novas, C., and Rose, N. (2000). "Genetic risk and the birth of the somatic individual." *Economy and Society* 29 (4):485–513.

Oatley, K. (2004). *Emotions: A brief history.* Malden, MA: Blackwell.

Oberman, L. M., Hubbard, E. M., McCleery, J. P., et al. (2005). "EEG evidence for mirror neuron dysfunction in autism spectrum disorders." *Cognitive Brain Research* 24 (2):190–98.

Oberman, L. M., and Ramachandran, V. S. (2007). "The simulating social mind: The role of the mirror neuron system and simulation in the social and communicative deficits of autism spectrum disorders." *Psychological Bulletin* 133 (2):310–27.

Ochsner, K. N. (2004). "Current directions in social cognitive neuroscience." *Current Opinion in Neurobiology* 14 (2):254–58.

Odgers, C. L., Moffitt, T. E., Poulton, R., et al. (2008). "Female and male antisocial trajectories: From childhood origins to adult outcomes." *Development and Psychopathology* 20:673–716.

Ogawa, S., Lee, T. M., Kay, A. R., et al. (1990). "Brain magnetic resonance imaging with contrast dependent on blood oxygenation." *Proceedings of the National Academy of Sciences* 87 (24):9868–72.

Olivier, B., Mos, J., Raghoebar, M., et al. (1994). "Serenics." *Progress in Drug Research* 42:167–308.

O'Malley, P. (1996). "Risk and responsibility." In A. Barry, T. Osborne, and N. Rose, eds., *Foucault and political reason*, London: UCL Press.

—— (1998). *Crime and the risk society.* Aldershot, UK: Dartmouth.

—— (2000). *Configurations of risk*. Abingdon, UK: Routledge.

Osler, W. (1882a). "On the brains of criminals: With a description of the brains of two murderers." *Canadian Medical and Surgical Journal* 10:385–98.

—— (1882b). "The brains of criminals (letter)." *Lancet* 2:38.

Owen, M. J., O'Donovan, M. C., Thapar, A., et al. (2011). "Neurodevelopmental hypothesis of schizophrenia." *British Journal of Psychiatry* 198 (3):173–75.

Packard, V. (1962). *The hidden persuaders*. Harmondsworth, UK: Penguin Books.

Pallen, M. (2006). "*Escherichia coli*: From genome sequences to consequences (or 'ceci n'est pas un éléphant . . .')." *Canadian Journal of Infectious Diseases and Medical Microbiology* 17 (2):114–16.

Panksepp, J. (2005). "Affective consciousness: Core emotional feelings in animals and humans." *Consciousness and Cognition* 14 (1):30–80.

Panofsky, A. L. (2006). "Fielding controversy: The genesis and structure of behavior genetics." PhD diss., New York University. Dissertations and Theses: AandI, publication no. AAT 3234171. http://www.proquest.com.

Pantelis, C., Velakoulis, D., Mcgorry, P. D., et al. (2003). "Neuroanatomical abnormalities before and after onset of psychosis: A cross-sectional and longitudinal MRI comparison." *Lancet* 361 (9354):281–88.

Parkin, A. J., and Hunkin, N. M. (2001). "British memory research: A journey through the 20th century." *British Journal of Psychology* 92 (pt. 1):37–52.

Parkin, H. E. (1969). *The social animal, an anthology for general and liberal studies*. London: Routledge and Kegan Paul.

Parton, N. (1991). *Governing the family: Child care, child protection and the state*. London: Macmillan Education.

Partridge, M. W. (1950). *Pre-Frontal Leucotomy: A survey of 300 cases personally followed over 1 1/2–3 years*. Oxford: Blackwell Scientific Publications.

Patton, P. (2003). "Language, power, and the training of horses." In C. Wolfe, ed. *Zoontologies: The question of the animal*, 83–99. Minneapolis: University of Minnesota Press.

Pauling, L., and Coryell, C. (1936). "The magnetic properties and structure of the hemochromogens and related substances." *Proceedings of the National Academy of Sciences* 22 (3):159.

Pavlov, P. (1941). *Lectures on conditioned reflexes*. Translated and edited by W. H. Gantt. Oxford: International Publishers.

Peay, J. (2010). *Mental health and crime*. New York: Routledge.

Pellow, S., Chopin, P., File, S. E., et al. (1985). "Validation of open: Closed arm entries in an elevated plus-maze as a measure of anxiety in the rat." *Journal of Neuroscience Methods* 14 (3):149–67.

Pellow, S., and File, S. E. (1986). "Anxiolytic and anxiogenic drug effects on exploratory activity in an elevated plus-maze: A novel test of anxiety in the rat." *Pharmacology, Biochemistry and Behavior* 24 (3):525–59.

Penfield, W. (1927). "The mechanism of cicatricial contraction in the brain." *Brain* 50 (3–4):499.

—— (1975). *The mystery of the mind: A critical study of consciousness and the human brain*. Princeton, NJ: Princeton University Press.

Penfield, W., and Erickson, T. C. (1941). *Epilepsy and cerebral localization*. London: Baillière, Tindall and Cox.

Penfield, W., Jasper, H. H., and MacNaughton, F. (1954). *Epilepsy and the functional anatomy of the human brain.* London: Churchill.

Penfield, W., and Rasmussen, T. (1950). *The cerebral cortex of man: A clinical study of localization of function.* New York,: Macmillan.

Perry, B. D. (1997). "Incubated in terror: Neurodevelopmental factors in the 'cycle of violence.'" In J. D. Osofsky, ed., *Children in a violent society*, 124–49. New York: Guilford Press.

—— (2002). "Childhood experience and the expression of genetic potential: What childhood neglect tells us about nature and nurture." *Brain and Mind* 3 (1):79–100.

—— (2008). "Child maltreatment: A neurodevelopmental perspective on the role of trauma and neglect in psychopathology." In T. P. Beauchaine and S. P. Hinshaw, eds., *Child and Adolescent Psychopathology*, 93–128. Hoboken, NJ: Wiley.

Petryna, A. (2009). *When experiments travel: Clinical trials and the global search for human subjects.* Princeton, NJ: Princeton University Press.

Phelan, J. C. (2002). "Genetic bases of mental illness—a cure for stigma?" *Trends in Neurosciences* 25 (8):430–31.

—— (2006). "Genes, mental illness and stigma." *American Journal of Medical Genetics Part B Neuropsychiatric Genetics* 141B (7):688.

Pincus, J. H. (1993). "Neurologists' role in understanding violence." *Archives of Neurology* 50 (8):867–69.

Pinel, P. (1801). *Traité médico-philosophique sur l'aliénation mentale, ou la manie.* Paris: Richard, Caille et Ravier.

Platt, A. (1977). *The child savers: The invention of delinquency.* Chicago: University of Chicago Press.

Platt, M. L., and Glimcher, P. W. (1999). "Neural correlates of decision variables in parietal cortex." *Nature* 400 (6741):233–38.

Pletscher, A., Shore, P. A., and Brodie, B. B. (1955). "Serotonin release as a possible mechanism of reserpine action." *Science* 122:374.

Poldrack, R. A. (2008). "The role of fMRI in cognitive neuroscience: Where do we stand?" *Current Opinion in Neurobiology* 18 (2):223–37.

Porter, R. (2002). *Madness: A brief history.* Oxford: Oxford University Press.

Porteus, S. D. (1950). *The Porteus maze test and intelligence.* Palo Alto, CA: Pacific Books.

Power, P., Iacoponi, E., Reynolds, N., et al. (2007). "The Lambeth early onset crisis assessment team study: General practitioner education and access to an early detection team in first-episode psychosis." *British Journal of Psychiatry* 191 (51):s133–s139.

Prasad, A. (2005). "Making images/making bodies: Visibilizing and disciplining through magnetic resonance imaging (MRI)." *Science, Technology and Human Values* 30 (2):291–316.

—— (2007). "The (amorphous) anatomy of an invention." *Social Studies of Science* 37 (4):533–60.

Pratt, J. (2000). "Dangerousness, risk and technologies of power." *Australian and New Zealand Journal of Criminology* 28 (1):3–31.

Premack, D., and Woodruff, G. (1978). "Does the chimpanzee have a theory of mind?" *Behavioral and Brain Sciences* 1 (4):515–26.

President's Council on Bioethics (U.S.), and Kass, L. (2003). *Beyond therapy: Biotechnology and the pursuit of happiness*. New York: Regan Books.

Pressman, J. D. (1998). *Last resort: Psychosurgery and the limits of medicine*. Cambridge: Cambridge University Press.

Pribram, K. H., and Gill, M. M. (1976). *Freud's 'Project' reassessed*. London: Hutchinson.

Proctor, R. (1988). *Racial hygiene: Medicine under the Nazis*. Cambridge, MA: Harvard University Press.

Provenzale, J. M., and Mukundan, S. (2005). "Getting small is suddenly very big: Review of the *Proceedings of the Third Annual Meeting of the Society for Molecular Imaging*." *American Journal of Roentgenology* 184 (6):1736.

Purves, D. (2010). *Brains: How they seem to work*. Upper Saddle River, NJ: Financial Times/Prentice Hall.

Putnam, H. (1981). *Reason, truth and history*. Cambridge: Cambridge University Press.

Quarton, G. C., Melnechuk, T., Schmitt, F. O., et al. (1967). *The neurosciences: A study program*. New York: Rockefeller University Press.

Quinn, N. (2003). "Cultural selves." *Annals of the New York Academy of Sciences* 1001 (1):145–76.

Rader, K. A. (1998). "'The mouse people': Murine genetics work at the Bussey Institution, 1909–1936." *Journal of the History of Biology* 31:327–54.

—— (2004). *Making mice: Standardizing animals for American biomedical research, 1900–1955*. Princeton, NJ: Princeton University Press.

Rafter, N. H. (1992). "Criminal anthropology in the United States." *Criminology* 30:525.

—— (2008). *The criminal brain: Understanding biological theories of crime*. New York: New York University Press.

Raine, A. (1993). *The psychopathology of crime: Criminal behavior as a clinical disorder*. San Diego: Academic Press.

Raine, A., Buchsbaum, M., and LaCasse, L. (1997). "Brain abnormalities in murderers indicated by positron emission tomography." *Biological Psychiatry* 42 (6):495–508.

Raine, A., Buchsbaum, M. S., Stanley, J., et al. (1994). "Selective reductions in prefrontal glucose-metabolism in murderers." *Biological Psychiatry* 36 (6):365–73.

Raine, A., Lencz, T., Bihrle, S., et al. (2000). "Reduced prefrontal gray matter volume and reduced autonomic activity in antisocial personality disorder." *Archives of General Psychiatry* 57 (2):119–27.

Rakic, P. (1985). "Limits of neurogenesis in primates." *Science* 227:1054–56.

Ramachandran, V. S. (2011). *The tell-tale brain: A neuroscientist's quest for what makes us human*. New York: W. W. Norton.

Ramachandran, V. S., and Oberman, L. M. (2006). "Broken mirrors: A theory of autism." *Scientific American* 295 (5):62–69.

Ramón y Cajal, S. (1928). *Degeneration and regeneration of the nervous system*. New York: Hafner.

—— (1937). *Recollections of my life*. Philadelphia: American Philosophical Society.

Rapport, R. L. (2005). *Nerve endings: The discovery of the synapse*. New York: W. W. Norton.

Reddy, W. M. (2001). *The navigation of feeling: A framework for the history of emotions*. Cambridge: Cambridge University Press.

Rees, J. R., Hill, D., and Sargeant, W. (1949). Special issue, "Anglo-American symposium on psychosurgery, neurophysiology and physical treatments in psychiatry." *Proceedings of the Royal Society of Medicine, Section of Psychiatry* 42
(suppl., September 12):1–93.

Regier, D. A. (2000). "Community diagnosis counts [commentary]." *Archives of General Psychiatry* 57:223–24.

Regier, D. A., Kaelber, C. T., Rae, D. S., et al. (1998). "Limitations of diagnostic criteria
and assessment instruments for mental disorders." *Archives of General Psychiatry* 55:109–15.

Regier, D. A., Narrow, W. E., Kuhl, E. A., et al. (2009). "The conceptual development of
DSM-V." *American Journal of Psychiatry* 166 (6):645–50.

Regier, D. A., and Robins, L. N. (1991). *Psychiatric disorders in America: The epidemiologic catchment area study*. New York: Free Press.

Relman, A. S., and Angell, M. (2002). "America's other drug problem." *New Republic*
227 (25):27–41.

Renvoisé, P., and Morin, C. (2007). *Neuromarketing: Understanding the "buy button" in
your customer's brain*. Nashville, TN: T. Nelson.

Restak, R. M. (2006). *The naked brain: How the emerging neurosociety is changing how
we live, work, and love*. New York: Harmony Books.

Rheinberger, H.-J. (1997). *Toward a history of epistemic things: Synthesizing proteins in
the test tube*. Stanford, CA: Stanford University Press.

——— (2000). "Gene concepts: Fragments from the perspective of molecular biology."
In R. F. Peter, J. Beurton, H.-J. Rheinberger, eds., *The concept of the gene in
development and evolution: Historical and epistemological perspectives*, 219–39.
Cambridge: Cambridge University Press.

——— (2006). "Vers la fin des organismes modèles?" In G. Gachelin, ed., *Les organismes
modèles dans la recherche médicale*, 275–77. Paris: Presses Universitaires de
France.

Richter, C. P. (1968). "Experiences of a reluctant rat-catcher: The common Norway
rat—friend or enemy?" *Proceedings of the American Philosophical Society* 112
(6):403–15.

Richter, J. (2000). "The Brain Commission of the International Association of Acad-
emies: The first international society of neurosciences." *Brain Research Bulletin*
52 (6):445–57.

Richter, S. H., Garner, J. P., and Würbel, H. (2009). "Environmental standardization:
Cure or cause of poor reproducibility in animal experiments?" *Nature Methods*
6 (4):257–61.

Rizzolatti, G. (2005). "The mirror neuron system and its function in humans." *Anatomy
and Embryology* 210 (5):419–21.

Rizzolatti, G., and Craighero, L. (2004). "The mirror-neuron system." *Annual Review of
Neuroscience* 27:169–92.

Rizzolatti, G., and Fabbri-Destro, M. (2008). "The mirror system and its role in social
cognition." *Current Opinion in Neurobiology* 18 (2):179–84.

Rizzolatti, G., Fadiga, L., Gallese, V., et al. (1996). "Premotor cortex and the recognition
of motor actions." *Cognitive Brain Research* 3 (2):131–41.

Rizzolatti, G., and Sinigaglia, C. (2008). *Mirrors in the brain: How our minds share actions and emotions*. Oxford: Oxford University Press.

Rodgers, R. J. (1997). "Animal models of 'anxiety': Where next?" *Behavioral Pharmacology* 8 (6–7):477–96; discussion 497–504.

Roepstorff, A. (2001). "Brains in scanners: An umwelt of cognitive science." *Semiotica* 134 (1):747–65.

——— (2002). "Transforming subjects into objectivity: An ethnography of knowledge in a brain imaging laboratory." *FOLK, Journal of the Danish Ethnography Society* 44:145–70.

——— (2004). "Mapping brain mappers: An ethnographic coda." In R. Frackowiak, ed., *Human brain function*. 1105–17. San Diego: Academic Press.

——— (2007). "Navigating the brainscape: When knowing becomes seeing." In C. Grasseni, ed., *Skilled visions: Between apprenticeship and standards*, 191–206. Oxford: Berghahn Books.

Rogers, J., Viding, E., Blair, R. J., et al. (2006). "Autism spectrum disorder and psychopathy: Shared cognitive underpinnings or double hit?" *Psychological Medicine* 36 (12):1789–98.

Rojas-Burke, J. (1993). "PET scans advance as tool in insanity defense." *Journal of Nuclear Medicine* 34 (1):13N–26N.

Roland, A. (1988). *In search of self in India and Japan: Toward cross-cultural psychology.* Princeton, NJ: Princeton University Press.

Romeo, R., Knapp, M., and Scott, S. (2006). "Economic cost of severe antisocial behaviour in children—and who pays it." *British Journal of Psychiatry* 188:547–53.

Romme, M. J., and Escher, S. (1993). *Accepting voices.* London: Mind (National Association for Mental Health).

Rose, F. C. (1999a). *A short history of neurology: The British contribution, 1660–1910.* Oxford: Butterworth-Heinemann.

Rose, N. (1979). "The psychological complex: Mental measurement and social administration." *Ideology and Consciousness* 5:5–70.

——— (1985). *The psychological complex: Psychology, politics and society in England, 1869–1939.* London: Routledge and Kegan Paul.

——— (1986). "Psychiatry: The discipline of mental health." In P. Miller and N. Rose, eds., *The power of psychiatry*, 43–84. Cambridge: Polity Press.

——— (1988). "Calculable minds and manageable individuals." *History of the Human Sciences* 1 (2):179–200.

——— (1989). *Governing the soul: The shaping of the private self.* London: Routledge.

——— (1994). "Medicine, history and the present." In C. Jones and R. Porter, eds., *Reassessing Foucault: Power, medicine and the body*, 48–72. London: Routledge.

——— (1996). *Inventing our selves: Psychology, power, and personhood.* New York: Cambridge University Press.

——— (1999). *Powers of freedom: Reframing political thought.* Cambridge: Cambridge University Press.

——— (2000). "The biology of culpability: Pathological identity and crime control in a biological culture." *Theoretical Criminology* 4 (1):5–43.

——— (2003a). "The neurochemical self and its anomalies." In R. V. Ericson and A. Doyle, eds., *Risk and morality*, 407–37. Toronto: University of Toronto Press.

——— (2003b). "Neurochemical selves." *Society* 41 (1):46–59.

——— (2004). "Becoming neurochemical selves." In N. Stehr, ed., *Biotechnology, Commerce and Civil Society*, 89–128. New Brunswick, NJ: Transaction.

Rose, N. (2006a). "Disorders without borders? The expanding scope of psychiatric practice." *BioSocieties* 1 (4):465–84.

—— (2006b). "Psychopharmaceuticals in Europe." In D. McDaid, M. Knapp, and G. Thornicroft, eds., *Mental health policy and practice in Europe*, Milton Keynes, UK: Open University Press.

—— (2007a). "Beyond medicalisation." *Lancet* 369 (9562):700.

—— (2007b). "Governing the will in a neurochemical age." In S. Maasem and B. Sutter, eds., *On willing selves: Neoliberal politics vis-à-vis the neuroscientific challenge*, 88–99. Basingstoke, UK: Palgrave Macmillan

—— (2007c). *The politics of life itself: Biomedicine, power, and subjectivity in the twenty-first century*. Princeton, NJ: Princeton University Press.

—— (2008). "Screen and intervene: Governing risky brains." Lecture given at Centre Koyre, Paris, May.

—— (2009). "Commerce vs. the commons: Conflicts over the commercialisation of biomedical knowledge." In S. Swee-Hock and D. Quah, eds., *The politics of knowledge*, 79–110. Singapore: ISEAS Press.

—— (2010a). "Normality and pathology in a biomedical age." In B. Carter and N. Charles, eds. *Nature, society and environmental crisis*, 66–83. Oxford: Wiley-Blackwell.

—— (2010b). " 'Screen and intervene': Governing risky brains." *History of the Human Sciences* 23 (1):79–105.

Rose, N., and Miller, P. (1992). "Political power beyond the state: Problematics of government." *British Journal of Sociology* 43 (2):173–205.

Rose, N. and Novas, C. (2004). "Biological citizenship." In A. Ong and S. J. Collier, eds., *Global assemblages: Technology, politics, and ethics as anthropological problems*, 439–63. Oxford: Blackwell.

Rose, S.P.R. (1986). *Molecules and minds: Essays on biology and the social order*. Philadelphia: Open University Press.

—— (2000). "God's organism? The chick as a model system for memory studies." *Learning and Memory* 7 (1):1.

Rosenberg, A. (1994). *Instrumental Biology, or, The disunity of science*. Chicago: University of Chicago Press.

Rosenberg, C. E. (2006). "Contested boundaries—psychiatry, disease, and diagnosis." *Perspectives in Biology and Medicine* 49 (3):407–24.

—— (2007). *Our present complaint: American medicine, then and now*. Baltimore: Johns Hopkins University Press.

Rosenhan, D. L. (1973). "On being sane in insane places." *Science* 179:250–58.

Rosner, R. (1999). "Dialogues in historiography—historiography and historians of neuroscience: Towards diversity in the ISHN." *Journal of the History of the Neurosciences* 8 (3):264–68.

Rothman, D. J. (1971). *The discovery of the asylum: Social order and disorder in the new republic*. Boston: Little, Brown.

Rowson, J. (2011). *Socialising with the brain*. London: Royal Society for the Encouragement of Arts, Manufactures and Commerce.

Roy, C. S., and Sherrington, C. (1890). "On the regulation of the blood-supply of the brain." *Journal of Physiology* 11 (1–2):85.

Royal Society of London. (2003). *Keeping science open: The effects of intellectual property policy on the conduct of science*. London: Royal Society.

———— (2011). *Brain Waves 4: Neuroscience and Law*. London: Royal Society.

Rubin, B. (2009). "Changing brains: The emergence of the field of adult neurogenesis." *BioSocieties* 4 (4):407–24.

Runciman, W. G. (1999). *The social animal*. London: Fontana.

Russell, T. A., Rubia, K., Bullmore, E. T., et al. (2000). "Exploring the social brain in schizophrenia: Left prefrontal underactivation during mental state attribution." *American Journal of Psychiatry* 157 (12):2040–42.

Rutter, M., Beckett, C., Castle, J., et al. (2007). "Effects of profound early institutional deprivation: An overview of findings from a UK longitudinal study of Romanian adoptees." *European Journal of Developmental Psychology* 4 (3):332–50.

Rutter, M., Kim-Cohen, J., and Maughan, B. (2006). "Continuities and discontinuities in psychopathology between childhood and adult life." *Journal of Child Psychology and Psychiatry* 47(3–4): 276–95.

Rutter, M., Sonuga Barke, E. J., and Castle, J. (2010). "I. Investigating the impact of early institutional deprivation on development: Background and research strategy of the English and Romanian adoptees (era) study." *Monographs of the Society for Research in Child Development* 75 (1):1–20.

Ryle, G. (1949). *The concept of mind*. London: Hutchinson.

Sachdev, P. S. (2002). "Neuropsychiatry—a discipline for the future." *Journal of Psychosomatic Research* 53:625–27.

———— (2005). "Whither neuropsychiatry?" *Journal of Neuropsychiatry and Clinical Neurosciences* 17 (2):140–44.

Sacks, O. W. (1976). *Awakenings*. New York: Vintage Books.

———— (1984). *A leg to stand on*. London: Duckworth.

———— (1985). *The man who mistook his wife for a hat*. London: Duckworth.

———— (1991). *Seeing Voices: A journey into the world of the deaf*. London: Picador.

Salomon-Bayet, C. (1978). *L'institution de la science et l'expérience du vivant: Méthode et expérience à l'Académie royale des sciences, 1666–1793*. Paris: Flammarion.

Salzer, H. M., and Lurie, M. L. (1953). "Anxiety and depressive states treated with isonicotinyl hydrazide (isoniazid)." *Archives of Neurology and Psychiatry* 70:317–24.

Sampson, E. E. (1988). "The debate on individualism: Indigenous psychologies of the individual and their role in personal and societal functioning." *American Psychologist* 43 (1):15.

Sanfey, A. G., Rilling, J. K., Aronson, J. A., et al. (2003). "The neural basis of economic decision-making in the ultimatum game." *Science* 300 (5626):1755–58.

Sani, S., Jobe, K., Smith, A., et al. (2007). "Deep brain stimulation for treatment of obesity in rats." *Journal of Neurosurgery: Pediatrics* 107 (4):809–13.

Santarelli, L., Saxe, M., Gross, C., et al. (2003). "Requirement of hippocampal neurogenesis for the behavioral effects of antidepressants." *Science* 301 (5634):805–9.

Sapolsky, R. M. (2004). "Is impaired neurogenesis relevant to the affective symptoms of depression?" *Biological Psychiatry* 56 (3):137–39.

Sargant, W., Hill, J., and Slater, E. T. (1948). *An Introduction to Physical Methods of Treatment in Psychiatry*. Edinburgh: E. and S. Livingstone.

Sargant, W., and Slater, E. T. (1944). *An introduction to physical methods of treatment in psychiatry*. Edinburgh: Livingstone.

Saunders, B. F. (2008). *CT suite: The work of diagnosis in the age of noninvasive cutting*. Durham, NC: Duke University Press.

Schaller, W. F. (1922). "The outlook in neuro-psychiatry." *California State Journal of Medicine* 20 (12):438–40.

Schatzman, M. (1973). *Soul murder: Persecution in the family*. London: Allen Lane.

Schildkraut, J. J. (1965). "The catecholamine hypothesis of affective disorders: A review of supporting evidence." *American Journal of Psychiatry* 122:509–22.

—— (2000). "The catecholamine hypothesis." In David Healy, ed., *The Psychopharmacologists*, 3:113–34. London: Arnold.

Schleim, S., and Roiser, J. P. (2009). "fMRI in translation: The challenges facing real-world applications." *Frontiers in Human Neuroscience* 3:1–7.

Schmitt, F. O. (1967). "Molecular biology among the neurosciences." *Archives of Neurology* 17 (6):561–72.

—— (1970). "Promising trends in neuroscience." *Nature* 227 (5262):1006–8.

—— (1985). "Adventures in molecular biology." *Annual Reviews of Biophysics and Biophysical Chemistry* 14:1–22.

—— (1990). *The never-ceasing search*. Philadelphia: American Philosophical Society.

Schuback, D. E., Mulligan, E. L., Sims, K. B., et al. (1999). "Screen for MAOA mutations in target human groups." *American Journal of Medical Genetics* 88:25–28.

Schultz, W. (2003). "Decisions, uncertainty, and the brain: The science of neuroeconomics." *Science* 300 (5626):1662–63.

Schwartz, J., and Begley, S. (2002). *The mind and the brain: Neuroplasticity and the power of mental force*. New York: Regan Books.

Scott, J. P. (1958). *Aggression*. Chicago: University of Chicago Press.

—— (1969). "Biological basis of human warfare: An interdisciplinary problem." In M. Sherif and C. W. Sherif, eds. *Interdisciplinary Relationships in the Social Sciences*, 121–36. New Brunswick, NJ: Aldine.

Scott, S., Knapp, M., Henderson, J., et al. (2001). "Financial cost of social exclusion: Follow up study of antisocial children into adulthood." *British Medical Journal* 323 (7306):191–94.

Scott, S., Sylva, K., Doolan, M., et al. (2010). "Randomised controlled trial of parent groups for child antisocial behaviour targeting multiple risk factors: The SPOKES project." *Journal of Child Psychology and Psychiatry* 51 (1):48–57.

Scull, A. T. (1979). *Museums of madness: The social organization of insanity in nineteenth-century England*. London: Allen Lane.

Searle, J. R. (1984). *Minds, brains and science: The 1984 Reith lectures*. Cambridge, MA: Harvard University Press.

—— (1992). *The rediscovery of the mind*. Cambridge, MA: MIT Press.

Senior C., Smythe H., Cooke R., et al. (2007). "Mapping the mind for the modern market researcher." *Qualitative Market Research* 10:153–67.

Serres, M. (1977). *La naissance de la physique dans le texte de Lucrèce: Fleuves et turbulences*. Paris: Editions de Minuit.

Shapin, S. (1994). *A social history of truth: Civility and science in seventeenth-century England*. Chicago: University of Chicago Press.

—— (2008). *The scientific life: A moral history of a late modern vocation*. Chicago: University of Chicago Press.

Shapin, S., and Schaffer, S. (1985). *Leviathan and the air-pump: Hobbes, Boyle, and the experimental life*. Princeton, NJ: Princeton University Press.

Sharma, T., Kumari, V., and Russell, T. (2002). "Exploring the social brain in schizophrenia." *Biological Psychiatry* 51 (8):104S.

Shephard, B. (2003). *A war of nerves: Soldiers and psychiatrists in the twentieth century.* Cambridge, MA: Harvard University Press.

Shizgal, P., and Conover, K. (1996). "On the neural computation of utility." *Current Directions in Psychological Science* 5 (2):37–43.

Shore, P. A., Silver, S. L., and Brodie, B. B. (1955). "Interaction of reserpine, serotonin, and lysergic acid diethylamide in brain." *Science* 122:284–85.

Shorter, E. (1997). *A history of psychiatry: From the era of the asylum to the age of Prozac.* New York: Wiley.

Shostak, S. (2007). "Translating at work: Genetically modified mouse models and molecularization in the environmental health sciences." *Science, Technology and Human Values* 32 (3):315–38.

Shultz, S., and Dunbar, R.I.M. (2007). "The evolution of the social brain: Anthropoid primates contrast with other vertebrates." *Proceedings of the Royal Society B: Biological Sciences* 274:2429–36.

Shweder, R. A., and Levine, R. A. (1984). *Culture theory: Essays on mind, self and emotion.* Cambridge: Cambridge University Press.

Silverman, C. (2004). "A disorder of affect: Love, tragedy, biomedicine and citizenship in American autism research, 1943–2003." PhD diss., University of Pennsylvania.

Simmel, G. (1903). *The metropolis and mental life.* In E. Shils, ed. and trans., *Social Sciences III: Selections and Selected Readings*, vol. 2. Chicago: University of Chicago Press.

Singer, T., and Lamm, C. (2009). "The social neuroscience of empathy." Special issue, "The Year in Cognitive Neuroscience 2009." *Annals of the New York Academy of Sciences* 1156:81–96.

Singer, T., Seymour, B., O'Doherty, J. P., et al. (2004). "Empathy for pain involves the affective but not sensory components of pain." *Science* 303 (5661):1157.

—— (2006). "Empathic neural responses are modulated by the perceived fairness of others." *Nature* 439 (7075):466–69.

Singh, I., and Rose, N. (2009). "Biomarkers in psychiatry." *Nature* 460 (7252):202–7.

Sinigaglia, C., and Rizzolatti, G. (2011). "Through the looking glass: Self and others." *Consciousness and Cognition* 20 (1):64–74.

Smieskova, R., Fusar-Poli, P., Allen, P., et al. (2010). "Neuroimaging predictors of transition to psychosis: A systematic review and meta-analysis." *Neuroscience and Biobehavioral Reviews* 34 (8):1207–22.

Smith, C.U.M. (1999). "Thomas Henry Huxley and neuroscience." *Physis: Rivista Internazionale di Storia della Scienza* 36 (2):355–65.

—— (2001). "*Renatus renatus*: The Cartesian tradition in British neuroscience and the neurophilosophy of John Carew Eccles." *Brain and Cognition* 46 (3):364–72.

Smith, J., and Berlin, L. (1999). "Informed consent when using medical devices for indications not approved by the Food and Drug Administration." *American Journal of Roentgenology* 173 (4):879.

Smith, R. (1992). *Inhibition: History and meaning in the sciences of mind and brain.* Berkeley: University of California Press.

—— (2002). "Representations of mind: C. S. Sherrington and scientific opinion, c. 1930–1950." *Science in Context* 14 (4):511–39.

Sourkes, T. L. (2006). "Introduction: Neuroscience in the Nobel perspective." *Journal of the History of the Neurosciences* 15:306–17.

Spence, S. A. (1996). "Free will in the light of neuropsychiatry." *Philosophy, Psychiatry and Psychology* 3:75–90.

———(2009). *The actor's brain: Exploring the cognitive neuroscience of free will.* Oxford: Oxford University Press.

Spitzer, R. L. (2001). "Values and assumptions in the development of *DSM-III* and *DSM-III-R*: An insider's perspective and a belated response to Sadler, Hulgus, and Agich's 'On values in recent American psychiatric classification.' " *Journal of Nervous and Mental Disease* 189 (6):351–59.

Spitzer, R. L., Endicott, J., and Robins, E. (1975). "Clinical criteria for psychiatric diagnosis and *DSM-III*." *American Journal of Psychiatry* 132 (11):1187–92.

Spitzer, R. L., and Wakefield, J. C. (1999). "*DSM-IV* diagnostic criterion for clinical significance: Does it help solve the false positives problem?" *American Journal of Psychiatry* 156 (12):1856–64.

Spitzer, R. L., and Williams, J.B.W. (1994). "American psychiatry transformation following the publication of *DSM-III*." *American Journal of Psychiatry* 151 (3):459–60.

Spitzer, R. L., Williams, J.B.W., and Skodol, A. E. (1980). "*DSM-III*: Major achievements and an overview." *American Journal of Psychiatry* 137 (2):151–64.

Spitzer, R. L., and Wilson, P. T. (1968). "A guide to the American Psychiatric Association's new diagnostic nomenclature." *American Journal of Psychiatry* 124 (12):1619–29.

Spurzheim, J. G., and Gall, F. J. (1815). *The physiognomical system.* London: Baldwin, Craddock and Joy.

Squire, L. R., ed. (1996). *The history of neuroscience in autobiography.* Washington DC: Society for Neuroscience.

Srole, L., Langner, T. S., Michael, M. K., et al. (1962). *Mental health in the metropolis: The midtown Manhattan study.* New York: McGraw-Hill.

Stahl, S. M. (1996). *Essential psychopharmacology: Neuroscientific basis and practical applications.* Cambridge: Cambridge University Press.

Star, S. (1999). "The ethnography of infrastructure." *American Behavioral Scientist* 43 (3):377.

Star, S. L., and Griesemer, J. R. ([1989] 1999). "Institutional ecology, 'translations,' and boundary objects: Amateurs and professionals in Berkeley's Museum of Vertebrate Zoology, 1907–39." In M. Biagioli, ed., *The Science Studies Reader,* 505–24. New York: Routledge.

Star, S., and Ruhleder, K. (1996). "Steps toward an ecology of infrastructure: Design and access for large information spaces." *Information Systems Research* 7(1): 111–34.

Stepnisky, J. N. (2007). "Narrative magic and the construction of selfhood in antidepressant advertising." *Bulletin of Science, Technology and Society* 27 (1):24–36.

Stoler, A. L. (1995). *Race and the education of desire: Foucault's* History of Sexuality *and the colonial order of things.* Durham, NC: Duke University Press.

Stranahan, A. M., Khalil, D., and Gould, E. (2006). "Social isolation delays the positive effects of running on adult neurogenesis." *Nature Neuroscience* 9 (4):526–33.

Suppes, P. (1962). "Models of data." In Ernest Nagel, Patrick Suppes, and Alfred Tarski, eds., *Logic, methodology and philosophy of science: Proceedings of the 1960 international congress.* 252–61. Stanford, CA: Stanford University Press.

Surman, T. (2008). "Working with C. S. Sherrington, 1918–24. Interview by E. M. Tansey." *Notes and Records of the Royal Society of London* 62 (1):123.

Swayze, V. W., II. (1995). "Frontal leukotomy and related psychosurgical procedures in the era before antipsychotics (1935–1954): A historical overview." *American Journal of Psychiatry* 152 (4):505–15.

Swazey, J. (1975). "Forging a neuroscience community: A brief history of the Neurosciences Research Program." In F. G. Worden, J. P. Swayze, and G. Adelman, eds., *The neurosciences: Paths of discovery*, 529–46. Cambridge, MA: MIT Press.

Szalavitz, M., and Perry, B. (2010). *Born for love: Why empathy is essential and endangered*. New York: HarperCollins.

Szyf, M., McGowan, P., and Meaney, M. J. (2008). "The social environment and the epigenome." *Environmental and Molecular Mutagenesis* 49:46–60.

Tallis, R. (2011). *Aping mankind: Neuromania, Darwinitis and the misrepresentation of humanity*. Durham, UK: Acumen.

Tancredi, L. R. (2005). *Hardwired behavior: What neuroscience reveals about morality*. New York: Cambridge University Press.

Tansey, E. M. (2006). "Henry Dale and the discovery of acetylcholine." *Comptes Rendus Biologies* 329 (5–6):419–25.

Taroni, F., Bozza, S., Biedermann, A., et al. (2010). *Data analysis in forensic science: A Bayesian decision perspective*. New York: Wiley.

Taylor, C. (1989). *Sources of the self: The making of the modern identity*. Cambridge: Cambridge University Press.

Taylor, J. S. (1995). "Neurolaw: Towards a new medical jurisprudence." *Brain Injury* 9 (7):745–51.

——— (1996). "Neurorehabilitation and neurolaw." *Neurorehabilitation* 7 (1):3–14.

Taylor, J. S., Harp, J. A., and Elliott, T. (1991). "Neuropsychologists and neurolawyers." *Neuropsychology* 5:293–305.

TenHouten, W. (1997). "Neurosociology." *Journal of Social and Evolutionary Systems* 20 (1):7–37.

Ter-Pogossian, M., Phelps, M., Hoffman, E., et al. (1975). "A positron-emission transaxial tomograph for nuclear imaging (PETT)." *Radiology* 114 (1):89.

Thagard, P., Magnani, L., and Nersessian, N. J., eds. (1999). *Model-based reasoning in scientific discovery*. New York: Kluwer Academic/Plenum.

Thaler, R. H., and Sunstein, C. R. (2008). *Nudge: Improving decisions about health, wealth, and happiness*. New Haven, CT: Yale University Press.

Tinbergen, N. (1968). "On war and peace in animals and man: An ethologist's approach to the biology of aggression." *Science* 160 (835):1411–18.

Toman, J. E. (1949). "The neuropharmacology of antiepileptics." *Electroencephalography and Clinical Neurophysiology* 1 (1):33–44.

Tonkonogy, J. M. (1991). "Violence and temporal lobe lesion: Head CT and MRI data." *Journal of Neuropsychiatry and Clinical Neuroscience* 3 (2):189–96.

Triggle, D. J. (2004). "Patenting the sun: Enclosing the scientific commons and transforming the university—ethical concerns." *Drug Development Research* 63 (3):139–49.

Trimble, M. R. (2007). *The soul in the brain: The cerebral basis of language, art, and belief*. Baltimore: Johns Hopkins University Press.

Tsakiris, M., and Haggard, P. (2005). "The rubber hand illusion revisited: Visuotactile integration and self-attribution." *Journal of Experimental Psychology: Human Perception and Performance* 31 (1):80.

Tully, J., ed. (1988). *Meaning and context: Quentin Skinner and his critics*. Cambridge: Polity.

——— (1993). *An approach to political philosophy: Locke in contexts*. Cambridge: Cambridge University Press.

Turk, D. J., Heatherton, T. F., Kelley, W. M., et al. (2002). "Mike or me? Self-recognition in a split-brain patient." *Nature Neuroscience* 5 (9):841–42.

Turk, D. J., Heatherton, T. F., Macrae, C., et al. (2003). "Out of contact, out of mind." *Annals of the New York Academy of Sciences* 1001 (1):65–78.

Valenstein, E. S. (1986). *Great and desperate cures: The rise and decline of psychosurgery and other radical treatments for mental illness*. New York: Basic Books.

——— (2005). *The war of the soups and the sparks: The discovery of neurotransmitters and the dispute over how nerves communicate*. New York: Columbia University Press.

Vallabhajosula, S. (2009). *Molecular imaging: Radiopharmaceuticals for PET and SPECT*. Berlin: Springer.

Valmaggia, L., McCrone, P., Knapp, M., et al. (2009). "Economic impact of early intervention in people at high risk of psychosis." *Psychological Medicine* 39 (10):1617–26.

van Os, J. (2009). "A salience dysregulation syndrome." *British Journal of Psychiatry* 194 (2):101–3.

van Os, J., and Kapur, S. (2009). "Schizophrenia." *Lancet* 374 (9690):635–45.

van Praag, H. M. (1991). "Serotonergic dysfunction and aggression control." *Psychological Medicine* 21 (1):15–19.

van Praag, H. M., Asnis, G. M., Kahn, R. S., et al. (1990). "Monoamines and abnormal behavior a multi-aminergic perspective." *British Journal of Psychiatry* 157:723–34.

van Praag, H. M., Brown, S. L., Asnis, G. M., et al. (1991). "Beyond serotonin: A multiaminergic perspective on abnormal behaviour." In S. L. Brown and H. M. van Praag, eds., *The role of serotonin in psychiatric disorder*, 302–32. New York: Brunner/Mazel.

Varela, F. (1992). "The Reenchantment of the Concrete." In J. Crary and S. Kwinter, eds., *Incorporations*, 320–38. New York: Zone.

Varela, F. J., Thompson, E., and Rosch, E. H. (1993). *The embodied mind: Cognitive science and human experience*. Cambridge, MA: MIT Press.

Verplaetse, J. (2004). "Moritz Benedikt's (1835–1920) localization of morality in the occipital lobes: Origin and background of a controversial hypothesis." *History of Psychiatry* 15 (3):305–28.

Vidal, F. (2002). "Brains, bodies, selves, and science: Anthropologies of identity and the resurrection of the body." *Critical Inquiry* 28 (4):930–74.

——— (2005). "Le sujet cérébral: Une esquisse historique et conceptuelle." *Psychiatrie, Sciences Humaines, Neurosciences* 3 (11):37–48.

——— (2009). "Brainhood, anthropological figure of modernity." *History of the Human Sciences* 22 (1):5–36.

Vidaltamayo, R., Bargas, J., Covarrubias, L., et al. (2010). "Stem cell therapy for Parkinson's disease: A road map for a successful future." *Stem Cells and Development* 19 (3):311–20.

Viding, E., Blair, R.J.R., Moffitt, T. E., et al. (2005). "Evidence for substantial genetic risk for psychopathy in 7-year-olds." *Journal of Child Psychology and Psychiatry* 46 (6):592–97.

Viding, E., Jones, A. P., Frick, P. J., et al. (2008). "Heritability of antisocial behaviour at 9: Do callous-unemotional traits matter?" *Development Science* 11 (1):17–22.

Viding, E., Larsson, H., and Jones, A. P. (2008). "Quantitative genetic studies of antisocial behaviour." *Proceedings of the Royal Society B: Biological Sciences* 363 (1503):2519–27.

Volavka, J., Martell, D., and Convit, A. (1992). "Psychobiology of the violent offender." *Journal of Forensic Sciences* 37 (1):237–51.

Volkow, N. D., and Tancredi, L. (1987). "Neural substrates of violent behaviour. A preliminary study with positron emission tomography." *British Journal of Psychiatry* 151 (5):668–73.

Vul, E., Harris, C., Winkielman, P., et al. (2009). "Puzzlingly high correlations in fMRI studies of emotion, personality, and social cognition." *Perspectives on Psychological Science* 4 (3):274.

Wade, N. J., and Bruce, V. (2001). "Surveying the seen: 100 years of British vision." *British Journal of Psychology* 92 (pt. 1):79–112.

Wagner, H. N. (2006). "From molecular imaging to molecular medicine." *Journal of Nuclear Medicine* 47 (8):13N–39N.

——— (2009). *Brain imaging: The chemistry of mental activity.* New York: Springer.

Wagner, H. N., Burns, H. D., Dannals, R. F., et al. (1983). "Imaging dopamine-receptors in the human brain by positron tomography." *Science* 221 (4617):1264–66.

Wakefield, J. C. (2007). "The concept of mental disorder: Diagnostic implications of the harmful dysfunction analysis." *World Psychiatry* 6 (3):149–56.

Waldeyer, H. von (1891). "Uber einige neuere Forschungen im Gebiete der Anatomie des Zentralnervensystems." *Deutsche Medizinische Wochenschrift* 17:1213–1356.

Wall, J., Kaas, J., Sur, M., et al. (1986). "Functional reorganization in somatosensory cortical areas 3b and 1 of adult monkeys after median nerve repair: Possible relationships to sensory recovery in humans." *Journal of Neuroscience* 6 (1):218.

Wang, P. S., Berglund, P., Olfson, M., et al. (2005). "Failure and delay in initial treatment contact after first onset of mental disorders in the national comorbidity survey replication." *Archives of General Psychiatry* 62 (6):603–13.

Wang, P. S., Lane, M., Olfson, M., et al. (2005). "Twelve-month use of mental health services in the United States: Results from the national comorbidity survey replication." *Archives of General Psychiatry* 62 (6):629–40.

Webster, A., Brown, N., and Rappert, B. (2000). *Contested futures: A sociology of prospective techno-science.* Aldershot, UK: Ashgate.

Wegner, D. M. (2002). *The illusion of conscious will.* Cambridge, MA: MIT Press.

Weiner, J. (1999). *Time, love, memory: A great biologist and his quest for the origins of behavior.* New York: Knopf.

Weisberg, D. S., Keil, F. C., Goodstein, J., et al. (2008). "The seductive allure of neuroscience explanations." *Journal of Cognitive Neuroscience* 20 (3):470–77.

Whitehouse, P., and George, D. (2008). *The myth of Alzheimer's: What you aren't being told about today's most dreaded diagnosis.* London: St. Martin's Press.

Wicker, B., Keysers, C., Plailly, J., et al. (2003). "Both of us disgusted in my insula: The common neural basis of seeing and feeling disgust." *Neuron* 40 (3):655–64.

Willingham, D. T., and Dunn, E. W. (2003). "What neuroimaging and brain localization can do, cannot do and should not do for social psychology." *Journal of Personality and Social Psychology* 85 (4):662–71.

Willis, T. ([1664] 1965). *The anatomy of the brain and nerves.* Edited by W. Feindel. Translated by S. Pordage. 2 vols. Montreal: McGill University Press.

Wills, A. (1999). "Herophilus, Erasistratus, and the birth of neuroscience." *Lancet* 354 (9191):1719–20.

Wilson, E. A. (2004a). "The brain in the gut." Chapter 2 in *Psychosomatic: Feminism and the Neurological Body.* Durham, NC: Duke University Press.

——— (2004b). "Gut feminism." *Differences: A Journal of Feminist Cultural Studies* 15 (3):66–94.

——— (2010). *Affect and artificial intelligence.* Seattle: University of Washington Press.

——— (2011). "Neurological entanglements: The case of paediatric depressions, SSRIs and suicidal ideation." *Subjectivity* 4(3): 277–97.

Wilson, M. (1993). "*DSM-III* and the transformation of American psychiatry: A history." *American Journal of Psychiatry* 150 (3):399–410.

Wilson, P., Minnis, H., Puckering, C., et al. (2009). "Should we aspire to screen preschool children for conduct disorder?" *Archives of Disease in Childhood* 94 (10):812.

Winch, P. (1990). *The idea of a social science and its relation to philosophy.* London: Routledge.

Winslow, J. T., Hastings, N., Carter, C. S., et al. (1993). "A role for central vasopressin in pair bonding in monogamous prairie voles." *Nature* 365 (6446):545–48.

Wittchen, H. U., and Jacobi, F. (2005). "Size and burden of mental disorders in Europe: A critical review and appraisal of 27 studies." *European Neuropsychopharmacology* 15 (4):357–76.

Wittchen, H., Jacobi, F., Rehm, J., et al. (2011). "The size and burden of mental disorders and other disorders of the brain in Europe 2010." *European Neuropsychopharmacology* 21 (9):655–79.

Wittgenstein, L. (1958). *Philosophical investigations.* Oxford: Basil Blackwell.

Woese, C. R. (2004). "A new biology for a new century." *Microbiology and Molecular Biology Reviews* 68 (2):173–186.

Wolfe, C. (2003). *Zoontologies: The question of the animal.* Minneapolis: University of Minnesota Press.

——— (2009). *What is posthumanism?* Minneapolis: University of Minnesota Press.

Woodruff, R. A., Goodwin, D. W., and Guze, S. B. (1974). *Psychiatric diagnosis.* New York: Oxford University Press.

Worden, F. G., Swazey, J. P., Adelman, G., eds. (1975). *The neurosciences: Paths of discovery.* Cambridge, MA: M.I.T. Press.

World Health Organization. (2001). *Mental health: New understanding, new hope.* Geneva: WHO.

Wright, H. W. (1921). "The service of neuropsychiatry to industrial medicine." *California State Journal of Medicine* 19 (12):464.

Young, A. (1995). *The harmony of illusions: Inventing post-traumatic stress disorder.* Princeton, NJ: Princeton University Press.

Young, J. Z. (1974). *An introduction to the study of man.* Oxford: Oxford University Press.

Young, R. M. (1970). *Mind, brain and adaptation in the nineteenth century: Cerebral localization and its biological context from Gall to Ferrier.* Oxford: Clarendon.

Yudofsky, S. C., and Hales, R. E. (2002). "Neuropsychiatry and the future of psychiatry and neurology." *American Journal of Psychiatry* 159 (8):1261–64.

Zahavi, D., and Roepstorff, A. (2011). "Faces and ascriptions: Mapping measures of the self." *Consciousness and Cognition* 20 (1):141–48.

Zeki, S. (1993). *A vision of the brain*. Oxford: Blackwell Scientific.

——— (1999). *Inner vision: An exploration of art and the brain*. Oxford: Oxford University Press.

Zerhouni, E. (2003). "The NIH roadmap." *Science* 302 (5642):63–72.

——— (2005). "Translational and clinical science: Time for a new vision." *New England Journal of Medicine* 353 (15):1621–23.

Zhu, J., Zhou, L., and Xingwu, F. G. (2006). "Tracking neural stem cells in patients with brain trauma." *New England Journal of Medicine* 355 (22):2376–78.

Zohar, D. (1997). *Rewiring the corporate brain: Using the new science to rethink how we structure and lead organizations*. San Francisco: Berrett-Koehler Publishers.

Zubin, J. (1952). "Abnormalities of behavior." *Annual Reviews of Psychology* 3:261–82.

Index

················

abnormality. *See* normality/abnormality

Academy of Molecular Imaging (AMI), 72

Adolphs, Ralph, 144

Adrian, Edgar, 12, 30, 32, 33, 70

aggression: animal models and studies of, 97–101, 183–86, 188–89, 255n40; brain imaging and studies of, 155, 175–76, 189; genomics and, 131, 156, 165, 181–89, 269n31; and intent, 184; intervention and control of, 188–92, 197; MAOA deficiencies and, 182–89; pharmacological interventions and, 183–85; social or cultural motives for, 98. *See also* antisocial behavior

Alexander, Franz, 35

Allen, Graham, 193, 195

Allport, Gordon, 160

Althusser, Louis, 209

Alzheimer's disease, 14, 15, 125, 138, 261n34

American Society for Neurochemistry, 32, 41

Andriezen, W. Lloyd, 170

animal models: anthropomorphism and, 84, 97–98, 103–4; and artificiality of laboratory as research setting, 85–92; behavioral studies, 83, 85, 92, 96–99; cerebral localization and, 64; computer simulations as replacement for, 108–9; and conceptual models, 92; and DSM categories, 105–8; ethics of animal-based research, 49, 247n37; generalization and, 94–95; and human psychopathologies, 92, 99–102, 138; and human sociality, 85, 150–51; and human specificity, 102–4; *vs.* mechanical or material models, 92–93; mereological fallacy and, 83–84; reverse translation, 106–7; self-vindication and laboratory research, 88–89, 92–93, 100, 105–9; "setup" and, 77, 85–89, 109, 226, 228; species selection for, 89–90; specific-

ity of individual organisms, 94–95; and stabilization of the experimenter, 85–86; stabilization (standardization) of animals used, 89–91; structures, functions, and disease models (models³), 92–95; "style of reasoning" and, 82–83; translation of findings, 82, 85, 95–96, 104–8

antisocial behavior: "bioprediction" of, 188; brain imaging and, 98, 165–66; and "callous and unemotional traits" (AS-CU), 191; costs of, 188, 192–93, 194, 197; diagnosis of antisocial personality disorders, 102–3, 126, 176–77; and environmental influences on brain development, 164, 185, 187–90; gene-environment interaction and, 187; genomics and, 131, 156, 180–90, 181–89, 261n31, 269n31; linked to physical abnormalities in the brain, 98, 164, 165, 172, 179; as neurobiological in origin, 179; parental input and, 193; preventive interventions, 11, 15–16, 188–96; "screen and intervene" control strategy, 15, 139, 166–67, 189–91, 196. *See also* criminality; criminal justice system

Aos, Steve, 193

Aronson, Elliot, 161

asylums: "asylum psychiatry" and treatment of mental disorders, 122; and clinical gaze, 55–59, 115, 123, 249n10, 249n12; as setting for research, 115, 170; as setting for treatment of mental disorders, 10, 34, 65, 124

Atkins v. Virginia, 268n27

autism spectrum disorders: and abnormal social cognition in, 148–50; as developmental, 15, 150, 261; diagnosis of, 129, 135, 240n29; mentalizing deficits and, 148–50, 216–17, 273n16; mirror neurons and, 273n16

autopsies, 61–65, 165–66, 168, 266n9

325